Physical Oceanography of the Baltic Sea

Matti Leppäranta and Kai Myrberg

Physical Oceanography of the Baltic Sea

Professor Matti Leppäranta
Department of Physics
University of Helsinki
Finland

Adjunct professor Kai Myrberg
Finnish Institute of Marine Research
Helsinki
Finland

Front cover illustrations: (Main image) Icebreaker *Fennica* in the Baltic Sea, courtesy of Riku Lumiaro. Small images are from the Baltic Sea, courtesy of Riku Lumiaro and the first author.

Back cover illustrations: Painting *On the Sea* by Albert Edelfelt (© Gothenburg Museum of Art) and photograph of Stockholm Archipelago by Leena Leppäranta.

SPRINGER–PRAXIS BOOKS IN GEOPHYSICAL SCIENCES
SUBJECT *ADVISORY EDITOR*: Philippe Blondel, C.Geol., F.G.S., Ph.D., M.Sc., Senior Scientist, Department of Physics, University of Bath, UK

ISBN 978-3-540-79702-9 Springer Berlin Heidelberg New York

Springer is part of Springer-Science + Business Media (springer.com)

Library of Congress Control Number: 2008933573

Apart from any fair dealing for the purposes of research or private study, or criticism or review, as permitted under the Copyright, Designs and Patents Act 1988, this publication may only be reproduced, stored or transmitted, in any form or by any means, with the prior permission in writing of the publishers, or in the case of reprographic reproduction in accordance with the terms of licences issued by the Copyright Licensing Agency. Enquiries concerning reproduction outside those terms should be sent to the publishers.

© Praxis Publishing Ltd, Chichester, UK, 2009
Printed in Germany

The use of general descriptive names, registered names, trademarks, etc. in this publication does not imply, even in the absence of a specific statement, that such names are exempt from the relevant protective laws and regulations and therefore free for general use.

Cover design: Jim Wilkie
Project management: Originator Publishing Services, Gt Yarmouth, Norfolk, UK

Printed on acid-free paper

Contents

Preface xi

List of figures xv

List of tables xxi

List of abbreviations and acronyms xxiii

List of symbols xxvii

1   Introduction 1
    1.1   The Baltic Sea 1

2   The Baltic Sea: History and geography 9
    2.1   Evolution of the Baltic Sea during the Holocene 10
    2.2   The present Baltic Sea 14
    2.3   Research history 16
          2.3.1   Birth of Baltic Sea research 16
          2.3.2   From World War I to 1950 18
          2.3.3   Period of rapid development—from the 1950s to 1980s . 20
          2.3.4   Modern research—from 1990 to the present 23
    2.4   Weather and climate in the Baltic Sea region 25
          2.4.1   General regional climate 25
          2.4.2   Solar and terrestrial radiation 27
          2.4.3   Air temperature 32
          2.4.4   Precipitation and humidity 33
          2.4.5   Wind conditions 37

3   Topography and hydrography of the Baltic Sea 41
    3.1   Topography 41

		3.1.1	General.	42
		3.1.2	The Kattegat and the southwestern Baltic	46
		3.1.3	The Gotland Sea	49
		3.1.4	The Gulf of Riga	51
		3.1.5	The Gulf of Finland	52
		3.1.6	The Gulf of Bothnia	53
	3.2	Brackish water		56
		3.2.1	Equation of state	56
		3.2.2	Adiabatic changes and stability	58
		3.2.3	Properties of brackish water	58
	3.3	Baltic Sea watermasses		60
		3.3.1	Watermasses, stratification, halocline, and thermocline	60
		3.3.2	Interaction between the Baltic Sea and the North Sea	61
		3.3.3	Vertical structure of the Baltic Sea water body	64
		3.3.4	Salinity and temperature in different basins	72
	3.4	Long-term variations		84
		3.4.1	Temperature	85
		3.4.2	Salinity	85
		3.4.3	Ice conditions.	86
4	**Water, salt, and heat budgets**			89
	4.1	Water budget		89
		4.1.1	Physical problem	90
		4.1.2	Precipitation and evaporation	91
		4.1.3	Drainage basin and river inflow	94
		4.1.4	Freshwater budget	96
		4.1.5	Water exchange through the Danish Straits	98
	4.2	Salt budget		101
		4.2.1	Salt exchange through the Danish Straits.	101
		4.2.2	Sea ice growth and melting.	103
		4.2.3	Analytic models for salinity	104
	4.3	Heat budget		105
		4.3.1	Heat transport in the water body	105
		4.3.2	Solar radiation	108
		4.3.3	Terrestrial radiation	109
		4.3.4	Turbulent heat exchange	111
		4.3.5	Heat input from precipitation	113
		4.3.6	Annual course	113
	4.4	Surface layer models		116
		4.4.1	Analytic models of temperature	116
		4.4.2	Mixed layer models	118
		4.4.3	Turbulence models	120
		4.4.4	Self-similarity approach	121
	4.5	Box models		124
		4.5.1	Basin systems	124

		4.5.2	Stochastic models	125
		4.5.3	Physics-based box models	125

5 Circulation … 131
5.1 Theoretical background … 131
5.1.1 Basic system of equations; scale analysis … 132
5.1.2 The dynamics of surface currents … 135
5.1.3 Important special cases in sea dynamics … 135
5.2 General circulation based on observations … 142
5.2.1 Surface currents … 142
5.2.2 Three-dimensional water circulation … 145
5.2.3 Examples of current dynamics … 146
5.3 Deepwater circulation and ventilation … 148
5.3.1 Internal water cycle … 148
5.3.2 Internal mesoscale dynamics and mixing … 152
5.4 Baltic inflows … 158
5.4.1 Characteristic features of Major Baltic Inflows … 158
5.4.2 Discharge index and characteristic numbers … 161
5.4.3 Major inflows in 1993 and 2003 … 168
5.4.4 Warm inflows and small-size and medium-size inflows … 171
5.4.5 Stagnation periods … 173
5.5 Numerical modeling … 174
5.5.1 General … 174
5.5.2 Modeling of general circulation dynamics … 176
5.5.3 Inflow modeling … 183
5.5.4 Renewal times of watermasses … 185

6 Waves … 189
6.1 Physics of wave phenomena … 189
6.2 Shallow-water waves … 193
6.2.1 General model … 193
6.2.2 Free waves in ideal basins and in the Baltic Sea … 194
6.2.3 Kelvin waves and Poincaré waves … 196
6.2.4 Topographic waves and eddies … 199
6.3 Tides … 201
6.4 Wind-generated waves … 203
6.4.1 Wave generation … 203
6.4.2 Wave statistics … 206
6.4.3 Similarity law of Kitaigorodskii … 207
6.4.4 Wave forecasting … 211
6.4.5 From linear to non-linear waves … 211
6.5 Internal waves … 212
6.5.1 Background … 212
6.5.2 Theory … 214
6.5.3 Observations … 215

viii Contents

7 The ice of the Baltic Sea 219
- 7.1 Ice season and its climatology 219
- 7.2 Small-scale ice structure and properties 225
  - 7.2.1 Crystal structure 225
  - 7.2.2 Ice salinity 227
  - 7.2.3 Sea ice impurities 231
- 7.3 Growth and decay of ice 232
  - 7.3.1 Growth of ice 232
  - 7.3.2 Melting of ice 235
  - 7.3.3 Numerical modeling of the ice growth and melting cycle . 236
- 7.4 Dynamics and morphology of drift ice 239
  - 7.4.1 Drift ice fields 239
  - 7.4.2 Dynamics of drift ice 245
  - 7.4.3 Analytical solutions of sea ice drift 252
- 7.5 Basin-scale sea ice models 255
  - 7.5.1 General 255
  - 7.5.2 Short-term models 255
  - 7.5.3 Sea ice climate models 257

8 Coastal and local processes 261
- 8.1 Coastal zone 262
  - 8.1.1 How best to define the coastal zone? 262
  - 8.1.2 Formation of fronts and river plumes 263
  - 8.1.3 Specific features of archipelago areas 265
- 8.2 Sea-level elevation 268
  - 8.2.1 Sea-level dynamics 268
  - 8.2.2 Long-term changes 269
  - 8.2.3 Observed variability 270
  - 8.2.4 Modeling 274
- 8.3 Upwelling 275
  - 8.3.1 What is upwelling? 275
  - 8.3.2 Theory 276
  - 8.3.3 Atmospheric forcing 278
  - 8.3.4 Early results based on *in situ* observations 279
  - 8.3.5 Analyses based on satellite measurements 280
  - 8.3.6 Observed regional features 281
  - 8.3.7 Ecosystem implications 282
  - 8.3.8 Modeling 282
- 8.4 Ice conditions 283
  - 8.4.1 Structure of the coastal ice zone 283
  - 8.4.2 Fast ice edge 286
- 8.5 Coastal weather 289
  - 8.5.1 The land–sea breeze 289
  - 8.5.2 Specific air–sea interactions 291

Contents ix

9 Environmental questions 295
   9.1 Light conditions 296
       9.1.1 Basic concepts 296
       9.1.2 Optical characteristics of Baltic Sea water 300
       9.1.3 Optical remote sensing. 303
   9.2 Oxygen budget of bottom water. 303
   9.3 Human impact 306
       9.3.1 Major constructions. 306
       9.3.2 Influence on coastal erosion 307
       9.3.3 Oils spills and sea traffic 309
       9.3.4 Eutrophication 310
       9.3.5 Winter traffic. 312
   9.4 Extreme situations 314
       9.4.1 Meaning of extreme cases. 314
       9.4.2 Sea-level 314
       9.4.3 Extremes of ice seasons 315
   9.5 Operational oceanography. 318
       9.5.1 General. 318
       9.5.2 Observational systems 319
       9.5.3 Modeling tools. 320
       9.5.4 Case study: Storm Gudrun, 2005 322

10 Future of the Baltic Sea 325
   10.1 Baltic Sea—unique basin. 326
   10.2 Climate change implications 328
      10.2.1 Climate change scenarios 328
      10.2.2 Overall climate change in the Baltic Sea region. 330
      10.2.3 Future changes in the Baltic Sea 331
   10.3 Environmental state 334
   10.4 Future of the physical oceanography of the Baltic Sea 335

APPENDICES

Useful constants and formulas 337

Study problems 341

References 345

Index 371

Preface

The Baltic Sea is a small intra-continental sea located in northeast Europe. It is a young basin, unveiled from beneath the Fennoscandian ice sheet at the end of the Weichselian glaciation 9,000–13,500 years ago. The area has been inhabited from the time of the retreating ice sheet, and presently there are altogether 85 million people living in the 14 countries of the drainage basin of the Baltic Sea. This sea has always been an important shipping route, a source for fishery, and an area of recreation. During the last one hundred years, however, human influence on the Baltic Sea has been harmful. Pollutants have been dumped into the basin and strong nutrient loading has led to severe eutrophication, and consequently the state of the Baltic Sea has become drastically worse. Research and protection strategies need to be further developed, as the Baltic Sea is maybe the most polluted sea in the world.

Scientific research of the Baltic Sea commenced in the 1800s. The main motivation was then, in addition to basic science, the knowledge required by fishery, shipping, and coastal conditions. One specific scientific and practical question was land upheaval in the region.

International science collaboration was intensive from the early stages on, leading to joint research programs and conference series. The role of the Baltic Sea countries was important in the foundation of the International Council for the Exploration of the Sea (ICES) in 1902 in Copenhagen. The strategic role of the Baltic Sea was important throughout the 20th century as told in the history of the World Wars. Due to efforts to take better care of these waters the Baltic Sea Protection Agreement was signed in 1974, and as a consequence the Helsinki Commission (HELCOM) was founded. This agreement was renewed in 2000 and deemed valid for the time being. The state of the Baltic Sea has, however, continued weakening, and as a response to this a number of actions have been undertaken and programmes have been initiated to "Save Our Sea" (the slogan adopted by people from the area).

Physics forms the fundamental basis for understanding the behavior of the Baltic Sea system. It is interesting to note that several classical findings in physical oceanography are due to the Baltic Sea research community. At the early stages of physical oceanography, Martin Knudsen worked in Copenhagen and developed his water and salt exchange laws for semi-enclosed basins. Torsten Gustafsson and Börje Kullenberg observed inertial oscillations in the Central Baltic Sea in the 1930s. Research on marine optics began in the 1930s, with Nils Jerlov becoming a worldwide leading scientist in the field. Since the Baltic Sea is a freezing basin with a dense coastal population, sea ice research began here early on. Erkki Palosuo made pioneering investigations about the structure and morphology of drift ice from the 1940s to the 1960s. More recently in climatological research, long oceanographic time-series (temperature, sea-level elevation, salinity, ice cover) from the Baltic Sea have provided very important information about the conditions in North Europe during the last 300 years.

There are a few general books about the geology, biology, chemistry, and physics of the Baltic Sea. The best-known examples are *Meereskunde der Ostsee* (1974), edited by L. Magaard and G. Rheinheimer; *The Baltic Sea* (1981), edited by A. Voipio; and *A System Analysis of the Baltic Sea* (2001), edited by F. Wulff, L. Rahm, and P. Larsson. Such books contain excellent reviews of various Baltic Sea research topics but they do not provide a general view or a systematic presentation of Baltic Sea physical oceanography starting from the basics. Two major handbooks *Baltiyskoe more: Gidrometeorologicheskie usloviya* [The Baltic Sea: Hydrometeorological Conditions] (1992), edited by F. S. Terzieva, V. A. Rožkova, and A. I. Smirnov, and *State and Evolution of the Baltic Sea, 1952–2005* (2008), edited by R. Feistel, G. Nausch, and N. Wasmund, have been published.

In consequence, a comprehensive textbook of the physics of the Baltic Sea has long been needed. The present authors have given a course on this topic in the University of Helsinki since 1995 and prepared an undergraduate textbook (in Finnish) *Itämeren fysiikka, tila ja tulevaisuus* [Physics, State and Future of the Baltic Sea] in 2006 with Harri Kuosa, a professor in marine ecology of the Baltic Sea. After this the present book—*Physical Oceanography of the Baltic Sea*—has been prepared with a focus on physics and targeted at the research community and as a course textbook for Ph.D. students. In addition, it is anticipated that the book would be very useful to environmental monitoring units, marine engineers, and decision-makers, who work both to exploit and protect the Baltic Sea, and to people who have a deep interest in learning about the Baltic Sea.

The book presents—as it is titled—a complete physical oceanography of the Baltic Sea including the physics of sea ice. Oceanography is presented from the basics, so that researchers coming from other fields such as physics, environmental science, or geography can use the book as a handbook or self-study material. This is also of great use for general oceanographers since the Baltic Sea is a brackish water or low-salinity basin, and because of that it has its own oceanographic characteristics. The leading principle is to present the physics of the Baltic Sea based on theoretical and observational material. Models are also presented as tools for analysis and practical applications. The introduction (Chapter 1) gives a general picture of the Baltic Sea. Chapter 2 presents the geological history and geography of the region. Chapters 3–6 present classical physical oceanography: hydrography; water, salt, and heat budgets; circulation; and waves, respectively. Chapter 7 treats the ice of the Baltic Sea, from its fine-scale structure to seasonal characteristics, and discusses the role of ice in the general oceanography of this sea. Chapter 8 is devoted to coastal processes. Chapter 9 introduces the interface between physical oceanography and environmental problems. Chapter 10, the closing chapter, discusses the future of the Baltic Sea and its research and poses the questions: What will be the role of physical oceanography in the future? and What will be the impact of the expected environmental and climate change on the Baltic Sea? There then follow two appendices entitled "Useful constants and formulas" and "Study problems". The book ends with an index, preceded by a list of references. There is no separate chapter for numerical modeling, as different models with their applications and main results are discussed in model sections in the appropriate chapters.

As our research progressed, we learned about the Baltic Sea from a large number of colleagues. Especially, we want to thank our teachers: former Director General of the Finnish Institute of Marine Research Pentti Mälkki, Professor Erkki Palosuo, and D.Sc. Rein Tamsalu. Estonian Academician Tarmo Soomere and Dr. Andreas Lehmann are greatly acknowledged for their careful review of the manuscript.

Professors Harri Kuosa, Peter A. Lundberg, and Anders Omstedt are thanked for their support in writing this book. We are furthermore deeply thankful for help from Lic.Phil. Pekka Alenius, Dr. Oleg A. Andrejev, D.Sc. Helgi Arst, Professor Jan Askne, M.Sc. Jan-Erik Bruun, Professor Jüri Elken, Professor Wolfgang Fennel, M.Sc. Maria Gästgifvars, Dr. Jari Haapala, Dr. Jari Hänninen, M.Sc. Riikka Hietala, Dr. Niels Kristian Højerslev, Dr. Hermanni Kaartokallio, Professor Kimmo Kahma, Dr. Seppo Kaitala, Dr. Harri Kankaanpää, Dr. Jouko Launiainen, B.Sc. Anna-Riikka Leppäranta, Dr. Brian MacKenzie, M.Sc. Kristine Madsen, Dr. Piotr Margonski, Dr.Habil. Wolfgang Matthäus, Dr. Markus Meier, Professor Alexei V. Nekrasov, M.Sc. Annu Oikkonen, M.Sc. Riitta Olsonen, M.Sc. Leena Parkkonen, Ms. Hilkka Pellikka, Dr. Heidi Pettersson, Professor Jan Piechura, Professor Markku Poutanen, Dr. Jouni Räisänen, D.Sc. Vladimir A. Ryabchenko, Professor Kunio Shirasawa, Dr. Artur Svansson, Dr. Marzenna Sztobryn, Professor Matti Tikkanen, M.Sc. Laura Tuomi, M.Sc. Jouni Vainio, Dr. Timo Vihma, Professor Markku Viitasalo, Dr. Ilppo Vuorinen, and Professor Zhang Zhanhai. The scientists and technicians who have participated in the research programs are gratefully acknowledged.

We want to thank the students and assistants of our Baltic Sea courses. M.Sc. Salla Jokela, Ms. Pirkko Numminen, and M.Sc. Hannu Linkola are thanked for several graphics products. In the collection of numerous illustrations we have received great help from M.Sc. Riku Lumiaro, Finnish Institute of Marine Research, and also from Dr. Reinhard Feistel, M.Sc. Vivi Fleming-Lehtinen, Dr. Nils Gustafsson, M.Sc. Andres Kask, M.Sc. Indrek Kask, Dr. Toshiyuki Kawamura, Professor Aarno Kotilainen, M.Sc. Ari O. Laine, Dr. Inga Lips, Professor Urmas Lips, Dr. Jouko Pokki, Mr. Tero Purokoski, Ms. Küllike Roovåli, Mr. Henry Söderman, Dr. Henry Vallius, and Dr. Victor Zhurbas.

Professor Juhani Keinonen, Director of the Department of Physics of the University of Helsinki, Professor Eeva-Liisa Poutanen, Director General of the Finnish

Institute of Marine Research, and Professor Hannu Savijärvi, the Department of Physics of the University of Helsinki are thanked for their support in book preparation. Financial support has been also provided by the Finnish Cultural Foundation, which is gratefully acknowledged. The authors want to thank the Praxis team—Clive Horwood, Dr. Philippe Blondel (University of Bath, U.K.), and Neil Shuttlewood—for professional guidance and help in getting this book completed.

We would like to thank our wives Leena Leppäranta and Riitta Kyynäräinen for their patience and understanding during the many years of this work.

Finally, we want to express our deep gratitude to *Merentutkimuslaitos* [Finnish Institute of Marine Research]. We both grew up with the physical oceanographers there and appreciate the great staff, high scientific profile, and honoured long traditions in science and culture in Finland. This year, its 90th anniversary, the Finnish Institute of Marine Research has become one of the leading centers of Baltic Sea research.

Matti Leppäranta and Kai Myrberg
Helsinki, August 10, 2008

Figures

1.1	The Baltic Sea and the Kattegat and their drainage basin.............	2
1.2	Postal route across the Åland Sea in 1638–1895....................	5
1.3	CTD-sounding (profiling electric conductivity and temperature as a function of depth) is about to commence onboard R/V *Aranda*.................	7
2.1	The location of the edge of the Weichselian ice sheet retreating north.....	11
2.2	The phases of the Baltic Sea, Fennoscandian ice sheet, and land uplift.....	12
2.3	The Baltic Sea and its sub-basins...............................	13
2.4	Lightship *Storkallegrund* in its site on the sea route to Vaasa..........	18
2.5	Sea ice chart on March 13, 1925 over the northern part of the Baltic Sea and Lake Ladoga...	19
2.6	Deployment of instrumentation in the Baltic Sea....................	22
2.7	Group photograph of the participants of the 5th Study Conference on BALTEX in Kuressaare, Estonia..................................	24
2.8	Sea-level pressure in 1901–2000 from the Hadley Centre simulation HadSLP2	26
2.9	The annual evolution of global radiation in Visby, Gotland in 1961–1990..	29
2.10	Mean seasonal cloudiness (unit 1/8) fields in the Baltic Sea region........	30
2.11	Daily and annual variation in cloudiness (in %) in Uppsala............	32
2.12	Mean monthly atmospheric temperatures in the Baltic Sea.............	34
2.13	Mean monthly precipitation at some Baltic stations..................	36
2.14	Mean annual relative humidity in the Baltic Sea....................	37
2.15	Mean wind speed and direction in the Baltic Sea area................	38
3.1	Fennoscandian land uplift relative to the center of the Earth...........	43
3.2	The main characteristics of the bottom topography of the Baltic Sea and related important geographical locations................................	44
3.3	A schematic view of sills between different basins in the Baltic Sea.......	45
3.4	The hypsographic curves of Baltic Sea depth.......................	47
3.5	Distribution of Quaternary deposits in the Baltic Sea................	color
3.6a	The bottom topography of the Skagerrak, Kattegat, and the southwestern Baltic Sea...	48
3.6b	The bottom topography of the Gotland Sea.......................	50

Figures

3.6c	The bottom topography of the Gulf of Riga and Gulf of Finland	52
3.6d	The bottom topography of the Gulf of Bothnia	54
3.7	The density of seawater, freezing point, and the temperature of maximum density	56
3.8	Climatological mean surface salinity in the Baltic Sea and North Sea in August	62
3.9	Cross-section of salinity in February and in August in Skagerrak	63
3.10	Schematic picture of the water exchange between the North Sea and the Baltic Sea	65
3.11	A schematic diagram of vertical stratification of Baltic Sea watermasses. Mean vertical gradient of salinity and temperature	66
3.12	Long-term mean annual course of salinity at different depths in Utö during 1911–2005	color
3.13a	T–S curves for Baltic Sea watermasses	68
3.13b	Hydrographic surveying aboard R/V *Nautilus* using Nansen bottles	69
3.14	Long-term mean annual course of temperature at different depths in Utö during 1911–2005	color
3.15	Time evolution of the different stages of thermocline development in the Bay of Bothnia in 1977	71
3.16	Typical profiles of temperature, salinity and σ_T (σ_T = density—1,000 kg/m^3) in the Gotland Deep in 2003	73
3.17	The salinity of the Baltic Sea water body in different seasons along different cross-sections: along the central axis Kattegat–Gulf of Finland, Kattegat–Eastern Gotland Basin and along the central axis of the Gulf of Bothnia	76
3.18	The temperature of the Baltic Sea water body in different seasons along different cross-sections: along the central axis Kattegat–Gulf of Finland, Kattegat–Eastern Gotland Basin and along the central axis of the Gulf of Bothnia	78
3.19	The mean distribution of salinity in the Baltic Sea in July	80
3.20	The mean distribution of temperature in the Baltic Sea in July	82
3.21	Temperature evolution in Christiansø, Denmark during 1880–1998 for different seasons and for the annual mean	85
3.22	Long-term variability in deepwater temperature in Bornholm Basin, Eastern Gotland Basin, Sea of Bothnia, and Bay of Bothnia in 1900–2000	86
3.23	Long-term variability in salinity in Bornholm Basin, Eastern Gotland Basin, Sea of Bothnia, and Bay of Bothnia in 1900–2000	87
3.24	Maximum annual ice extent in the Baltic Sea during 1900–2008	88
4.1	The annual Baltic Sea water balance	90
4.2	The distribution of annual precipitation in the Baltic Sea	92
4.3	Monthly values of evaporation in the Gotland Sea	93
4.4	The saturation water vapor pressure of air as a function of temperature	94
4.5	The Neva in St. Petersburg is by far the largest river flowing into the Baltic Sea	97
4.6	Monthly river inflows into Baltic Sea basins	98
4.7	Mean freshwater budget in the Baltic Sea	99
4.8	Annual discharge from the Vistula, Tornio, and Kalix, and Neva for the period 1950–1993	99
4.9	Baltic Sea salt balance and idealization of the Baltic Sea as a two-layer box	101
4.10	Surface heat balance in the Baltic Sea	107
4.11	Comparison of the mixed layer model output with observed data from June to November, 1974	119

Figures xvii

4.12	Simulation of surface layer cooling by the $k - \varepsilon$ model for October–November 1979 in the Bay of Bothnia	122
4.13	Self-similarity profile of salinity	124
4.14	The distribution of particles after a 5-year random walk starting from Helsinki	126
4.15	Model simulations for the sea-level and temperature for the Arkona Basin and for the Åland Sea from the 13-box Baltic Sea model in winter 1986–1987	129
5.1	Inertial oscillations in the Baltic Sea between August 17 and 24, 1933	136
5.2	The spectral density of current velocity at a depth of 13 m in the Gulf of Bothnia	137
5.3	The Ekman profile of current velocity in shallow seas	139
5.4	The wind-driven motion of ice and currents at different depths in the Ekman layer in the Bay of Bothnia, April 1975	140
5.5	Upwelling off the northern coast of the Gulf of Finland on September 2, 2002	color
5.6	The tilt of the sea surface across the Great Belt calculated according to sea-level differences between Korsör and Slipshavn and the relation between tilt and surface currents	142
5.7	Long-term mean surface circulation in the Baltic Sea according to lightship measurements	143
5.8	Current roses based on measurements outside Rauma, Sea of Bothnia during the International Gulf of Bothnia Year 1991	144
5.9	A schematic of the large-scale internal water cycle in the Baltic Sea	color
5.10	The movement of drifting buoys in the Gulf of Bothnia between July 7 and 18, 1991	146
5.11	An example of an instantaneous strait flow in the Southern Quark at Station F33	147
5.12	Surface and bottom currents, and near-bottom oxygen, salinity, and temperature in the Archipelago Sea between September 12 and December 14, 2002	149
5.13	Direction diagram of near-bottom currents south of Utö. Bottom topography at the site	152
5.14	A cross-section of temperature in the Gotland Sea	154
5.15	Intrusions in the Baltic Sea proper	156
5.16	A sketch of the inflow into the Baltic Sea and determining processes	159
5.17	Main phases of a major inflow: precursory period, main inflow period, and post-inflow period, and the related changes in sea-level	160
5.18	Mean sea-level pressure patterns during a typical Major Baltic Inflow event	162
5.19	Main routes followed by Major Baltic Inflows, main sills, and channels	164
5.20	Major Baltic Inflows shown in accordance with the Q_{96}-index between 1880 and 2005 and their seasonal distribution	167
5.21	Long-term variation of temperature, salinity, oxygen, and hydrogen sulfide at 200-meter depths in the Gotland Deep between 1870 and 2004	168
5.22	Longitudinal transects of salinity in the Darss Sill area during the Baltic Major Inflow in January 1993	170
5.23	Potential temperature and salinity on February 16–18, 2003 along the axis Arkona Basin–Bornholm Gate–Bornholm Deep–Stolpe Channel–Gdańsk Deep	color
5.24	Bornholm Deep–Stolpe Channel-Gdańsk Deep temperature transects measured by R/V *Oceania* in November 2003, January 2004, February 2004, and May 2004 showing the eastward propagation of warm inflow water over Stolpe Sill into the southeastern Gotland Basin	color

xviii **Figures**

5.25	Long-term variations in river runoff into the Baltic Sea compared with salinity in central Baltic deepwater.	173
5.26	Baroclinic circulation in the Baltic Sea based on a diagnostic model	177
5.27	Annual mean transport per unit length for 1981–2004: upper layer and lower layer.	179
5.28	Simulated mean surface circulation in the Gulf of Finland. The stability of currents. Calculations are for the years 1987–1992	color
5.29	Section of salinity through the Belt Sea, Arkona Sea, and Bornholm Sea, March 1993.	184
5.30a	Mean age for the last 5 years of the 96-year spin-up associated with tracer-marking of inflowing water at the Darss and Drogden Sills.	color
5.30b	As in Figure 5.30a, but ages are associated with tracer-marking of freshwater from all rivers.	color
6.1	Waves are a common feature on the sea surface.	190
6.2	Schematic illustration of shallow-water and deepwater waves	192
6.3	Sea-level variations at Ystad, southern Sweden during one week in December 1975.	197
6.4	Dispersion relation of Kelvin and Poincaré waves	198
6.5	Theoretically calculated coastal profile of Kelvin waves	198
6.6	Energy spectra of current measurements off Pori, Sea of Bothnia, August–September 1976.	202
6.7	Energy spectrum of sea-level elevation in Helsinki mareograph from the period 1980 to 1999.	203
6.8	Co-tidal lines of the M_2 tide in the Kattegat and Baltic Sea; and co-tidal lines of the K_1 tide in the Kattegat and Baltic Sea	204
6.9	One-dimensional wave spectra from the Baltic Sea showing local waves and swell.	205
6.10	Histograms of significant wave height and period of the peak wave based on buoy measurements in the Northern Gotland Basin 1997–2004	208
6.11	Scatter diagram of wave heights and periods at Almagrundet in the Western Gotland Basin southeast of Stockholm 1978–1995	209
6.12	Significant wave height as a function of wind speed, duration, and fetch	210
6.13	Wave growth as a non-dimensional diagram	color
6.14	Convergence bands due to internal waves on the sea surface in the Bay of Bothnia seen in an ERS-1 SAR image; isotherms for 16-day period in July–August 1978 in the Sea of Bothnia	213
6.15	Vertical distribution of the Brunt–Väisälä frequency N in the Gotland Deep	216
7.1	Drift ice fields in the Baltic Sea make up its winter landscape	220
7.2	The probability of annual ice occurrence in the Baltic Sea	222
7.3	Baltic Sea ice chart on March 20, 2006	224
7.4	Ice crystal structure in Baltic Sea ice and lake ice.	color
7.5	Evolution of mean vertical salinity of ice in the Baltic Sea	228
7.6	Sea ice phase diagram with the proportions of ice, main ions, brine, and solid salts as a function of temperature	229
7.7	Flexural strength of sea ice as a function of brine content.	230
7.8	The mean thickness of congelation ice, snow ice, and snow at Kemi, 1979–1990	238
7.9	Drift ice at Nahkiainen lighthouse, Bay of Bothnia	239
7.10	Mapping the topography of ridged ice in the Bay of Bothnia	242
7.11	Ice thickness distribution in the Bay of Bothnia.	245

7.12	Ship under ice pressure in the Baltic Sea	246
7.13	Schematic illustration of the rheology of drift ice	247
7.14	Ice drift and wind in the central Bay of Bothnia, March 1977	249
7.15	Ice field displacement March 5–8, 1994	250
7.16	Free drift solution as a vector sum of wind-driven ice drift and ocean current in deepwater	253
7.17	Ice drift speed relative to free drift speed as a function of ice compactness in the Baltic Sea	253
7.18	The sea-level difference between Kemi and Vaasa in open sea and ice conditions as a function of wind speed	254
7.19	A bay model for a compact ice field	254
7.20	Convoy of ships assisted by an icebreaker in the Baltic Sea	256
7.21	Velocity field calculated by the sea ice dynamics model and compared with observations	258
7.22	Calibration of a Baltic Sea ice climate model	259
8.1	Coastal landscape in the Baltic Sea from the Curonian Spit	262
8.2	The quasi-permanent salinity front at the mouth of the Gulf of Finland	264
8.3	A river plume (Luleå River) at the Swedish coast of the Bay of Bothnia observed by ERS-1 SAR	266
8.4	The island of Seili in the Archipelago Sea	267
8.5	Detrended annual sea-level at Helsinki and the annual mean NAO-index	269
8.6	Annual mean sea-level, sea surface temperature, and air temperature along with linear trends in Klaipėda, Lithuania in 1961–2002	271
8.7	Sea-level probability distributions at Hanko for the 20-year periods of 1897–1916 and 1980–1999	274
8.8	Principal response of an elongated basin to constant wind in the length direction of the basin	277
8.9	Common upwelling zones in the Baltic Sea and typical atmospheric conditions favorable for upwelling	279
8.10	Tvärminne Zoological Station of the University of Helsinki	280
8.11	Monthly mean composites of sea surface temperature for September 2003 and September 2005	color
8.12	Measured temperature cross-section in the Gulf of Finland in summer 2006 (Estonia on the left side, Finland on the right side)	color
8.13	Comparison of sea surface temperature maps of the Gulf of Finland obtained from satellite imagery and simulated	color
8.14	The stability of the ice cover in the archipelago as a function of ice thickness, fetch, and wind speed	284
8.15	Grounded ridge at the fast ice edge, eastern side of the Bay of Bothnia	285
8.16	Schematic diagram on the bearing capacity of ice	286
8.17	Boundary zone configuration	287
8.18	Steady state solution of wind-driven zonal flow, northern hemisphere	288
8.19	Schematic diagram of sea breeze and land breeze	290
8.20	NOAA satellite image of snow bands over the Baltic Sea	292
9.1	Solar radiation spectrum on top of the atmosphere and on sea surface	297
9.2	Description of radiance and irradiance	297
9.3	Secchi depth in the Baltic Sea 1900–2000	300
9.4	Attenuation spectra of light in Gdansk Bay and the Gotland Sea compared with general coastal seawater and pure water	301

xx Figures

9.5	Chlorophyll *a* maximum in July 2005 in the Baltic Sea based on MODIS data from the Terra/Aqua satellite	color
9.6	Long-term oxygen variation in deepwater and bottom water at Stations F 9, F 26, and F 64 and long-term oxygen and hydrogen sulfide variation in deepwater and bottom water at Stations BY 5, BY 15, and BY 31	305
9.7	Near-bottom oxygen conditions in the Baltic Sea in 1993, 1994, 2006, and 2007	color
9.8	The Öresund bridge, seen from the east looking toward Copenhagen	308
9.9	The shore of the island of Aegna in 2000 and in 2002 from the same location	color
9.10	The operation area of the clean-up vessel *Hylje* during March 15-18, 2006	311
9.11	Ice road between the island of Hailuoto and mainland at Oulu, eastern coast of the Bay of Bothnia	313
9.12	Cumulative distribution of the maximum annual ice extent in the Baltic Sea	316
9.13	The cumulative distributions of freezing and melting dates	317
9.14	The wave forecast for January 9, 2005 at 08:00 GMT provided by the Finnish Institute of Marine Research	color
9.15	Flooding in Pärnu during Storm Gudrun in January 2005	323
10.1	The Baltic Sea landscape	326
10.2	The Baltic Sea herring	327
10.3	The Finnish research vessel *Aranda* and Sea Captain Riitta Kyynäräinen keeping watch on the bridge	329
10.4	Mean thickness of level ice in the control simulation 1961–1990 and by 2100	332
10.5	Forecast global sea-level changes in the Baltic until 2100	333

Tables

1.1	Comparison between the Baltic Sea and other intra-continental seas and large lakes	3
2.1	The principal characteristics of the Baltic Sea	14
2.2	Climatological data from the normal period 1961–1990 for three Baltic Sea weather stations	28
3.1	The dimensions of the Baltic Sea basins and the Kattegat	46
3.2	Physical properties of Baltic Sea brackish water compared with freshwater and normal oceanic water	60
3.3	The main characteristics of Baltic Sea stratification	70
3.4	Typical values of salinity in different basins of the Baltic Sea. In the southwestern Baltic Sea and in the Gulf of Finland	74
3.5	Mean values of surface and bottom temperatures in the Baltic Sea for different seasons	84
4.1	Estimations of mean annual precipitation and evaporation in the Baltic Sea	91
4.2	Mean annual river inflow (cubic kilometers) into the Baltic Sea	96
4.3	The largest rivers in the drainage basin of the Baltic Sea and the Kattegat	97
4.4	Mean monthly solar radiation in Jokioinen, southern Finland	110
4.5	Annual course of magnitudes of heat budget components in the Baltic Sea, assuming that the sea surface is ice-free all year	114
4.6	Sea surface heat balance at Finngrundet, southwestern Sea of Bothnia in March 1961–February 1962	115
5.1	Typical values for different terms of the equations of motion	133
5.2	Characteristic numbers of the 35 most intensive Major Baltic Inflows between 1880 and 2005	165
6.1	Theoretical uninodal seiche periods in Baltic Sea basins	195
6.2	Oscillation modes of the whole Baltic Sea based on absence of a Coriolis effect and with a Coriolis effect	196
7.1	Statistics on annual maximum ice thickness and the proportion of superimposed ice during 1961–1990 for Baltic Sea landfast ice	227
7.2	Ice and snow thickness in 1961–1990 in Kemi	233

Tables

7.3	Ice field classification based on ice compactness	240
7.4	General and common Baltic Sea ice types	241
7.5	Atmospheric and oceanic boundary layer parameters for sea ice dynamics	251
9.1	Extremes of the physical characteristics of the Baltic Sea and their environmental impact	314
9.2	Extremes of the physical characteristics of the Baltic Sea and their impact on human activities	315
9.3	Satellite systems used in operational oceanography in the Baltic Sea	320

Abbreviations and acronyms

AVHRR	Advanced Very High Resolution Radiometer
BACC	Baltex Assessment of Climate Change
BALTEX	Baltic Sea Experiment
BASIS	Baltic Air–Sea–Ice Study
BASYS	Baltic Sea System Study
BED	Baltic Environmental Database
BEERS	Baltic Experiment for ERS-1
BEPERS	Bothnian Experiment in Preparation for ERS-1
BMB	Baltic Marine Biologists
BMG	Baltic Marine Geologists
BMP	Baltic Monitoring Program
BOOS	Baltic Operational Oceanographic System
BOSEX	Baltic Open Sea Experiment
BP	Before present
BSH	*Bundesamt für Seeschifffahrt und Hydrographie* [Federal Maritime and Hydrographic Agency, Germany]
BSHcmod	Model code developed at *Bundesamt für Seeschifffahrt und Hydrographie* (BSH)
BSI	Baltic Sea Index
BSSC	Baltic Sea Science Congress
CBO	Conference of Baltic Oceanographers
CDOM	Colored dissolved organic matter
COHERENS	Coupled Hydrodynamical Ecological Model for Regional Shelf Seas
CTD	Conductivity–Temperature–Depth
DAS	Data Assimilation System
DHI	Danish Hydraulic Institute
DIAMIX	DIApycnal MIXing

Abbreviations and acronyms

DIN	Dissolved inorganic nitrogen
DIP	Dissolved inorganic phosphorus
DMI	Danish Meteorological Institute
DR	Drogden Sill
DS	Darss Sill
ECMWF	European Centre for Medium-range Weather Forecasting
EDIOS	European Directory of the Ocean-observing System
ERS-1	European Remote Sensing Satellite-1
ESA	European Space Agency
EU	European Union
FIMR	Finnish Institute of Marine Research
FinEst	Finnish–Estonian model
FMI	Finnish Meteorological Institute
FRESCO	Finnish–Russian–Estonian Cooperation
GCM	General circulation model
GFDL	Geophysical Fluid Dynamics Laboratory
GMES	Global Monitoring of Environment and Security of the European Area
GMT	Greenwich Mean Time
GOOS	Global Ocean Observing System
GPS	Global Positioning System
HBV	*Hydrologiska Byråns Vattenbalansmodell* [Water balance model of the Hydrological Office, SMHI, Sweden]
HELCOM	Helsinki Commission
HIRLAM	High Resolution Limited Area Model
HIROMB	High Resolution Operational Model for the Baltic Sea
I/B	Icebreaker
IBY	International Baltic Year
ICES	International Council for the Exploration of the Sea
IOW	*Leibniz-Institut für Ostseeforschung Warnemünde* [Leibniz Institute for Baltic Sea Research, Germany]
IPCC	International Panel on Climate Change
MBI	Major Baltic Inflow
MERIS	Medium Resolution Imaging Spectrometer Instrument
MODIS	Moderate Resolution Imaging Spectroradiometer
MOM	Modular Ocean Model
NAO	North Atlantic Oscillation
NATO	North Atlantic Treaty Organization
NOAA	National Oceanographic and Atmospheric Administration
NOAMOD	Northeastern Atlantic 2D Model
OAAS	Oleg Andrejev–Alexander Sokolov
OAS	Optically active substance
OCCAM	Ocean Circulation Climate Advanced Modeling
PAR	Photosynthetically active radiation
PCB	Polychlorinated biphenyl

PEX-86	Baltic Sea Patchiness Experiment
POM	Princeton Ocean Model
PSU	Practical salinity unit
R/V	Research vessel
S/S	Steam ship
SAR	Synthetic Aperture Radar
SCOR	Scientific Committee on Oceanic Research
SEGUE	Searching efficient protection strategies for the eutrophied Gulf of Finland: the integrated use of experimental and modeling tools
SMHI	Swedish Meteorological and Hydrological Institute
SOOP	The ship of opportunity
SPBIO	St. Petersburg Institute of Oceanology
SYKE	Finnish Environment Institute
TS	Temperature–Salinity
UNESCO	United Nations Educational, Scientific and Cultural Organization
WAM	Wave Prediction Model
WAMDI	Wave Model Development and Implementation
WMO	World Meteorological Organization

Symbols

A	Surface area; ice compactness
A_H, A_v	Eddy diffusion coefficient, horizontal and vertical
c	Wave phase speed
c_g	Group speed of waves
c_p	Specific heat at constant pressure
C_D, C_H, C_E	Turbulent exchange coefficients for momentum, heat, and moisture
D	Ekman depth
E	Evaporation; stability
f	Coriolis parameter
F	Freshwater budget
g	Acceleration due to gravity on Earth's surface
h	Ice thickness
H	Sea depth
i	$\sqrt{-1}$
K	Bulk modulus of seawater; Eddy diffusion coefficient
$\boldsymbol{K}$	Wave vector
L	Latent heat of freezing
L_E	Latent heat of evaporation
N	Cloudiness; Brunt–Väisälä frequency
p	Pressure
P	Precipitation
q	Specific humidity; complex velocity vector
Q	Power (rate of change of energy)
Q_b	Heat flux from sea bottom
Q_c	Sensible heat flux
Q_e	Latent heat flux
Q_{La}	Atmospheric thermal radiation

Symbols

Q_{Lo}	Outgoing thermal radiation
Q_n	Net heat flux
Q_P	Heat flux from precipitation
Q_R	Radiation balance
Q_s	Incoming solar radiation
Q_{sc}	Solar constant
Q_r	Outgoing solar radiation
r	Sun–Earth distance
R	Relative humidity; Rossby radius of deformation
S	Salinity; spectral density of surface waves
t	Time
T	Temperature (°C); timescale
$\underline{T}$	Temperature (K)
T_f	Freezing point temperature
T_m	Temperature of maximum density
$\mathbf{u}$	Three-dimensional velocity
u	East velocity component
u^*	Friction velocity
U	Horizontal velocity
v	North velocity component
V_i	Inflow into Baltic Sea at the mouth
V_o	Outflow into Baltic Sea at the mouth
V_r	River discharge into the Baltic Sea
w	Vertical velocity
x	East coordinate
y	North coordinate
z	Vertical coordinate
z_0	Surface roughness
Z	Zenith angle
α	Albedo; temperature expansion coefficient of seawater
β	Salinity coefficient for density of seawater
ε	Emissivity
γ	Fraction of visible light in solar radiation
Γ	Adiabatic temperature gradient
κ	von Kármán constant; light attenuation coefficient
λ	Wavelength
ω	Frequency
Ω	Angular velocity of the Earth
ρ	Density of water
ρ_a	Density of air
σ	Stefan–Boltzmann constant
τ	Reynolds stress
τ_a	Wind stress
θ	Potential temperature
ξ	Sea-level elevation

Subscripts

a	Air
a	Ice; inflow
H	Horizontal
o	Surface; outflow
V	Vertical

1

Introduction

Rolf Witting (1879–1944) was the founder of physical oceanography in Finland and one of the leading authorities in Baltic Sea research during the first half of the 20th century. He developed measurement techniques and worked in such areas as the sea-level, hydrography, circulation, and waves. After Finland gained independence in 1917, Rolf Witting became the first Director of the Institute of Marine Research (1918–1936). He also became a Member of Parliament in Finland and was the Foreign Minister during a period of the Second World War (1940–1943). Courtesy of the Finnish Institute of Marine Research.

1.1 THE BALTIC SEA

The Baltic Sea[1] is a unique basin of the World Ocean. It is small and shallow, rather a series of basins, and connected to the main Atlantic Ocean only via the Danish Straits (Figure 1.1). The exchange of water through these straits is quite limited, and as a consequence of the positive freshwater balance the Baltic Sea water mass is brackish, with the mean salinity about 7‰—one-fifth of the salinity of normal ocean waters.

[1] The origin of the word "Baltic" is not clear. It may come from "belt" (*belt* in Germanic languages), a sea shaped like a belt, or "white" (*baltas/balts* in Lithuanian/Latvian languages).

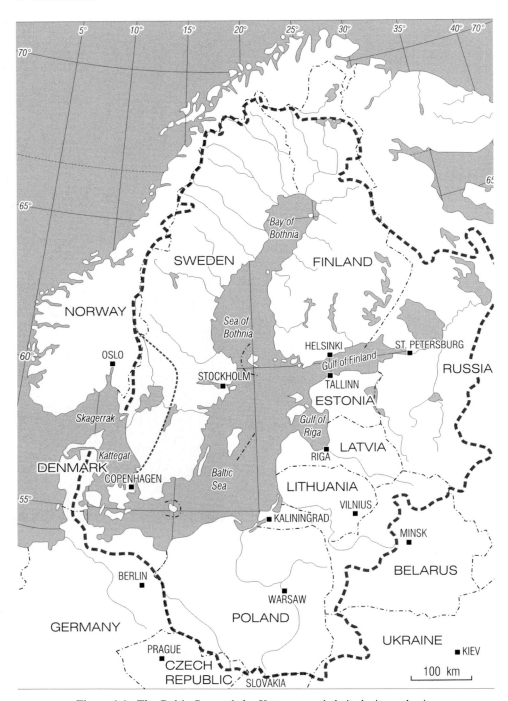

Figure 1.1. The Baltic Sea and the Kattegat and their drainage basin.

Table 1.1. Comparison between the Baltic Sea and other intra-continental seas and large lakes.

Basin	Area (10³ km²)	Mean depth (m)	Mean salinity (‰)	Freshwater budget	Ice cover on average	Location centre, (lat., long.)
Baltic Sea	393	54	7	+	Half	60°N 20°E
Black Sea	436	1,197	20	+	Northeast	43°N 35°E
Gulf of Ob	41	12	5	+	All	73°N 74°E
Chesapeake Bay	12	6	15	+	Shores	38°N 76°W
Hudson Bay	1,232	128	30	+	All	58°N 85°W
Red Sea	438	491	40	−	None	22°N 38°E
Persian Gulf	239	25	40	−	None	27°N 52°E
Caspian Sea	374	211	12	0	North	43°N 50°E
Lake Superior	82	149	<0.1	0	All	48°N 88°W

This elongated sea lies between maritime temperate and continental sub-Arctic climate zones. In winter it is partly ice-covered and during the most severe winters it is completely frozen over. The variable coastal geomorphology and the wide archipelago areas give the Baltic Sea its individual appearance.

There are four major brackish water[2] basins in the World Ocean. Listed from the largest to the smallest they are the Black Sea (Kostianoy and Kosarev, 2008) located between Europe and Asia Minor, the Baltic Sea, the Gulf of Ob in the Kara Sea (Volkov *et al.*, 2002) and Chesapeake Bay, surrounded by Virginia and Maryland on the east coast of the United States of America. They all developed into brackish water basins during the Holocene. During the Weichselian glaciation the Black Sea was a freshwater lake, the Baltic Sea and the Gulf of Ob were under the ice sheet, and Chesapeake Bay was a river valley. The mean depth of the Black Sea is 1,197 m and, due to the strong salinity stratification and extremely slow deepwater renewal, the water masses below 200 m are anoxic. The Sea of Azov in the northeastern part of the Black Sea is often frozen during the winter. The Gulf of Ob is a long (800 km) and narrow estuary of Ob River in the Kara Sea in the Russian Arctic, ice-covered in winter. Chesapeake Bay is a small and very shallow basin. It is a ria or a drowned valley in a humid subtropical climate zone with hot summers and with ice formation in river mouths in some winters.

Table 1.1 gives some basic information about brackish water seas and other basins comparable with the Baltic Sea. Closest to the brackish water seas is Hudson Bay. It is an oceanic, semi-enclosed basin with a positive freshwater budget, and

---

[2] Salinity less than 24.7‰ (and more than about 0.5‰).

its salinity is about 30‰. In contrast, small intra-continental seas with a negative freshwater budget have salinity above the oceanic level. Such seas exist in the tropical zone; for example, the Red Sea and Persian Gulf, where salinities are above 40‰. Very large lakes are comparable in size with the Baltic Sea, and the Caspian Sea is even larger in volume.

The Baltic basin is a very old depression in the bedrock. Prior to the Weichselian glaciation the basin was covered by the Eem Sea, which extended from the North Sea to the Barents Sea leaving Fennoscandia as an island. At the end of the Weichselian glaciation, 13,000–14,000 years ago, the Baltic Ice Lake was born from glacier meltwater, and then during the Holocene there were fresh and brackish phases dictated by glacier melting, land uplift, and eustatic changes in global sea-level. The present brackish state commenced 7,000 years ago, and since about 2,000 years ago salinity has been close to the present level. Land uplift is slowly changing the Baltic Sea landscape. Here it is possible to observe how land rises from the sea and how terrestrial life gradually takes over. During the history of the Baltic Sea, people living in the region have adapted to this gradual, long-term change.

In this book we examine the present brackish Baltic Sea. It is a shallow basin, with a mean depth of 54 meters, and its fundamental feature is the permanent salinity stratification. This limits vertical convection, and consequently oxygenation of deep-water masses is weak and anoxic bottom areas are found. The salinity field shows a continuous decrease from oceanic levels at the North Sea boundary to freshwater in river mouths. The permanent circulation is weak, and sea ice is found for 5–7 months annually in the basin. The renewal time of the entire water mass is around 50 years.

The Baltic Sea has been central to North European nations throughout its lifetime (Figure 1.2). It has nine coastal countries, and at present (2008) there are about 85 million people living in its drainage basin. All of its coastal countries except Russia are members of the European Union, and three—Denmark, Finland and Sweden—are members of the Nordic Council. The Baltic Sea is one of the most actively and systematically investigated sea areas in the world. The fact that the marine environment of the Baltic Sea is highly vulnerable has been known for several decades. This has also activated close co-operation between the coastal countries, despite their different historical, political, and cultural backgrounds, to monitor the state of the sea and to find ways of protecting it. For this reason—and due to initiative of Finnish Academician Ilmo Hela, former Director General of the Finnish Institute of Marine Research—the Baltic Sea coastal countries signed the Baltic Sea Protection Agreement in 1974. According to this agreement all these countries are responsible for carrying out national monitoring programs, bringing the data to a common database, and enhancing protection of the Baltic Sea. These activities are co-ordinated by the Baltic Sea Commission or Helsinki Commission (HELCOM). In spite of the agreement the state of the Baltic Sea has become worse, the loading of nutrients and pollutants has increased, and the protection of "Our Sea" (the term used by countries bordering the sea) has become even more important.

Today the Baltic Sea faces major environmental problems. The human impact has increased the nutrient load and led to eutrophication, which has further increased oxygen consumption and worsened the state of the Baltic Sea. Fish populations are

Figure 1.2. Postal route across the Åland Sea in 1638–1895. Peasants in Eckerö, Åland designed a mail boat for the difficult drift ice conditions, to cross leads, new ice, thick ice, and ridged ice. This system ceased when steamboats were introduced to provide the service. (Watercolor painting by J.A.G. Acke, 1909, reproduced by permission of the Åland Museum of Art, Finland.)

accumulating heavy metals, and restrictions are now imposed on the use of the fish stock (in particular, the Baltic herring, cod, and salmon, which have been very important sources of nutrition). Ship traffic is increasing (in particular, oil transport), which means higher risks to the vulnerable Baltic Sea environment. The state of the Baltic Sea highlights an evident and immediate pollution problem. The impact of climate change is an additional potential threat on a centennial timescale. The predicted global climate change would increase the temperature of the Baltic Sea water and decrease the area and thickness of ice, and south of Finland the eustatic sea-level rise would overrun the opposite influence of land uplift. Other implications for the physical oceanography of the Baltic Sea (in particular, the effects on salinity and circulation) are uncertain.

The scientific literature on Baltic Sea physics is numerous and multilingual, but textbooks and monographs are few in number. Very recently a textbook was written in Finnish (Myrberg *et al.*, 2006). There are a few general books about Baltic Sea geology, biology, chemistry, and physics (Magaard and Rheinheimer, 1974; Voipio, 1981; Wulff *et al.*, 2001); although these books present most useful reviews, knowledge of the actual physics of the Baltic Sea remains quite limited, and how the laws of physical oceanography lead to an understanding of the physics of the Baltic Sea is not explained in a consistent way from the basics to modern achievements. In addition, one major Baltic Sea handbook was published in Russian (Terzieva *et al.*, 1992) and an extensive 50-year (1952–2005) time-slice handbook of Baltic Sea conditions was published in 2008 (Feistel *et al.*, 2008). Mälkki and Tamsalu (1985) presented a monograph on the dynamics of the Baltic Sea and Fonselius (1995) wrote an excellent monograph of the classical oceanography of the Baltic Sea. The impacts of global climate change on the Baltic Sea are presented by the BACC Author Group (2008). The textbook approach taken here is complementary to these reviews, monographs, and handbooks.

For the general theory of physical oceanography and sea ice geophysics, the main background references to the present work are Dietrich *et al.* (1963), Neumann and Pierson (1966), Holton (1979), Csanady (1982), Gill (1982), Apel (1987), Kowalik and Murty (1993), Cushman-Roisin (1994), Leppäranta (1998), Curry and Webster (1999), and Kundu and Cohen (2004).

In this book the principal goal and the overall idea is to present the physical oceanography of the Baltic Sea based on physical theory and observations (Figure 1.3). Field observations have been to some degree acquired by our own research teams, but most data have been taken from publications and public databases. In addition, satellite remote-sensing imagery is used to illustrate physical phenomena in the Baltic Sea. Mathematical models form a very powerful tool in physical oceanography and a very large variety of problems have been successfully examined by model simulations, such as circulation, vertical mixing, wind-generated waves, and sea ice dynamics. There is no modeling chapter in the book, but models are used in most chapters in specific modeling sections. Having models in separate sections emphasizes the value of fundamental theory and data. These stay but new models replace old models, basically allocating just a historical value to the old ones.

The book presents—as it is titled—the physical oceanography of the Baltic Sea including the physics of sea ice. It is targeted at the research community, Ph.D. students, and environmental protection authorities. The oceanography is presented from the basics, so that researchers coming from other fields such as physics, environmental sciences, ecology, or geography can use the book as a handbook or self-study material. The book may also be useful for general oceanographers since the Baltic Sea is a brackish water or low-salinity basin with peculiar, interesting oceanographic characteristics.

There are ten text chapters, a reference list, a list of study questions, an appendix containing a summary of the properties of brackish seawater and some useful constants and formulae, and an index. The first chapter (i.e., the current chapter) gives an introduction to our topic, and the second chapter describes the evolution of the Baltic Sea during the Holocene, a brief history of the research, and the geography of the region. The classical oceanography of the Baltic Sea is presented in Chapters 3–6: topography and hydrography[3] in Chapter 3; water, salt, and heat budgets in Chapter

---

[3] Temperature and salinity characteristics of water masses.

Figure 1.3. CTD-sounding (profiling electric conductivity and temperature as a function of depth) is about to commence onboard R/V *Aranda*. Field observations form the basis of knowledge in physical oceanography. (Reproduced with the permission from Ari O. Laine, M.Sc.)

4; and circulation and waves in Chapters 5–6. The theoretical background is also provided, introducing brackish water, air–sea interaction, mixing of water masses, ocean dynamics, and mathematical modeling methods. The physical system is described with observations. Model simulations are also included, trying to clarify the distinction between what is known from observations and where models can help us in filling gaps in our empirical knowledge. The wave chapter (Chapter 6) includes shallow and deepwater waves, internal waves, and also a brief section on tidal waves, which are however weak in the Baltic Sea.

Chapter 7 presents the ice of the Baltic Sea, a unique brackish ice, which forms every winter and is a major factor in the physics and ecology as well as in the practical life in this region. The structure, growth and melting, and drift of the ice are included. Chapter 8 focuses on coastal processes: upwelling, sea-level elevation, weather, and ice. Chapter 9 steps into the interface of physics and environmental questions, which is an extremely important research topic in the highly polluted Baltic Sea. The main questions are light conditions, the oxygen budget of deepwaters, extreme situations, and human impact. In addition, operational oceanography in the Baltic Sea is introduced. Chapter 10 concludes the book discussing the future of the Baltic Sea and the challenges facing future research. The list of references is extensive though it cannot be considered complete. There is also a set of study problems at the end of the book, largely based on our educational efforts in the University of Helsinki.

2

The Baltic Sea: History and geography

Stepan Osipovich Makarov (1848–1904) was born in Nikolaev (now Mykolayiv, Ukraine) in 1849 (1848 according to the Julian calendar). After Marine Academy he joined the Russian Navy in 1865 to serve in the Baltic Sea fleet. He became vice-admiral and commander of the Russian Navy, and he died in the Russo-Japanese war as his flagship *Petropavlovsk* struck a mine and sunk. Admiral Makarov directed two round-the-world oceanographic expeditions on the corvette *Vityaz* and designed and commanded the world's first icebreaker *Yermak* in two Arctic expeditions. He examined the hydrography and circulation in the Gulf of Finland and in connection with *Yermak* cruises also made ice observations in this basin. Admiral Makarov was a pioneering Russian oceanographer who was honored by the Russian Academy of Sciences. (© Reproduced from the Collections of the National Board of Antiquities, Finland, with permission.)

This chapter introduces the reader to the Baltic Sea starting with a short life history in Section 2.1. In 13,500 years the basin has undergone several different phases. The treatment in this book, however, focuses on the present brackish water Baltic Sea introduced in Section 2.2. Section 2.3 gives a brief history of Baltic Sea research, which first commenced in the late 1800s making it one of the oldest research basins in oceanography. The last section (Section 2.4) describes the climate and weather conditions in the Baltic Sea region, where influences are felt from the North Atlantic via westerly winds, from continental conditions on its eastern side, and from polar airmasses in the north.

2.1 EVOLUTION OF THE BALTIC SEA DURING THE HOLOCENE

The Baltic basin is an old depression in the Fennoscandian bedrock. It was covered by ice sheets during ice ages and filled with water in warmer periods. The water body formed after the Weichselian glaciation is referred as the Baltic Sea (Figure 2.1). The massive ice sheet induced a large further depression into the bedrock and the slow recovery has resulted in land uplift, which is still growing by up to 9 mm/year in the north. In the course of its history the Baltic Sea has undergone several different freshwater and brackish water phases, dictated by the land uplift and the eustatic rise in global sea-level (e.g., Winterhalter *et al.*, 1981; Eronen, 1990; Donner, 1995; Tikkanen and Oksanen, 2002). Many of the Baltic Sea phases have been named for species of mussels or mollusks that existed in that period.

The history of the Baltic Sea began 13,000–13,500 BP[1] (Tikkanen and Oksanen, 2002), when the ice sheet had retreated to the southern edge of the Baltic basin. First, a glacial lake was born, named the *Baltic Ice Lake* (12,600–10,300 BP). About 10,600 BP the ice edge had retreated to the southern coast of Finland, and the lake had grown to its maximum extent with outflow in the south (Figure 2.2) at the present-day sound of Öresund (see Figure 2.3). The retreat of the ice sheet slowed down for almost 600 years resulting in the formation of the long Salpausselkä moraine eskers across southern Finland (Figure 2.2). At that time there were people already living by the Baltic Ice Lake.

Due to climate warming the retreat of the ice sheet became faster about 10,300 BP, and land uplift continued. The southern outlet closed up and a new outlet, the Billingen gateway, across central Sweden formed. This gateway opened into the Närke strait (Figure 2.2) with a major natural catastrophe taking place: the surface of the Baltic Ice Lake dropped in 1–2 years by about 26 metres to meet the level of the North Sea (BACC Author Group, 2008).

The *Yoldia Sea* period (10,300–9500 BP) had begun (Figure 2.2). Oceanic water flowed into the Baltic Sea and the salinity increased. The Yoldia Sea became a brackish water basin, and new marine species came in from the North Sea (e.g., the *Yoldia arctica* mussel, which gave its name to this phase). The northernmost part of the present Baltic Sea, the Gulf of Bothnia (which consists of the Bay of Bothnia, the Sea of Bothnia, the Archipelago Sea, and the Åland Sea, see Figure 2.3) was still beneath the ice sheet. Then after about 800 years the channel to the North Sea closed up, because land uplift was faster than eustatic sea-level rise. Then Lake Vänern was

---

[1] BP is years before present (1950) based on radiocarbon dating. Calibrated, actual years are denoted by cal. years BP. These timescales overlap 3,000 years ago and 11,000 cal. years BP ≈ 10,000 years BP (abbreviated throughout to BP).

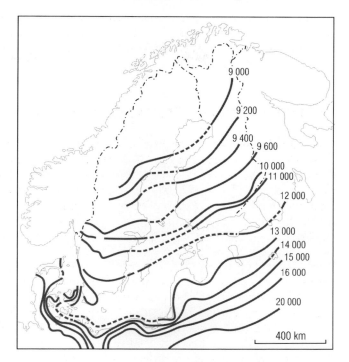

Figure 2.1. The location of the edge of the Weichselian ice sheet retreating north (years BP).

formed from the Närke Strait, with River Göta Älv as its outlet to the Kattegat (see Figures 1.1 and 2.3).

A lake phase followed (9500–8000 BP), called *Ancylus Lake*, the name coming from the *Ancylus fluviatilis* mollusk. In the beginning the Bay of Bothnia, the northernmost basin of the Baltic Sea, was still under the shrinking ice sheet, but during this phase the ice sheet retreated to the northern Scandinavian mountains. The glacial meltwater and river inflow provided a positive water balance. Land uplift was stronger in the north than in the south, and therefore a new outlet opened in the south. This was River Dana flowing along the route of what is now Darss Sill–Great Belt, the main channel between the North Sea and the present Baltic Sea. In this region land uplift had almost ceased, while the eustatic sea-level was rising by about 10 mm per year. Consequently, River Dana developed into a strait, and saline water could again enter the Baltic Sea. The salinity of the water started to increase slowly, and a transition period, now termed the *Mastogloia Sea* (named for *Mastogloia* diatoms), followed around 8000–7500 BP.

The transition period ended in the *Litorina Sea* phase (7500–4000 BP) named for the *Littorina* mollusk, a saline water species. During this phase and later the evolution of the Baltic Sea was less catastrophic, slowly changing salinity being dictated by water exchange with the North Sea and slowly decreasing surface area due to land uplift. The salinity of the Litorina Sea was somewhat higher than at present, 13‰ in the central region and 8‰ in the north. Land uplift was strong in the Gulf of Bothnia,

12 The Baltic Sea: History and geography   [Ch. 2

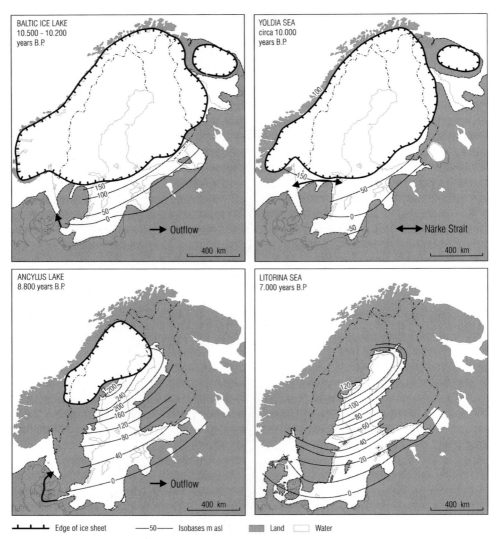

Figure 2.2. The phases of the Baltic Sea, Fennoscandian ice sheet, and land uplift: Baltic Ice Lake, Yoldia Sea, Ancylus Lake, and Litorina Sea (according to Eronen and Haila, 1990). The present Baltic Sea is shown by the dotted line in all phase charts. The curves with numbers are isobaths showing the current altitudes with respect to the sea-level.

and at the start of the Litorina Sea phase the coastline in the Finnish lowlands was 100 km inland from the present location.

The *Limnea Sea* phase (named for the *Limnea* mollusc) covers the period from about 4000 BP to 2000 BP, and since then the water body has been what is now the *present Baltic Sea*. An exception is the Finnish western lowlands, where land uplift of

Sec. 2.1]   Evolution of the Baltic Sea during the Holocene   13

Figure 2.3. The Baltic Sea and its sub-basins (with small modifications from Fonselius, 1995).

10 m to 20 m in 2,000 years has created large areas of new land as reflected by the locations of pre-historical coastal villages, which are now far inland. In the warm period at about 4000 BP nearly all glaciers disappeared from the Scandinavian Mountains, but thereafter grew again, achieving their maximum extent about 400 years ago.

In the text below the name "Baltic Sea" refers to this "present Baltic Sea" unless otherwise explicitly expressed. The salinity of the present Baltic Sea is slightly lower than in the *Limnea Sea* and the surface area is slightly smaller. In its ecosystem can be found relict species from both fresh and more saline earlier phases. The surface area of the Gulf of Bothnia is decreasing by close to 10,000 km^2 (10%) every 2,000 years, while that of the whole Baltic Sea is shrinking by 2% every 2,000 years. The corresponding volume decrease of the whole Baltic Sea would be about 4% per 2,000 years. In physical oceanography the timescale of interest is normally less than 100 years; so, for the last 100 years or for corresponding future projections the size of the Baltic Sea can be considered as fixed.

Table 2.1. The principal characteristics of the Baltic Sea (the open ocean boundary is taken as the southern edge of the Kattegat).

Quantity	Value
Surface area	392,978 km^2
Mean depth/maximum depth	54.0/459 m
Volume	21,205 km^3
Surface area of drainage basin	1,633,290 km^2
Age	7,500 years as a brackish basin, 2,000 years at the present salinity level
Actual land uplift[a]	0 to 9 mm/year
Water renewal time	50 years
Mean salinity[b]	7.4‰ (1/5 of normal ocean salinity)
Salinity range	0–32‰
Salinity in the Gotland Sea (surface/bottom)	6–7‰/11–13‰
Annual ice extent	12.5%–100% of the total area of the Baltic Sea
Annual ice season	5–7 months
Primary production	30–250 g C m^{-2} yr^{-1}
Number of coastal countries	9
Number of countries in the drainage basin	14
Population in the drainage basin	85 million people

[a] Observed (apparent) land uplift = actual land uplift minus eustatic rise in global mean sea-level.
[b] The level has varied within 7‰–8‰ during the last 100 years (Winsor *et al.*, 2001); Meier and Kauker (2003a) give 7.4‰ based on model calculations.

2.2 THE PRESENT BALTIC SEA

The present Baltic Sea is classified as a small, intra-continental sea[2] of the Atlantic Ocean (Figure 2.3; Table 2.1). Its outermost part consists of three parallel straits: the Little Belt, the Great Belt, and the Öresund (sometimes called the "Sound" in English

---

[2] In older literature intra-continental seas have also been called "small mediterranean seas" (see Neumann and Pierson, 1966).

language texts), and together they constitute the *Danish Straits*. After the boundary comes the Kattegat,[3] which is the transition zone between the Baltic Sea and the North Sea and between brackish and oceanic watermasses. In the north, the Kattegat extends to Skagerrak, which is the outermost part of the North Sea and has an oceanic watermass.

The geographical division of the Baltic Sea is based on coastal morphology, sills, and other topographical formations. There are 14 principal parts (Figure 2.3). Starting from the oceanic side, the first parts are the *Öresund Strait* and the *Belt Sea*, which consists of the Great Belt, the Little Belt, Mecklenburg Bight, and Kiel Bight. Then in the south there are two small basins: the *Arkona Basin* between Sweden and Germany and the *Bornholm Basin* between Sweden and Poland. These basins with the Belt Sea and Öresund constitute the southwestern Baltic Sea. There then follows the *Gotland Sea*, which contains half of the Baltic Sea watermass and which is subdivided into the *Western*, *Eastern*, and *Northern Gotland Basins* and *Gdansk Bay*. The deepest point in the Baltic Sea is the Landsort Deep (459 m) in the Western Gotland Basin, southeast of Stockholm. The Gotland Sea, Arkona Basin, and Bornholm Basin form the Baltic Sea proper.

East and north of the Gotland Sea there are three major gulfs: the *Gulf of Riga*, the *Gulf of Finland*, and the *Gulf of Bothnia*. The Gulf of Bothnia is the northernmost part of the Baltic Sea, surrounded by Finland and Sweden. It is a large water body consisting of four basins: the *Åland Sea*, the *Archipelago Sea*, the *Sea of Bothnia*, and the *Bay of Bothnia*. The Gulf of Riga is surrounded by Estonia and Latvia and is relatively isolated from the Gotland Sea. The Gulf of Finland is an eastward elongated basin with no sill toward the Gotland Sea surrounded by Estonia, Finland, and Russia.

The surface area of the Baltic Sea is 392,978 km^2 and the mean depth is 54 m. The North Sea is considerably larger (750,000 km^2) and deeper (95 m). In the Atlantic Ocean the surface area is 88 million km^2 and the mean depth is 3,300 m, while in the (European) Mediterranean Sea these numbers are 2.5 million km^2 and 1,500 m. Thus, the surface area and mean depth of the Baltic Sea are just 0.5% and 1.7% of the Atlantic dimensions or 16% and 3.6% of those of the Mediterranean Sea. Land uplift is still rather strong in the Baltic Sea region, non-existent in the south but up to 9 mm per year in the north. The freshwater budget is strongly positive due to river discharge, and the watermass is consequently brackish. Salinity decreases from 25‰ in the Danish Straits to zero in river mouths, and the water renewal time is some 50 years. Occasionally, during strong inflow events, salinity has been as high as 32‰ in the Danish Straits. The location of the Baltic Sea is at the edge of the seasonal sea ice zone of the World Ocean, and climate variations show up strongly in the ice season. The length of the ice season is 5–7 months, and the maximum annual ice extent has ranged within 12.5%–100% of the surface area of the Baltic Sea. The corresponding averages are 6.4 months and 45%.

---

[3] Sometimes the Kattegat is included as part of the Baltic Sea: the surface area, mean depth, and volume of the Baltic Sea would then be 415,265 km^2, 52 m, and 21,715 km^3, respectively. The drainage area of the Kattegat is 86,980 km^2.

Nine countries (Denmark, Estonia, Finland, Germany, Latvia, Lithuania, Poland, Russia, and Sweden) have coastlines in the Baltic Sea, and apart from Russia all belong to the European Union. The drainage basin area covers about four times the area of the Baltic Sea and includes five more countries (Belarus, Czech Republic, Norway, Slovakia, and Ukraine) in addition to the coastal countries. The population of the drainage basin totals some 85 million.

2.3 RESEARCH HISTORY

2.3.1 Birth of Baltic Sea research

The oldest regular oceanographic measurements were made for sea-level elevation in the Baltic Sea. The first records began in 1703 in St. Petersburg, but the longest continuous time-series began in 1774 in Stockholm (Ekman, 1999). Accordingly, quantitative knowledge of land uplift in the Baltic Sea has been available for a long time. In addition, at several sites a time-series on ice was commenced at much the same time, providing valuable data for present-day historical climate investigations.

Observational programs grew gradually in the 1800s. At that time only four countries surrounded the Baltic Sea: Denmark, Germany, Russia, and Sweden. In 1834 the first documented scientific observation of upwelling in the Baltic Sea was carried out by Alexander von Humboldt (Kortum and Lehmann, 1997). The first hydrographic survey was made in 1871 on the German R/V *Pommerania*, a paddle-steamer (Meyer *et al.*, 1873; Smed, 1990). The cruise covered the Baltic Sea as far north as Stockholm. In summer 1877 Swedish chemist F. L. Ekman[4] conducted an extensive expedition on the gunship *Alfhild* to the Baltic Sea, the Kattegat, and Skagerrak collecting about 1,800 records of temperature and salinity from different depths. The results were later analyzed by the Swedish chemist Otto Pettersson (Ekman and Pettersson, 1893). The expedition covered nearly the entire Baltic Sea area providing the first comprehensive view of its hydrography. It was found, for example, that there is a layer of minimum temperature just above the halocline in summer (Smed, 1990; Fonselius, 2001), referred to today as the dicothermal layer. This expedition was the most important and detailed oceanographic investigation in the Baltic Sea in the 19th century. About the same time K. E. von Baer and M. G. C. C. Braun conducted hydrographical and biological investigations in Estonia.

In the late 19th century (Dr.Habil. Wolfgang Matthäus, pers. commun.), a center for Baltic Sea research developed in Kiel owing to the interdisciplinary cooperation of the well-known German marine scientists Gustav Karsten (1820–1900), Kurt Brandt (1854–1931), Otto Krümmel (1854–1912), Karl Möbius (1865–1908), and Victor Hensen (1835–1924). The pioneer of progressive oceanography in Germany, Otto Krümmel, summarized the status of knowledge on the physics of the Baltic Sea in his book (Krümmel, 1907).

---

[4] Father of V. W. Ekman, a famous physical oceanographer.

Regular ice observations and hydrographic station data collection started to expand in the 1880s. Ice monitoring was closely connected with the development of winter navigation in ice-covered waters, made possible by steamships. More oceanographical cruises were performed late in the 19th century, and in 1890–1893 Swedish, German, and Danish oceanographers started to carry out monitoring cruises four times a year—a standard procedure still today. In 1894 Admiral Stepan Makarov from St. Petersburg published a study of the hydrography and circulation of the Gulf of Finland. The first Finnish cruise was done in 1898 on the S/S *Suomi*. In the same year an agreement was made at a Nordic conference about a hydrographic measurement program in the Baltic Sea, initiated by Theodor Homén (1858–1923) from Finland, Martin Knudsen (1871–1949) from Denmark, and Otto Pettersson (1848–1941) from Sweden (Simojoki, 1978).

In 1899 the Swedish government invited a number of European governments to the Preparatory Conference for the Exploration of the Sea, to be held in Stockholm in 1899. This was due to the activity of Otto Pettersson and Gustaf Ekman who had gone to visit King Oscar II of Sweden and Norway and asked him to sign the invitation to the Conference. The response of the King was positive. The Second Preparatory Conference was held in Kristiania (since 1925 called Oslo) in 1901 and finally the International Council for the Exploration of the Sea (ICES) was founded in Copenhagen in 1902 by its eight founder member countries (Smed, 1990). Finland, then an autonomous grand duchy of Russia, could participate as one member. Through ICES, co-operation in hydrographic and marine fishery investigations increased considerably (Dybern and Fonselius, 2001). The founding of ICES stimulated regular reporting and publishing of oceanographic observations, and data were collected at fixed stations, including some in the Baltic Sea where there were the most important deep stations (Bornholm Deep, Gotland Deep, Landsort Deep, Åland Deep, Ulvö Deep, Luleå Deep, etc.). Measurements started again after World War I, but with only a few sampling stations (Dybern and Fonselius, 2001).

With the onset of oceanographic research of the Baltic Sea, several outstanding pioneering oceanographers worked in the region, the first being Swedish professor Otto Pettersson. Rolf Witting (1879–1944) made hydrographical surveys and collected current measurements on lightships (Figure 2.4). A step forward in the early theoretical considerations was taken when Martin Knudsen presented in 1899 his famous model of the water exchange between the Baltic Sea and the North Sea and reported the irregular saltwater inflows into the Baltic Sea, known presently as "Major Baltic Inflows" (Knudsen, 1900). Knudsen also set up the relationship between wind, air pressure, and water exchange in the transition area between the North Sea and the Baltic Sea. Knudsen's theories are still used today. In 1910 Vagn Walfrid Ekman moved from Kristiania to the University of Lund in Sweden and had an important influence on Baltic Sea research there (e.g., by developing oceanographic measurement techniques). The research station Bornö, in Skagerrak, was built in 1902 by Otto Pettersson and Gustaf Ekman and became a major site for Swedish oceanographical research.

18   The Baltic Sea: History and geography   [Ch. 2

Figure 2.4. Lightship *Storkallegrund* in its site on the sea route to Vaasa 2.5 nautical miles from Judastenarna shoal. This ship served until 1957. The first current charts of the Baltic Sea were constructed on the basis of lightship measurements (Witting, 1910). (Reproduced with permission from the Nautical Museum of Finland.)

2.3.2   From World War I to 1950

After the First World War the political geography in the Baltic Sea region was much as it is now. Oceanographic research was steadily expanding. The observational time-series had become several decades long, and a realistic picture of the Baltic Sea system could be developed. Time-series analysis of the hydrography was published (Granqvist, 1938b), and a thorough analysis of the circulation of the northern part was made by Palmén (1930). Erik Palmén (1898–1985) was a well-known meteorologist, who worked in the Finnish Institute of Marine Research between 1922 and 1947. In 1914 the Oceanographic Institution at the University of Gothenburg (Göteborg) was founded and in 1938 it was moved to Stigbergstorget in Gothenburg. In this institute many studies have been carried out but especially so at the entrance area of the Baltic Sea. The son of Otto Pettersson, Hans Pettersson (1888–1966) was the first professor there, and he initiated research into the marine optics of the southwestern Baltic Sea. Nils Jerlov (1910–1990) continued this work and became a lead scientist in this field.

Sec. 2.3] **Research history** 19

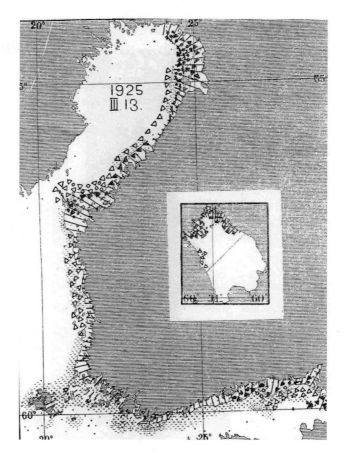

Figure 2.5. Sea ice chart on March 13, 1925 over the northern part of the Baltic Sea and Lake Ladoga. The insert shows Lake Ladoga which is located east of the Gulf of Finland (see Figure 1.1). In the 1920s there was no access to observe the ice conditions farther out from the boundary of landfast ice and the drift ice fields had unknown properties Granqvist (1926). (Printed with permission from the Finnish Institute of Marine Research.)

Pioneering general findings have been made during Baltic Sea oceanographic research. Gustafsson and Kullenberg (1936) published their observations of inertial oscillations in the Gotland Sea; this was the first time observations of inertial oscillations were reported. Between 1925 and 1938, the German R/V *Poseidon* was very active in the Baltic Sea and basic observations were carried out during ten cruises throughout the whole Baltic (Schulz, 1956). The first German center for Baltic Sea research—the Institute of Marine Sciences at the University of Kiel—was founded in 1937 (Krauss, 1990).

Sea ice charting developed into an operational system, a necessity for winter traffic (Figure 2.5). The Baltic Ice Code and chart symbols were standardized in the 1920s, and Risto Jurva (1888–1953) developed a time-independent cartographic method for description of the evolution of ice seasons (Jurva, 1937a). A major international program "Baltic Ice Week" was planned for the winter 1938 (Granqvist, 1938a), but unfortunately this winter was very mild with ice only in the very north and east.

In World War II the Baltic Sea was strategically a key area on the eastern front. In this period two very severe winters occurred with the whole Baltic Sea frozen over. Erkki Palosuo (1912–2007), a pilot in the Finnish Air Force, performed ice reconnaissance flights across the Baltic Sea proper; he later used the data for his Ph.D. thesis (Palosuo, 1953). In 1947 the Baltic Sea once again froze over, but thereafter this situation has so far not been repeated. There was some progress in research during the 1940s concerning the transparency of the waters, circulation, sea-level (Hela, 1944), and energy balance. Brogmus (1952) published the outcome of his fundamental research of the hydrography of the Baltic Sea, one of the main references in this topic.

After World War II the political geography of the Baltic Sea experienced a major change. The eastern side (except Finland) was under Warsaw Pact control and the Baltic countries (Estonia, Latvia, and Lithuania) were under Soviet Union control. The southwestern corner of the Baltic Sea was in the North Atlantic Treaty Organization (NATO), while Finland and Sweden in the north formed a neutral territory. These political boundaries made gaining access to oceanographic stations difficult.

2.3.3 Period of rapid development—from the 1950s to 1980s

In all Baltic Sea countries, research activity steadily increased in the 1950s, but due to political restrictions there were difficulties in carrying out basin-wide investigations. Dietrich *et al.* (1963) showed that the geostrophic method is applicable in the Baltic Sea despite the shallow depths. The first numerical circulation models were constructed (e.g., Svansson, 1959; Uusitalo, 1960). Automatic measurement systems became available that could obtain a fine-resolution time-series of hydrographical and dynamical characteristics. Marine optics research continued in the southern Baltic Sea by Danish, Polish, and Swedish oceanographers (Jerlov, 1976; Højerslev, 1988; Dera, 1992).

As a result of developments after World War II a further center for investigation of the Baltic Sea developed in Warnemünde in the 1950s—the Institute of Marine Research of the Academy of Sciences, German Democratic Republic, today called the Baltic Sea Research Institute (IOW, *Leibniz-Institut für Ostseeforschung Warnemünde*) (Brosin, 1996). During the Cold War, the border between the two political and military blocs limited Baltic Sea research. Despite this complicated situation, ICES took the first step to resume co-operation in Baltic Sea research after World War II. In 1957, the CBO was initiated by the Finnish oceanographer Ilmo Hela (1915–1976) with the assistance of ICES (Matthäus, 1987). During the 3rd CBO in 1962, Erich Bruns (1900–1978) proposed an international investigation, and in August 1964 the first joint research program in the Baltic Sea after World War II was carried out. This joint research program was the starting point for the successful collaboration of the marine scientific community in the Baltic region resulting finally in close cooperation within the framework of the Helsinki Convention for the Protection of the Baltic Marine Environment. In the 1950s and 1960s, the German contribution mainly focused on the water exchange between the North Sea and the

Baltic Sea (e.g., Dietrich, 1950, 1951; Wyrtki, 1953, 1954) and internal movements of the Baltic Sea (e.g., Krauss, 1966, 1972).

In 1969 Finland and Sweden decided to keep the main harbors open all year, right up to Kemi and Luleå on the northern coast of the Gulf of Bothnia. This meant the construction of powerful icebreakers and major investment in research into Baltic Sea ice. The Finnish–Swedish Winter Navigation Research Board was founded to coordinate and finance sea ice science and technology. Much of the science funding was allocated to the development of sea ice–forecasting models and remote-sensing methods and to investigations of sea ice ridges. As a result, real-time ice forecasting based on a numerical sea ice model was commenced in 1976 (Leppäranta, 1981a). Remote-sensing methods were extensively investigated in the Sea Ice '75 experiment in 1975 (Blomqvist *et al.*, 1976), and NOAA (National Oceanographic and Atmospheric Administration) satellite imagery was taken into operational use for ice charting around 1980. Sea ice ridges, the most difficult obstacles for winter shipping in the Baltic Sea, were examined in the field from the late 1960s (Palosuo, 1975), and a detailed ice atlas was published based on digitized ice charts (SMHI and FIMR, 1982).

In 1969–1970 the International Baltic Year (IBY) was organized. This planned for national cruise programs to be co-ordinated so that a series of 39 stations, called BY-stations, in the Baltic Sea proper and the Gulf of Finland were sampled every month by one participating ship. Altogether 12 research vessels participated. In 1974 the ICES/SCOR (Scientific Committee on Oceanic Research) Working Group formulated the future research tasks that were considered important for better understanding of the Baltic Sea system in relation to pollution. The Baltic Open Sea Experiment (BOSEX) was performed in September 1977 in the Baltic Sea proper, south of Gotland. Measurements were carried out in a 30 km square with 11 research ships and scientists from 14 institutions representing all the Baltic countries (ICES, 1984). An important topic in the ICES/SCOR Working Group was the spatial heterogeneity ("patchiness") of hydrographic and biological quantities with related effects on monitoring and trend investigations. In 1986 the Baltic Sea Patchiness Experiment (PEX-86) was carried out south of Gotland by 12 ships. The project was followed by a series of publications (ICES, 1994).

The state of Baltic Sea waters had become worse over the course of the years due to pollutants and nutrient loads. To prevent further worsening of the situation, the Baltic Sea countries signed the Agreement for the Protection of the Baltic Sea in 1974 in Helsinki. The Baltic Sea Commission or the Helsinki Commission (HELCOM) was established in 1980, with Professor Aarno Voipio as its first Secretary General (*http://www.helcom.fi/*). This body runs a continuous Baltic monitoring program (BMP), in which all Baltic Sea countries participate.

Air–sea interaction investigations became an important topic in the 1970s (Mälkki, 2001). Professor Sergey A. Kitaigorodskii visited Finland frequently and had a major impact in this field (e.g., Kitaigorodskii and Mälkki, 1979). Atmospheric boundary layer studies were carried out by Launiainen (1979) in the Gulf of Finland and by Joffre (1981) over the ice cover in the Gulf of Bothnia. Kahma (1981a) studied the growth of the wave spectrum in fetch-limited conditions in the Sea of Bothnia.

Figure 2.6. Deployment of instrumentation in the Baltic Sea aboard R/V *Aranda* ("Old Aranda"), which served Baltic Sea oceanographic research between 1953 and 1988. (Courtesy of the Finnish Institute of Marine Research.)

Along with the development of measurement technology, knowledge of the dynamics of the Baltic Sea advanced to a new level (Figure 2.6). Fonselius (1962, 1967, 1969) published a number of papers related to the hydrography of the Baltic Sea. For example, Krauss (1981) investigated the processes influencing thermocline erosion, while Aitsam and Elken (1982) studied the meso-scale[5] variability of the hydrophysical fields of the Baltic Sea. In addition, studies into Major Baltic Inflows advanced (e.g., Matthäus and Franck, 1988).

The development of mathematical models progressed in leaps. Vertically integrated circulation models for sea-level simulations were studied by, for example, Wübber and Krauss (1979) and Häkkinen (1980) and further two-layer models were developed by Welander (1974). Diagnostic models were examined by Sarkisyan *et al.* (1975) and three-dimensional fine-resolution models by Simons (1978) and Kielmann (1981). A model for the exchange of water and salt between the Baltic and the Skagerrak was developed by Stigebrandt (1983) and a model of the vertical circulation of Baltic deepwater was presented by Stigebrandt (1987). A one-dimensional second-order turbulence model (presently well-known as the program "PROBE") was used as a tool to examine mixing and temperature structure in the water column (Svensson, 1979; Omstedt *et al.*, 1983).

---

[5] Meso-scale variability is 10 km to 100 km in space and a few days in time.

2.3.4 Modern research—from 1990 to the present

We do not attempt to write the history of the time in which we are currently living. This section here lists the main research topics and programs in the Baltic Sea since 1990 without judging their significance to the progress of science. The years around 1990 marked an important milestone, during which the former research of single processes or phenomena expanded to complex studies of multidisciplinary problems, and the question of climate change impact spread to all geosciences. Additionally, the political geography of the Baltic Sea region once more experienced a major change, computer-based forecast of oceanographic fields became feasible and reliable, and cost analyses started to impact on the direction and overall control of Baltic Sea research.

Characteristic of modern time research work in the Baltic Sea is the presence of extensive programs often involving several countries and institutions. The SKAGEX-project in the Skagerrak–Kattegat area was carried out in its many phases in the early 1990s by means of 17 participating ships (Danielssen *et al.*, 1997). The Gulf of Bothnia Year was carried out in 1991 by Finland and Sweden (Perttilä and Ehlin, 1993) and the Gulf of Finland Year in 1996 by Estonia, Finland, and Russia (Sarkkula, 1997). The Gulf of Riga project—an environmental research program—took place in the years 1993–1997 (Anon., 1999). EU funding of Baltic Sea research has also grown in modern times as the EU has expanded more and more to the Baltic Sea region.

The BALTEX (Baltic Sea Experiment) program (*http://www.baltex.de/*) commenced in 1993 and is still ongoing in 2008; its focus is in water and heat cycles and consequently has a strong meteorological component (Raschke *et al.*, 2001; Omstedt *et al.*, 2004). The BALTEX community also organized several scientific workshops from 1995 onwards (Figure 2.7) and recently published a comprehensive analysis of our present understanding of climate change in the Baltic Sea area (BACC Author Group, 2008). Between 1994 and 1998 the Baltic Sea System Study (BASYS, see *http://www.io-warnemuende.de/Projects/Basys/*) was carried out, an umbrella study for a large number of mainly national projects in climatology, hydrology, marine biology, paleobiology, etc. with participating institutes from all the Baltic Sea countries. The BASIS project (Baltic Air–Sea–Ice Study, see, e.g., Launiainen *et al.*, 2001) took place between 1997 and 2000 and involved extensive field studies to improve coupled air–sea–ice models. An experimental study (DIAMIX) was devoted to studying mixing processes: how energy input to the sea propagates down and dissipates. The study revealed small-scale dynamic features such as internal waves, coastal dynamics, and deepwater eddies (Stigebrandt *et al.*, 2002).

Three-dimensional models were further developed in the 1990s (see, e.g., Lehmann, 1995) and came up with results concerning circulation and transport with relatively high spatial and vertical resolution. Model investigations of the impact of climate change on conditions in the Baltic Sea were performed by Meier (2006) among others. Eutrophication of the Baltic Sea became a strong motivation for multidisciplinary research in the 1990s. Three-dimensional hydrodynamic–ecological models were developed and nutrient load scenario runs were carried out (Neumann *et*

Figure 2.7. Group photograph of the participants of the 5th Study Conference on BALTEX in Kuressaare, Estonia, 4–8 June 2007. The BALTEX conferences are held every third year in Baltic Sea islands and form an important forum for international research collaboration. (Photograph by Dr. Marcus Reckermann, BALTEX Secretariat, printed by permission.)

al., 2002). Recently, present knowledge of ecosystem modeling in a wide perspective was summarized by Fennel and Neumann (2004). The operational algae-monitoring program Alg@line based on automatic ship measurements started and became an important tool in monitoring algal blooms (Rantajärvi, 2003). Our existing thorough knowledge of the dynamics of the Baltic Sea has made it possible to identify and to some extent quantify the contribution of ship wakes to hydrodynamic activity (Erm and Soomere, 2004; Soomere, 2006).

Sea ice research progressed by dint of climatological modeling and remote sensing. Coupled ice–ocean models were constructed to facilitate long-term simulation to evaluate the influence of climate change on the ice season (Haapala and Leppäranta, 1996, 1997). Major remote-sensing programs BEPERS (Bothnian Experiment in Preparation for ERS-1) and BEERS (Baltic Experiment for ERS-1) were performed between 1986 and 1994; they were organized jointly by Finland, Germany, and Sweden (Askne *et al.*, 1992; Leppäranta *et al.*, 1992). Sea ice ecology and the connections between physics and ecology resulted in extensive research activity, as the brackish sea ice in the Baltic Sea has biota very similar to polar sea ice (Ikävalko, 1997).

Pan-European co-operation is ongoing and active in operational oceanography under EuroGOOS—an association of agencies—founded in 1994 to reach the goals set by GOOS (the Global Ocean Observing System) and in particular the development of operational oceanography in European seas and adjacent ocean basins. In the Baltic Sea the BOOS (Baltic Operational Oceanographic System) is an association of institutes from all the Baltic Sea countries, each one taking national responsibility for operational collection of marine observations, forecasts, and dissemination of information and services including warnings (see, e.g., Gorringe and Håkansson, 2005). Real-time data collection using weather and wave buoys commenced in the 1990s, as did operational wave forecasting. Many of the efforts in operational oceanography today are supported by forcing from the HIRLAM (High Resolution Limited Area Model).

Research into Baltic Sea oceanography is advancing steadily. Extensive international research programs continue and new ones are born (e.g., Global Monitoring of Environment and Security of the European Area, GMES). The health of the Baltic Sea is of great concern to all the countries in the drainage basin. Co-operation between scientists is carried out today through many international projects and through scientific meetings organized by ICES, as well as through the Baltic Sea Science Congress (BSSC) series, which is jointly organized by Baltic Marine Geologists (BMG), Baltic Marine Biologists (BMB), and the Conference of Baltic Oceanographers (CBO). The history of the CBO has been summarized by Matthäus (1987) and by Dybern and Fonselius (2001).

2.4 WEATHER AND CLIMATE IN THE BALTIC SEA REGION

2.4.1 General regional climate

The Baltic Sea is located between maritime temperate and continental sub-Arctic climate zones, in the latitude–longitude box 54°N–66°N × 9°E–30°E. The climate has large variability due to the opposing effects between the moist and relatively mild marine air flows from the North Atlantic and the Russian continental climate. The Baltic Sea climate varies depending on the exact location of the polar front[6] and the strength of westerlies, and both seasonal and interannual variations are considerable. Westerlies are important particularly in winter, when the temperature difference between marine and continental airmasses is large.

The intensity of westerlies is described by the NAO (North Atlantic Oscillation) index. The NAO is defined as the air pressure difference between the Azores (Ponta Delgada) and Iceland (Stykkisholmur). The NAO index is positive when there is high pressure in the south and low pressure in the north, at which time relatively mild westerly winds prevail and winters are typically much warmer than on average over most of Europe (Figure 2.8). The Baltic Sea lies along the path of the North Atlantic storm track and therefore low-pressure systems frequently come into the Baltic Sea region bringing warm airmasses and reducing the temperature difference between northern and southern latitudes. When the NAO index is negative there is high pressure in the north and low pressure in the south (Figure 2.8), at which time the winds blow from northerly and easterly directions and mean wintertime temperatures are much below normal.

---

[6] Separating the polar air of the north and the warmer air of mid-latitudes.

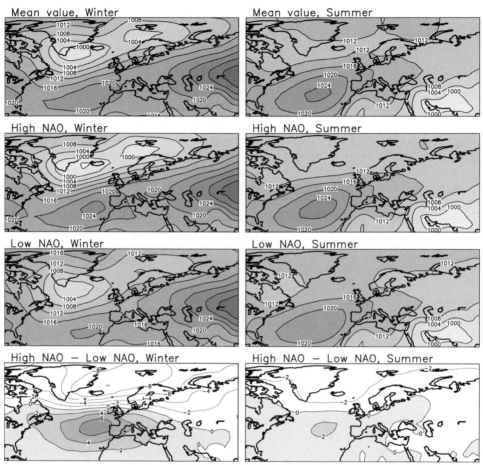

Figure 2.8. Sea-level pressure in 1901–2000 from the Hadley Centre simulation HadSLP2 based on Allan and Ansell (2006). Top row: Mean sea-level pressure in winter (December–February) and summer (June–August). Second and third rows: Composite pressure fields for winters and summers with high and low NAO indexes. Bottom row: the difference between high-NAO and low-NAO cases. The NAO index was calculated as the difference between normalized seasonal mean pressure anomalies at (40°N, 20°W) and (65°N, 20°W). The high-index and low-index composites include cases that deviated more than one standard deviation from the mean; the number of cases was 1,619. (Prepared by Dr. Jouni Räisänen, Department of Physics, University of Helsinki, printed with permission.)

The southern and western parts of the Baltic Sea belong to the central European mild climate zone and have a westerly circulation. The northern part locates at the polar front where the winter climate is cold and dry due to outbreaks of cold.[7] In

---

[7] Sometimes a very cold, Arctic airmass flows into the Baltic Sea area from the Arctic Ocean. The Arctic front is located between polar and Arctic airmasses.

terms of classical meteorology, during winter the polar front fluctuates over the Baltic Sea region, but during summer it is located farther to the north. The central part of the Baltic Sea can be in the mild or cold side of the polar front depending on the particular year. The temperature difference between winter and summer is much larger in the north. During summer the sea surface temperature is 15°C–20°C in the whole Baltic Sea; during winter in normal years the Gulf of Bothnia, the Gulf of Finland, and the Gulf of Riga are ice-covered with maximum ice thickness more than half a meter whereas the surface temperature is 2°C–3°C in the south. During warm summers and cold winters the air pressure field is smooth and winds are weak, and blocking high-pressure situations are a common feature. During such a period the weather can be very stable for several weeks.

Table 2.2 presents climatological data from three stations in the Baltic Sea: Kemi on the northern coast, the island of Utö in the southern Archipelago Sea at middle latitudes of the Baltic Sea, and Świnoujście on the shore of the Szczecin lagoon in Poland. These data present an overall picture to help understand the annual cycle of weather conditions in the Baltic Sea and will later be used in the discussion of the heat budget of the Baltic Sea.

The thermal "memory" of the Baltic Sea is some 2–3 months. Although the water volume is small, the strong seasonal cycle allows the basin to store much heat in the summertime for release in winter. Under fall and winter conditions, high-latitude seas are characterized by strong sensible and latent heat fluxes to the atmosphere as long as the surface is ice-free and by the transport of warm air from lower latitudes through intense cyclone activity. This transport is especially strong for central and northern Europe since the North Atlantic Current brings warm water and adds heat to the atmosphere. As a result the mean atmospheric temperature in the Baltic Sea region is much higher than anywhere else at corresponding latitudes. There are no other areas in the world where agriculture is possible at latitudes higher than 60°.

Wind conditions in the Baltic Sea region are determined by the general atmospheric circulation of Northern Europe. However, the vertical structure of the atmospheric surface layer over the Baltic Sea modifies the air–sea transfer of momentum, heat, and moisture. This layer has a clear seasonal dependence. In autumn and winter the sea is warmer than the air, and the marine boundary layer is unstable and winds are strong. Turbulent fluxes are strong and the sea cools rapidly. In spring and early summer the air is warmer than the sea, while winds usually remain weak. Turbulent fluxes are weak and directed downwards. Cloud coverage has an important role in the radiation balance and consequently in the level of the atmospheric temperature. Especially during winter, if the sky is clear, air temperatures can in the north reach very low values down to −40°C.

2.4.2 Solar and terrestrial radiation

The radiation budget is the governing background forcing on the seasonal course of physical conditions in the Baltic Sea. The monthly mean solar radiation peaks to around $200\,W\,m^{-2}$–$250\,W\,m^{-2}$ in June but almost vanishes in December, while the

Table 2.2. Climatological data from the period 1961 to 1990 for three Baltic Sea weather stations: Kemi airport (65°47′N, 24°35′E), Utö (59°47′N, 21°23′E), and Świnoujście (53°52′N, 14°10′E). Kemi and Utö data are from FMI (1991) and Świnoujście sources from Mietus and Owczarek (1994) for temperature, precipitation, and wind speed, Kwiecień (1987) for cloudiness (1951–1970) and humidity (1956–1980), and Mietus (1998) for wind direction.

Kemi	Temperature	Precipitation	Cloudiness	Humidity	Wind	
	(°C)	(mm)	(1/8)	(%)	(m/s)	(direction)
January	−12.4	32.3	5.6	86	3.7	165
February	−11.5	25.8	5.7	86	3.7	171
March	−6.9	28.7	5.5	84	3.8	168
April	−0.5	26.1	5.3	77	3.9	153
May	6.1	30.3	5.1	70	3.8	163
June	12.5	39.2	5.0	68	3.8	235
July	15.2	52.2	5.0	72	3.7	201
August	13.0	63.4	5.4	79	3.7	150
September	7.9	58.1	5.8	84	3.9	194
October	2.3	59.9	5.9	86	4.2	194
November	−4.5	44.8	5.9	89	3.9	164
December	−9.9	33.0	5.5	88	3.8	173

Utö	Temperature	Precipitation	Cloudiness	Humidity	Wind	
	(°C)	(mm)	(1/8)	(%)	(m/s)	(direction)
January	−2.1	33.1	6.3	85	7.9	244
February	−3.5	21.5	6.0	87	6.9	198
March	−1.7	23.5	5.4	86	6.5	193
April	1.9	28.9	5.2	85	5.8	237
May	6.7	28.3	4.3	81	5.3	205
June	12.6	30.1	4.1	81	5.3	258
July	15.9	46.7	4.6	82	5.3	254
August	15.7	63.8	5.0	81	5.7	266
September	11.9	60.1	5.6	82	6.7	257
October	7.9	59.9	5.9	83	7.6	242
November	3.8	62.8	6.4	83	8.1	245
December	0.4	47.9	6.4	83	8.3	252

Świnoujście	Temperature	Precipitation	Cloudiness	Humidity	Wind	
	(°C)	(mm)	(1/8)	(%)	(m/s)	(direction)
January	0.6	39	5.8	87	4.2	205
February	0.0	27	5.6	86	4.1	240
March	2.7	37	4.8	82	4.3	210
April	6.2	39	4.8	81	4.3	150
May	11.3	45	4.6	79	4.2	60
June	15.1	53	4.3	79	3.9	225
July	16.9	52	4.6	80	3.6	280
August	16.9	55	4.6	80	3.7	260
September	13.8	52	4.1	82	3.6	240
October	9.6	45	5.0	86	3.7	200
November	4.7	51	5.9	88	4.2	245
December	1.1	48	6.0	88	4.1	250

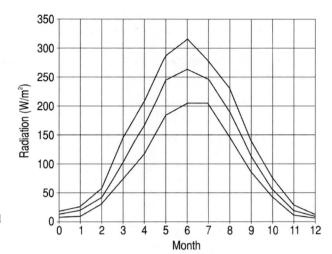

Figure 2.9. The annual evolution of global radiation in Visby, Gotland in 1961–1990. The curves show the monthly means, maxima, and minima. (Redrawn from Bergström et al. (2001).)

terrestrial net loss is around 50 W m^{-2} throughout the year. The balance is normally positive from April to September.

The coverage of direct radiation measurements is quite poor in the Baltic Sea region. Traditionally, for long-term investigations only observations of global radiation[8] and sunshine duration are available. Variations in global radiation can be primarily explained by solar elevation, which causes yearly and daily cycles, and cloudiness. A secondary influence on the radiation level comes from water vapor, other greenhouse gases, and aerosols. Albedo also has an important role in the radiation balance due to the presence of ice and snow in winter.

Figure 2.9 shows the global radiation in Gotland Island. Large annual variability is clearly seen. For terrestrial radiation there are no long-term records available, and it is normally estimated indirectly from the surface temperature, air temperature, cloudiness, and humidity. The atmospheric and surface components of terrestrial radiation undergo similar annual cycles to those of temperature, but their sum or the net terrestrial radiation at the surface shows little seasonal variability. The mean daily level is about −50 W m^{-2} and ranges between 0 W m^{-2} and −100 W m^{-2} depending mainly on cloudiness. Solar radiation comes as direct and diffuse components, and their proportion depends on the altitude of the Sun and cloudiness. In summer the direct part is on average 60% but in winter only 25%.

Thus, cloudiness is the key factor in the radiation balance as well as in its solar and terrestrial components. Average cloudiness is 4/8–6/8 in the Baltic Sea region (Table 2.2, Figure 2.10). The high level is observed in autumn and winter (5/8–7/8), while in summer the level is lower (3/8–6/8). In the north the variability in monthly averages is less (1/8) than in other parts of the Baltic (2/8). There is a significant difference in cloudiness between open sea stations and land stations reflecting the

---

[8] The total downwelling solar irradiance on a horizontal plane.

30 The Baltic Sea: History and geography [Ch. 2

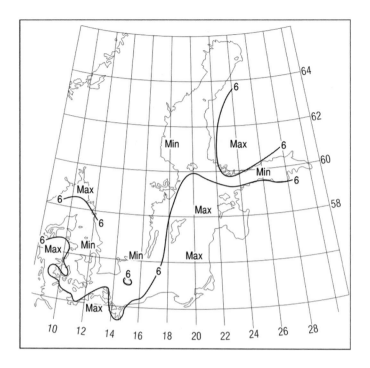

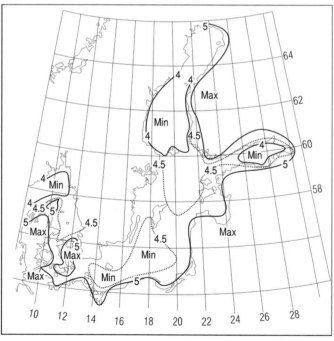

Figure 2.10. Mean seasonal cloudiness (unit 1/8) fields in the Baltic Sea region. This page, top: winter; bottom: spring. Opposite page, top: summer; bottom: autumn. (Redrawn from Mietus, 1998.)

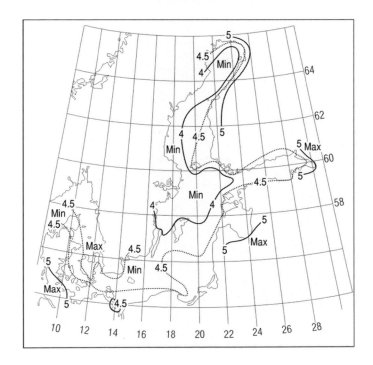

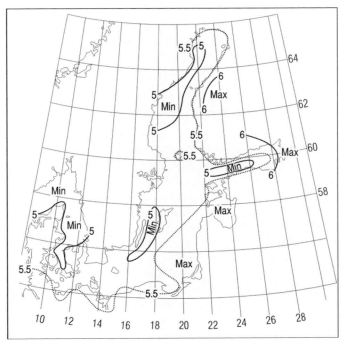

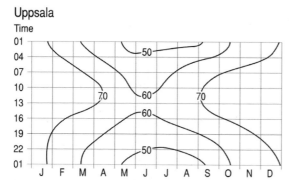

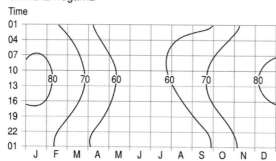

Figure 2.11. Daily and annual variation in cloudiness (in %) in Uppsala (59°54'N, 17°35'E) and Svenska Högarna (59°27'N, 19°30'E) for the period 1961–1990 (modified from Bergström et al., 2001). Uppsala represents land area and Svenska Högarna represents sea area.

differences in, for example, convective conditions (Figure 2.11). In winter, cloudiness is higher above the sea and *vice versa* in summer. It is seen that the day–night difference in cloudiness is much stronger over land than over the sea.

2.4.3 Air temperature

The radiation balance guides the annual cycle of air temperature in the Baltic Sea (Figure 2.12). In general, the amplitude of the annual cycle is larger in land areas than in the open sea. In winter the range is in a north–south direction from −12°C to 0°C due to the strong influence of westerlies (i.e., the NAO index, and the location of the polar front). In the north, ice formation weakens air–sea coupling and thus weakens the heat flow from the water body to the atmosphere. There are also large air temperature gradients perpendicular to the coastline. If the sea is open, atmospheric temperature at the coast is several degrees higher than in inland areas due to the warming influence of the sea.

In April the radiation balance becomes positive, air temperature rises, and its horizontal variability decreases. In the northernmost area the mean air temperature is about 0°C due to the presence of ice, in the Gotland Basin it is 3°C–5°C, and in the south it is 6°C–7°C. The warming of the land areas in the Baltic Sea region starts after the melting of the snow. Therefore, in April the atmospheric temperature is over land a few degrees higher than over the sea in the southern and central parts, but in the north the land areas are still snow-covered and temperature is not far from 0°C.

In the summertime air temperature differences are much smaller than in winter over the Baltic Sea. In July the average air temperature has a small range: 14°C–17°C over the whole basin. This means that in the north the amplitude of the annual air temperature cycle is much larger than in the south. During summer the lowest air temperatures are observed in the open seas and the highest ones occur over inland areas. The warm air loses heat to the cold sea. By September the situation has reversed, as the sea is warmer than the cooling atmosphere. In October the mean monthly air temperature is between 3°C in the north and 11°C in the south. In fall, when the sea surface is warmer than the air and winds are strong, the turbulent heat loss from the sea becomes large.

The role of short-term "transient" cyclones in bringing warm and moist air is pronounced in northern latitudes in winter. The cloud cover also strongly affects the air temperature. During dry and cloud-free summer periods the temperature reaches its maximum, while in winter cloud-free and dry conditions lead to large terrestrial radiation losses from the surface and consequently to low temperatures. During cloudy weather, temperature variations are usually much smaller than during clear sky weather conditions.

2.4.4 Precipitation and humidity

Monthly precipitation amounts usually to 20 mm–100 mm in the Baltic Sea region (Figure 2.13). The largest monthly values are reached in late summer and in autumn, while the smallest values are observed in winter and spring. In autumn and winter, precipitation is usually connected to intense cyclones, but in the summer half the precipitation comes down as showers connected to the convection of air due to daytime surface heating. In spring and early summer, showers are much more frequent over land areas than over the sea due to the low sea surface temperature and the related stable surface layer over the sea. In addition, the location of the polar front is strongly related to cyclonic activity due to baroclinic instability (Holton, 1979) and has a strong influence on the distribution of precipitation.

The spatial distribution of precipitation is related to geographical location and topography. According to Bergström et al. (2001) the largest amount of precipitation in the Baltic Sea drainage basin is estimated to be 1,500 mm–2,000 mm/year in the Scandinavian Mountains due to their orographic influence on airmasses that are passing by. In southwestern Sweden precipitation levels of 1,000 mm–1,200 mm/year have been measured. Precipitation decreases eastward over Sweden reaching a minimum at the island of Öland, where it is some 500 mm/year. Along the east coast of the central basin of the Baltic Sea, precipitation is 700 mm–800 mm/year in a narrow zone due to the combined effects of several factors: prevailing southwesterly winds, friction caused by the coasts, forced lifting of the air due to orographic effects, and thermal convection close to the coast due to land warming.

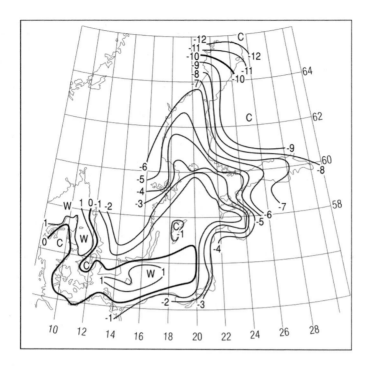

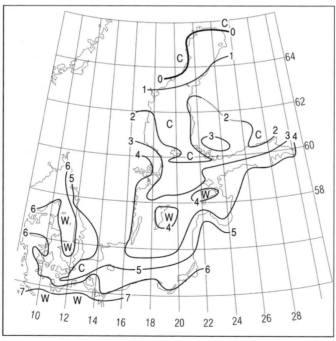

Figure 2.12. Mean monthly atmospheric temperatures in the Baltic Sea. This page, top: January; bottom: April. Opposite page, top: July; bottom: October. W = warm, C = cold. (Redrawn from Mietus, 1998.)

Weather and climate in the Baltic Sea region

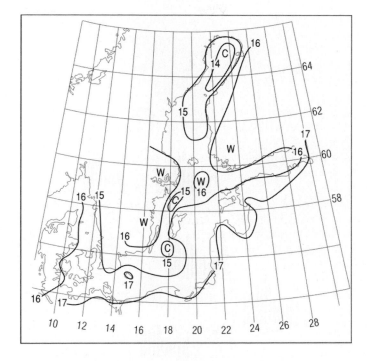

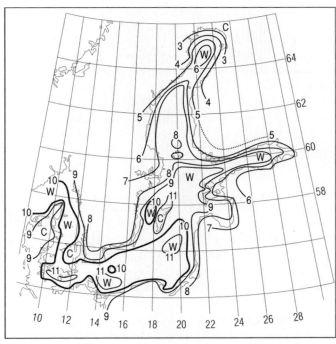

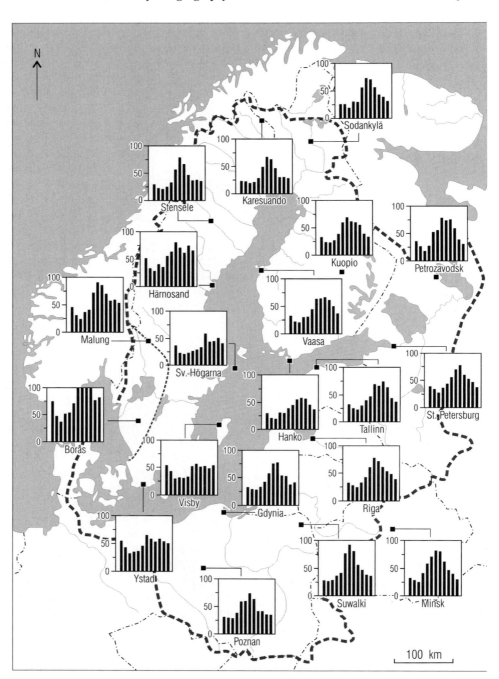

Figure 2.13. Mean monthly precipitation (in millimeters) at some Baltic stations. The dashed line shows the boundary of the drainage basin. (Redrawn from Bergström *et al.*, 2001.)

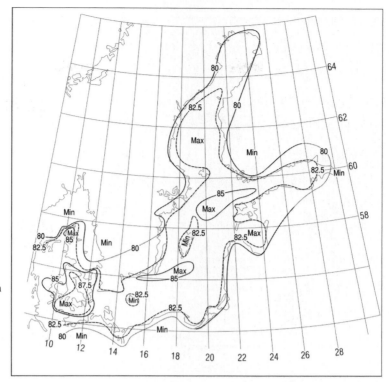

Figure 2.14. Mean annual relative humidity (in percentages) in the Baltic Sea. (Redrawn from Mietus, 1998.)

In the drainage basin of the Gulf of Finland, precipitation is lowest in the eastern part (550 mm/year). In Finland its maximum is observed in the southwestern part (700 mm/year) and its minimum in the north (400 mm/year). In offshore areas of the Baltic Sea, annual precipitation equals 450 mm–500 mm/year, but estimates for the open sea are still uncertain. The lowest values of the whole drainage basin are recorded in some valleys of the Scandinavian Mountains, where precipitation can be as low as 300 mm/year, after much of the atmospheric water content has been lost to orographic precipitation (see Bergström et al., 2001 for details). The distribution of precipitation over the Baltic Sea itself is discussed in Section 4.1.

Mean relative humidity is 80%–90% over the Baltic Sea (Figure 2.14). Maximums are found in the central areas of large basins, while over islands and at the coast the values are lowest. The annual cycle shows large values in fall and winter and low values in summer (Table 2.2). Seasonal variability is small, in summer at coastal sites minimum mean humidity is 70%.

2.4.5 Wind conditions

Wind conditions in the Baltic Sea region are dominated by westerlies (Figure 2.15). The winds are strongest between October and February and weakest between April

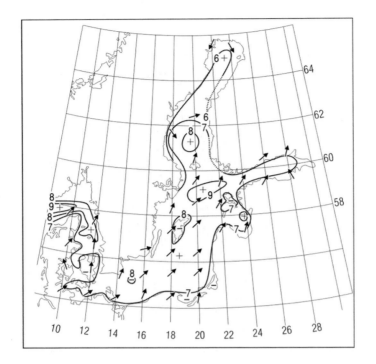

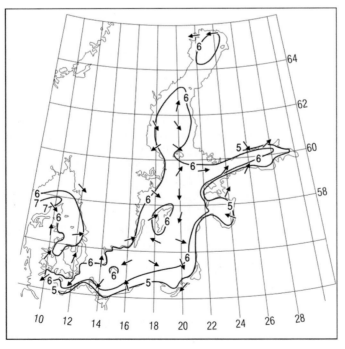

Figure 2.15. Mean wind speed (meters per second) and direction (shown by arrows) in the Baltic Sea area. This page, top: January; bottom: April. Opposite page, top: July; bottom: October. (Redrawn from Mietus, 1998.)

Sec. 2.4] **Weather and climate in the Baltic Sea region** 39

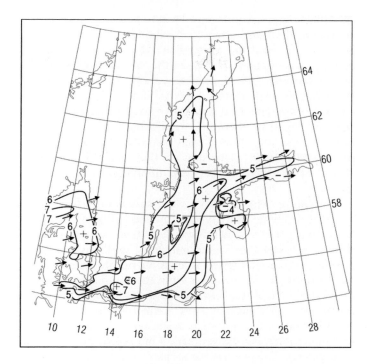

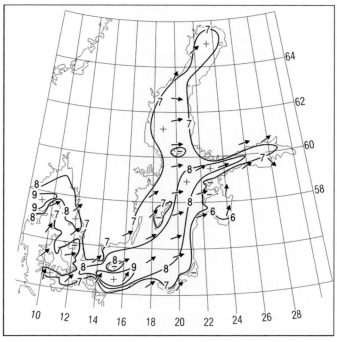

and June. The distribution of wind speed and direction is determined by cyclonic activity in the Northern Europe–Baltic Sea area and by related changes in the air pressure gradient. In wintertime (December–February) the westerly circulation, called zonal circulation, causes the winds to blow on average from the south to the west sector. Occasionally there are northerly winds when a cyclone passes the Baltic Sea. In such winters the NAO index is positive. In cold winters northerly winds advect cold Arctic airmasses down to the southern Baltic Sea, and the NAO index becomes negative. Maximum wind speeds are observed in wintertime. As an example Niros et al. (2003) studied wind speeds in the northern and central Baltic Sea for the period 1991–1999. They showed that mean wintertime wind speed was $8\,\mathrm{m\,s^{-1}}$–$10\,\mathrm{m\,s^{-1}}$ and variability between different weather stations was highest during wintertime.

In spring (March–May), wind speeds are lower than in winter. This is caused by weak cyclonic activity and by the stable atmospheric boundary layer, which furthermore reduces wind speeds. Mean wind speeds are around $6\,\mathrm{m\,s^{-1}}$. Wind distribution still shows predominantly southerly and westerly directions. However, northerly winds connected to cold-air outbreaks and southeasterly winds related to advection of warm airmasses from the south also occur. In summer (June–August), cyclonic activity is weakest and the winds blow on average from between the west and northwest with a speed of $5\,\mathrm{m\,s^{-1}}$–$6\,\mathrm{m\,s^{-1}}$. Then the role of local sea and land breezes is important in coastal areas resulting in a strong diurnal cycle in wind speed. In autumn (September–November), wind speeds once more increase due to intensifying cyclogenesis. The average values are $7\,\mathrm{m\,s^{-1}}$–$8\,\mathrm{m\,s^{-1}}$ and the predominating direction is between the south and west. Mean wind speed gives an overall view of wind forcing of the Baltic Sea (Figure 2.15). The highest wind speeds ever recorded in the open Baltic are more than $30\,\mathrm{m\,s^{-1}}$, observed during November–January. However, the share of strong winds (speed greater than $17\,\mathrm{m\,s^{-1}}$) is less than 5%. More than 50% of winds have a speed of $5\,\mathrm{m\,s^{-1}}$–$9\,\mathrm{m\,s^{-1}}$ and in about 15% of cases they are very weak ($0\,\mathrm{m\,s^{-1}}$–$3\,\mathrm{m\,s^{-1}}$).

It has been shown by Soomere and Keevallik (2003) that the most frequently occurring wind directions are southwest and north to northwest in the entire Baltic Sea basin. However, the directions of the strongest winds do not necessarily coincide with the most frequent wind directions. The reason is the mismatch of the geometry of the sea basin and the direction of predominating winds (e.g., as has been observed in the Gulf of Finland).

The Baltic Sea was introduced in this chapter, beginning with geology. Oceanographic research in the Baltic Sea has a long history, commencing in the late 1800s with investigations into hydrography, sea-level, and ice conditions. The present Baltic Sea is a small and shallow brackish water basin with a mean salinity of only 7.4‰. The regional climate has very large seasonal and interannual variability as a result of its location in the boundary zone of the North Atlantic, Russian continental, and polar climates. Climatic conditions are strongly controlled by the North Atlantic Oscillation (NAO). Chapter 3 investigates the fundamental properties of the Baltic Sea, its topography and hydrography, and its subdivision into different basins.

3

Topography and hydrography of the Baltic Sea

Professor Otto Pettersson (1848–1941) was a world-famous oceanographer and today he is considered as the father of oceanography in Sweden. He remained active until the very end of his life in 1941 at the age of 92 years. He was a professor in marine chemistry at the University of Stockholm. He played an active part in investigations of the Skagerrak, the Kattegat, and the Baltic Sea. In summer 1877 Swedish chemist F. L. Ekman conducted an extensive expedition to these waters collecting about 1,800 records of temperature and salinity from different depths. The results were later analyzed by Professor Pettersson and the main characteristics of Baltic Sea hydrography were then determined. Professor Pettersson was also very active at the early stages of the development of ICES, and he was also one the initiators in organizing the first Nordic conference proposing a hydrographic measurement program in the Baltic Sea in 1898. (© Archives of Dr. Arthur Svansson, Gothenburg, Sweden.)

3.1 TOPOGRAPHY

This section starts with a detailed description of the topographic features of the Baltic Sea. The dimensions of the main basins are listed and a detailed description of each basin is given. In Section 3.2 the basic physical characteristics of brackish water are given. A discussion follows on how brackish waters differ from oceanic waters. In Section 3.3, watermasses and their specific stratification conditions are described in detail and the interaction of the Baltic Sea with the North Sea is discussed. Section 3.4 gives a summary of long-term changes in the hydrographic conditions of the Baltic Sea during the last 100 years, a period in which systematic measurements have been carried out.

3.1.1 General

The outermost part of the Baltic Sea in the southwest is the narrow region between the Danish mainland and Sweden. The normal definition is to take the boundary as the line between the Kattegat and the Danish Straits, which consist of the Öresund and the Belts. This definition is also employed here. As discussed in Chapter 2, sometimes the Kattegat is included in the Baltic Sea and sometimes the Danish Straits are left out, the resulting range becoming about ±5% in surface area. The present Baltic Sea consists of different basins, where the character of the coastline and the bottom topography vary considerably. This variability is one of the key background factors of the complex physics of the Baltic Sea. The geometry, morphology, deeps, and depressions have their own specific features. The coastal areas are often characterized by archipelagos and modified by the effects of waves.

A characteristic feature of the Baltic Sea is its shallowness. It has a mean depth of 54.0 meters and the greatest depth of only 459 meters is in Landsort Deep between Stockholm and the island of Gotland. The area of the Baltic Sea is also small, 392,978 km^2, while its volume equals 21,205 km^3.

Land uplift has been ongoing in the Baltic area throughout the Holocene (Figure 3.1). Its magnitude ranges from 0 mm/year in the south up to 9 mm/year in the northern Gulf of Bothnia. The location of the shoreline is influenced by the balance between eustatic sea-level rise, the hydrography of the Baltic watermass, and land uplift pattern. At present, eustatic sea-level rise is 2 mm–3 mm/year (IPCC, 2007). When all these effects are put together, one number representing apparent land uplift in relation to mean sea-level is obtained, this being largest in the Bay of Bothnia. In Kemi it equals 7.20 ± 0.27 mm/year while in the eastern Gulf of Finland (Hamina) it is only 1.62 ± 0.26 mm/year. The zero line of apparent uplift (i.e., where eustatic sea-level rise compensates the influence of actual land uplift) extends across Estonia, and in the southern Baltic Sea the land is apparently dropping. The latest results (Johansson et al., 2004) show that apparent uplift, from the long-term perspective, will eventually cease in most of the Gulf of Finland. This is due to increase in eustatic sea-level rise, which is in turn caused by the recent warming trend in global climate. However, on top of eustatic sea-level rise, the mean annual sea-level of the Baltic Sea is also positively correlated with the strength of westerlies.

The topographic features of the Baltic Sea (Figure 3.2) will now be discussed according to the division of the Baltic Sea into the 14 basins given in Chapter 2 (Figure 2.3). This presentation is largely based on the detailed description of bottom topography by Fonselius (1995).

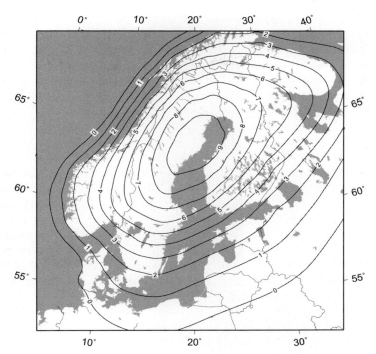

Figure 3.1. Fennoscandian land uplift relative to the center of the Earth (mm/year). The model is based on leveling, tide gauges, GPS observations, and geophysical models in Vestøl (2006), Ågren and Svensson (2007), and Lidberg (2007). (Prepared by Prof. Markku Poutanen, Finnish Geodetic Institute, printed with permission.)

Water exchange between the sub-basins is largely governed by the sills between them. Sill depth is the maximal depth at which water can freely or horizontally flow between two basins. So, water exchange is effective down to sill depth. In Figure 3.3 a schematic view is given concerning the inflow of water over sills to inner parts of the Baltic Sea.

An overall view of the topographic features can be obtained by taking a look at the areas, volumes, and depths in different basins (Table 3.1). The largest basin is the Gotland Sea with an area of 151,920 km^2 or 39% of the total area of the Baltic Sea, whereas its volume 10,824 km^3 is 51% of total volume. Not surprisingly, the mean depth of the Gotland Sea (71 m) is larger than the mean depth of the entire Baltic Sea. The second largest basin is the Gulf of Bothnia, which has an area of 115,516 km^2, volume of 6,369 km^3, and mean depth of 55 m. The next in size is the Bornholm Basin. Its area is 38,942 km^2, volume 1,780 km^3, and mean depth 46 m. The Gulf of Finland covers an area of 29,498 km^2 (8% of the Baltic Sea) and has a volume of 1,098 km^3 (5% of the Baltic Sea). Its mean depth is 37 m, clearly less than that of the entire Baltic Sea. The Danish Straits, the Arkona Basin, and the Gulf of Riga have relatively small volumes.

Depth distribution can be investigated from hypsographic curves (Figure 3.4). The shallowness of the sea becomes clearly visible. Only about 12% of the total area has a depth more than 100 m and only 2.7% more than 150 m. So, a large part of the Baltic Sea belongs to the shallow, coastal-like area. As a rule of thumb it can be said that about 50% of the area of the Baltic has a depth of 50 meters or less.

44 Topography and hydrography of the Baltic Sea [Ch. 3

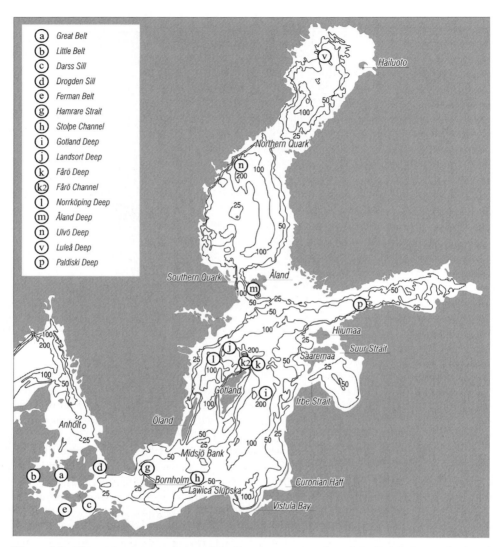

Figure 3.2. The main characteristics of the bottom topography of the Baltic Sea and related important geographical locations. (Based on Fonselius, 1995.)

The Weichselian ice sheet removed much of the older sediments from the sea floor. During the Holocene, while the Baltic Sea underwent evolutionary phases, particle transport and sedimentation again accumulated material to the bottom of the sea at varying intensities, depending on location (Figure 3.5, see color section). The bottom substrate is hard in areas where sedimentation has been weak (areas of erosion). In some parts of the Baltic Sea the sedimentation process has continued without any external disturbance. The oldest sedimentation basins are found in the

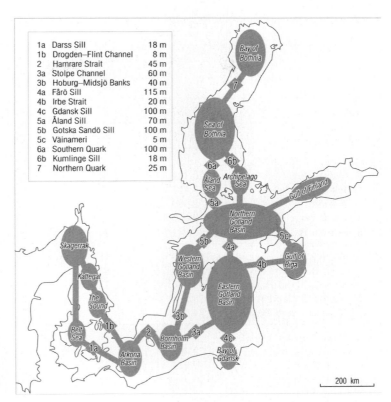

Figure 3.3. A schematic view of sills between different basins in the Baltic Sea.

southern Baltic from where the continental ice sheet first retreated. The depth of sediments varies between different areas. In the Bay of Bothnia post-glacial sediments are about 10 m thick, in the Sea of Bothnia they reach a thickness of 40 m–50 m, in the Gotland Basin about 30 m. Close to the edge of the ice sheet, layering was fast, even several centimeters per year. When the ice sheet retreated, sedimentation velocity reduced and is presently about 1 mm/year in the deepest basins.

Approximately 30% of the seabed is covered by modern soft (organic-rich) sediments or mud (Dr. Harri Kankaanpää, pers. commun.). In areas where the water depth is less than 50 meters, no permanent sedimentation takes place (normally at least) because the waves affect the bottom, inducing erosion. Despite their general low intensity, local bottom currents also contribute to sediment erosion (lateral transport). As the pattern of areas with different sedimentation properties mimics wind properties and wind–wave activity (Danielssen et al., 1997), the sedimentation rate may have varied in the past together with temporal variation in wave conditions.

In the following sub-sections the bottom topography of the Baltic Sea is described in detail, going from the Kattegat to the southwestern Baltic Sea, to the Gotland Sea, and finally to the large gulfs. For the geographical nomenclature see Figures 3.2 and 3.3, and the basin boundaries are given in Figure 2.3.

Table 3.1. The dimensions of the Baltic Sea basins and the Kattegat (modified from Fonselius, 1995).

	Area (km^2)	Mean depth (m)	Maximum depth (m)	Volume (km^3)
Kattegat	22,287	23	130	515
Baltic Sea	392,978	54	459	21,205
Southwestern Baltic Sea				
Danish Straits	20,121	14	81	287
Belt Sea	17,821	15	81	260
Öresund	2,300	12	53	27
Arkona Basin	19,068	23	53	442
Bornholm Basin	38,942	46	105	1,780
Gotland Sea	151,920	71	459	10,824
Gdansk Bay	25,234	57	114	1,439
Eastern Gotland Basin	63,478	77	249	4,911
Northern Gotland Basin	28,976	71	150	2,056
Western Gotland Basin	34,232	71	459	2,418
Eastern and northern gulfs				
Gulf of Finland	29,498	37	123	1,098
Gulf of Riga	17,913	23	51	405
Gulf of Bothnia	115,516	55	293	6,369
Åland Sea	5,477	75	301	411
Archipelago Sea	8,893	19	104	169
Sea of Bothnia	64,886	66	293	4,308
Bay of Bothnia	36,260	41	146	1,481

3.1.2 The Kattegat and the southwestern Baltic

The Skagerrak has a mean depth of 230 meters with no sill towards the Kattegat. The border between the Kattegat and the Skagerrak is the line between Skagen and Marstrand. The Kattegat is the transition zone between the Baltic Sea and the North Sea (Figure 3.6a). Normally, salty Atlantic water penetrates the Baltic Sea through the bottom layer from the North Sea side, Skagerrak, and the Kattegat. The Baltic Sea water flows as the Baltic Current in the surface layer through the Kattegat to the Skagerrak, where it turns to the west as the Norwegian Coastal Current. This water exchange pattern is typical for estuarine circulation; for that reason the Baltic Sea is sometimes considered a large estuary. Baltic Sea watermasses can be tracked over long distances along the Norwegian coast due to their low-salinity signals.

In the Kattegat strong mixing of Baltic Sea and North Sea waters takes place, and the water entering the Baltic Sea has a salinity that only ranges between 20‰ and 30‰, which depends on local topographic features. There are local deeps in the

Sec. 3.1] Topography 47

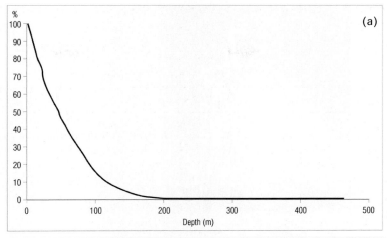

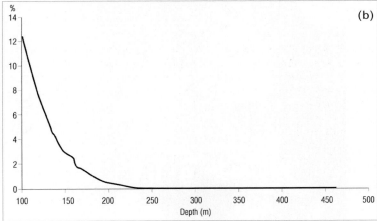

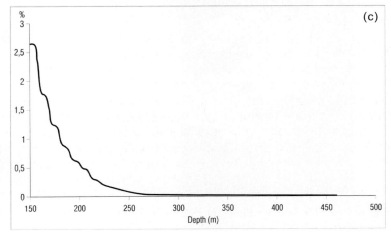

Figure 3.4. The hypsographic curves of Baltic Sea depth: (a) all depths; (b) depths greater than 100 m; (c) depths greater than 150 m.

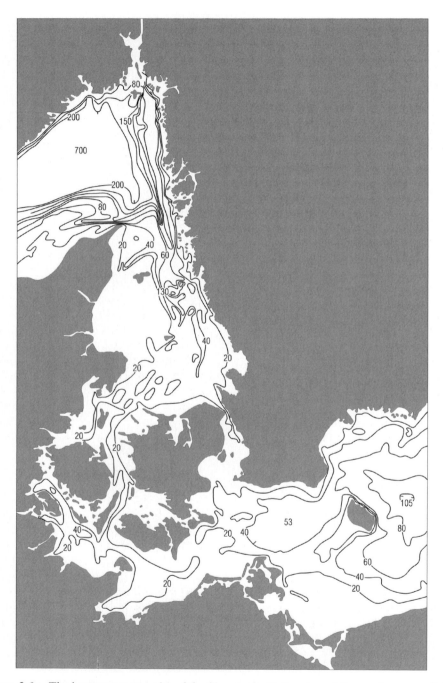

Figure 3.6a. The bottom topography of the Skagerrak, Kattegat, and the southwestern Baltic Sea. (Redrawn from Fonselius, 1995.)

Kattegat, the deepest one is close to Anholt (100 m). There is a deep channel along the Swedish side where the depth is between 75 m and 100 m, getting shallower in the south. In the Danish Straits—at the gate to the Baltic Sea—the depth is mostly less than 20 meters. The boundary line between the Kattegat and the Danish Straits goes through Hasenør–Sjællands Odde from Jylland to Sjælland and Gilleleje–Kullen from Sjælland to Skåne. The former part is the boundary to the Belt Sea and the latter part is the boundary to the Öresund.

The Danish Straits consist of the Belt Sea and the Öresund. The fragmented Belt Sea consists of the following water areas in the Danish and German archipelago: Great Belt between Sjælland and Fyn, Smålands-farvandet, Kiel Bight, Fehmarn Belt, Mecklenburg Bight, Århus Bight, Little Belt, and the water areas south and north of Fyn. The depth conditions are variable in the Belt Sea. In the deepest channels the depth can reach 60 m, but normally depths are around 20 m; in Little Belt they are less than 15 m. The southern boundary line of the Danish Straits follows the line from Darsser Ort to the Gedser Peninsula. Here can be found the most important topographic sill in relation to the water exchange between the North Sea and the Baltic Sea: the Darss Sill with a depth of 18 meters. The main water exchange channel between the North Sea and the Baltic Sea goes through the Great Belt. The narrow channel extends from the Kattegat to the Darss Sill where the maximum draught for incoming ships is determined.

The Öresund is a narrow strait (5 km wide at its northern entrance) between Denmark and Sweden. Its southern border goes through the line from Stevns Klint to Falsterbo Peninsula. In the northern part of Öresund, sea depth is typically between 23 m and 26 m, but in Landskrona Deep, south of the island of Ven, the depth reaches 53 m. About 70%–75% of the water exchange goes through the Belts and 25%–30% through the Öresund. This is a long-term average, but for certain situations it can be quite different (Jacobsen, 1980). Water transport through the Great Belt is one order of magnitude larger than that through the Little Belt (Jacobsen and Ottavi, 1997). In the southern Öresund on the Swedish side the sill depth in Flint Channel is 8.6 m, whereas on the Danish side the sill depth of Drogden Channel is 8 m. From the sill area farther southward, on the Baltic side, depth increases to 20 m.

In the southwestern Baltic Sea (Figure 3.6a) two basins can be distinguished: the Arkona Basin (maximum depth 53 m), followed by the Bornholm Basin east of the island of Bornholm. The sill depth between these two basins is 45 m in Hamrare Strait between Skåne (southern Sweden) and Bornholm. The typical depth in the Bornholm Basin is 60 m–80 m and the maximum depth in the Bornholm Deep equals 105 m. In the area southwest of Bornholm the depth is 20 m–25 m only. The deepest passage farther into the Baltic Sea, reaching the Eastern Gotland Basin, is the Stolpe Channel. In the north the Bornholm Basin is bounded by the Swedish coast and the Midsjö Bank, where the sill toward the Western Gotland Basin is shallow.

3.1.3 The Gotland Sea

The Gotland Sea consists of the Gdansk Bay and the Gotland Basin, which is subdivided into the Eastern, Western, and Northern Gotland Basins (Figure 3.6b).

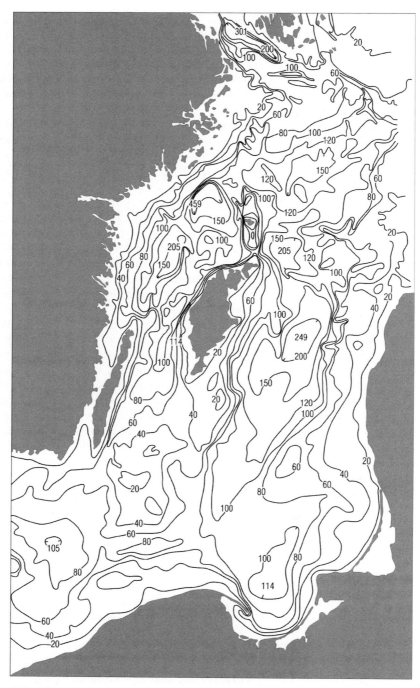

Figure 3.6b. The bottom topography of the Gotland Sea. (Redrawn from Fonselius, 1995.)

It contains about half of the Baltic Sea watermass (10,824 km^3 out of 21,205 km^3). In the south it is separated from the Bornholm Basin by a broad underwater ridge stretching from Öland to Poland. The ridge consists of several shallow areas: Northern Midsjö Bank (9 m), Southern Midsjö Bank (13 m), and Lawica Slupska (8 m). The Stolpe Channel penetrates through this ridge. A northern extension of this ridge, Hoburg Bank (minimum depth 10 m) goes to Gotland and separates the Eastern and Western Gotland Basins.

Gdansk Bay has a maximum depth of 114 m. Its outer limit is the line between Briusterort and Rozewie (Prof. Jan Piechura, pers. commun.), and the sill depth to the Eastern Gotland Basin equals 100 m. There are two remarkable water areas along the coasts of Poland, the Kaliningrad region, and Lithuania—namely, the Vistula Lagoon and the Curonian Lagoon, the latter being well known as an amber deposit.

The Eastern Gotland Basin is an extensive, bowl-shaped deep with a maximum depth of 249 m, the Stolpe Channel is its southern mouth. The maximum depth of Stolpe Channel is 90 m and the sill depth is 80 m towards the Gotland Sea and 60 m towards the Bornholm Basin. There are two depressions in the Eastern Gotland Basin: Gotland Deep (249 m) east of Gotland and Fårö Deep (205 m) northeast of Gotland. The sill depth between these depressions is 140 m. The Fårö Channel locates between the island of Fårö and the island of Gotland. The sill depth between the Eastern Gotland Basin and the Northern Gotland Basin is 115 m, and the boundary line between these basins is slightly north of the island of Gotland along the line Fårö–Gotska Sandö–Ristna.

The Northern Gotland Basin extends up to the Gulf of Finland and to the Gulf of Bothnia. From a hydrographic viewpoint the Gulf of Finland is a continuation of the Gotland Basin because there is no sill in between the basins. The topography of the Northern Gotland Basin is very variable and there are large areas with depths more than 150 m. The northern boundary of the basin is the Åland Sill with a depth of 70 m. In the west the boundary line to Western Gotland Basin is the line Fårö–Kopparstenarna–outer Archipelago of Stockholm.

Landsort Deep (459 m), the deepest place in the whole Baltic Sea, is located in the Western Gotland Basin. It is a cleft shaped like a half-moon with very steep walls. The Western Gotland Basin contains additionally the Norrköping Deep (205 m). Between Västervik (Swedish mainland) and the town of Visby, Gotland there are areas with depths more than 150 m. The Western Gotland Basin becomes southwards shallower and at the Hoburg Bank and at Midsjö Banks the depth ranges from 20 m to 40 m only. The island of Öland belongs to this basin as well. The depth between the Swedish mainland and Öland is less than 20 m.

3.1.4 The Gulf of Riga

The Gulf of Riga is an almost regular-shaped bowl-like water body situated between Estonia and Latvia (Figure 3.6c). It is largely isolated from the Gotland Sea due to presence of the islands of Hiiumaa and Saaremaa. The water exchange with the

52   **Topography and hydrography of the Baltic Sea**   [Ch. 3

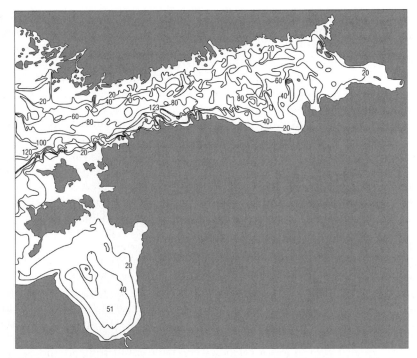

Figure 3.6c. The bottom topography of the Gulf of Riga and Gulf of Finland. (Redrawn from Fonselius, 1995.)

Gotland Sea takes place through two straits south and east of Saaremaa.[1] The broader one is the Irbe Strait in the western corner, where the boundary is the line between Sõrve, Saaremaa and Ovisi, Latvia. The width of Irbe Strait is 27 km, its cross-sectional area is 0.27 km^2, and its sill depth is 21 m. East of Saaremaa and Hiiumaa there is the shallow Väinameri Sea (mean depth less than 10 m). It is connected with other sea areas by a network of straits between the main basin of the Gulf of Riga and the Gotland Basin, starting from Suur Strait (Suur Väin in Estonian, sometimes Virtsu Strait in Soviet maps) between the island of Muhu and the Estonian mainland. The maximum depth of the Gulf of Riga is 51 m and the mean depth is 23 m. The coasts are to a great extent sandy dunes and far offshore there are only two islands: Kihnu and Ruhnu.

3.1.5 The Gulf of Finland

The Gulf of Finland is the easternmost basin in the Baltic Sea, surrounded by Finland, Russia, and Estonia (Figure 3.6c). This elongated and narrow basin has

---

[1] Hiiumaa and Saaremaa are called Dagö and Ösel on many old maps. These names originate from the time when Estonia belonged to Sweden during the 17th century (until 1721).

no sill towards the Gotland Sea. Anyway, the western boundary is defined as the line Pöösapea–Osmussaar–Hanko town. The Gulf of Finland becomes shallower toward its eastern end. The deepest areas (80 m–100 m) are located in the western and southern parts, the maximum depth (Paldiski Deep[2]) being 123 m. The Estonian coast has rather steep and locally (on the scale of a few kilometers) almost straight walls and more or less regular bathymetry, and there are only a few islands whereas the Finnish coast is shallow (20 m–40 m), has an extremely irregular coastline and ragged bathymetry, and contains extensive archipelago areas. Adjacent to the deep connection of the Gulf of Finland with the Baltic Sea proper are located extensive very shallow areas close to Hiiumaa that almost reach the entrance to the Gulf of Finland. In the eastern part there are a few major islands far offshore: Suursaari (Gogland) between Kotka and Narva, and Tytärsaari and Seiskari farther east. East of Seiskari the depth is less than 40 m and in the mouth of the River Neva the depth is usually below 5 m (except for the dredged waterway to Saint Petersburg).

North of the Hanko Peninsula there is a specific, stratified fjord-like inlet in the Gulf of Finland: Pojo Bay (see *http://luoto.tvarminne.helsinki.fi/english/*). It has been sometimes considered as a miniature model of the Baltic Sea. This bay is a 40 km long estuary of River Mustionjoki, connected to the outer archipelago by a series of small basins with variable depths. The 6 m deep sill of Pojo Bay is located in the town of Ekenäs. From there the water depth gradually increases to a maximum of 42 m in Sällvik, *ca.* 5 km north of Ekenäs. Freshwater outflow from the River Mustionjoki forms an oligohaline (generally less than 4‰) surface layer on top of more saline deepwater. This vertical stratification leads to regular summertime stagnation and gradual oxygen decrease in the deepwater. Deepwater renewal takes place only under special meteorological conditions, mainly during autumn and winter, when hydraulic control at the basin entrance, Ekenäs Sill, becomes weaker and allows saline inflows from the archipelago.

3.1.6 The Gulf of Bothnia

The Gulf of Bothnia is an embedded semi-enclosed Baltic Sea basin whose hydrography is quite different from that in other parts of the Baltic Sea. This is because sills and archipelagos in the southern section largely isolate the basin from the Gotland Sea (Figure 3.6d). The Gulf of Bothnia consists of the Archipelago Sea, the Åland Sea, the Sea of Bothnia, and the Bay of Bothnia. The Archipelago Sea covers the island area from the southwest coast of the Finnish mainland to Åland, and the Åland Sea is the sea area between Åland and Sweden. The Sea of Bothnia is separated from the Åland Sea by the Southern Quark and toward the Archipelago Sea the boundary follows the outer northern islands. The boundary between the Sea of Bothnia and the Bay of Bothnia is the Northern Quark. The Southern Quark and the Northern Quark are not treated as separate basins, but the former belongs to the Åland Sea and the latter is divided between the Sea of Bothnia and the Bay of Bothnia.

---

[2] Unofficial name, located just outside the town of Paldiski, Estonia.

54 Topography and hydrography of the Baltic Sea [Ch. 3

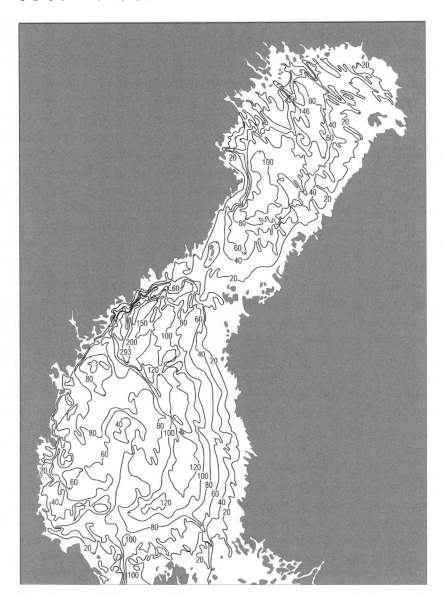

Figure 3.6d. The bottom topography of the Gulf of Bothnia. (Redrawn from Fonselius, 1995.)

The Åland Sea consists of two basins and three sills. The Southern Åland Sill (70 m) is a very narrow channel located toward the Northern Gotland Basin. The maximum depth in the southern Åland Sea basin is 220 m (Lågskär Deep). The Åland Sea Middle Sill (also 70 m) between Söderarm and Lågskär separates the two Åland Sea basins from each other. The deepest place in the Åland Sea is 301 m (see Figure

3.6b) in the northern basins, between Eckerö and Grisslehamn, called the Åland Deep (Ålandsdjup). From this depression several channels with depths of 150 m lead water through the Southern Quark to the Sea of Bothnia, forming the Northern Åland Sill. However, the sill in the south governs the salinity of waters that can enter the Sea of Bothnia. In some sources it has been suggested that the Southern Åland Sill should be 45 m, but 70 m seems to be the right value (see Fonselius, 1995). The boundary line between the Åland Sea and the Sea of Bothnia follows the island chain: Söderö (the northeast peninsula)–Ormö–Norrskär–Märkeskällan–Märket–Långö–Eckerö.

The Archipelago Sea is a mosaic of sea and land, including thousands of islands. The boundary line towards the Sea of Bothnia goes from the northernmost peninsula of Saltvik in Åland to the Finnish mainland via Landtö–Kauritsala–Lokalahti. The Archipelago Sea is very shallow. There are north–south channels with depths ranging from 30 m to 40 m through the Archipelago Sea, and they form a part of the water exchange system between the Sea of Bothnia and the Gotland Sea. The sill depth towards the Sea of Bothnia is about 18 m. The main route goes through the Kihti Strait. The mean depth of the Archipelago Sea is 19 m, which restricts the inflow of deepwater from the Gotland Sea to the Sea of Bothnia.

The Sea of Bothnia is separated from the Bay of Bothnia by the boundary line: Iskmo–Raippaluoto–Björkö–Lappören–Valassaaret–Hadding Peninsula. The Sea of Bothnia is asymmetric in the east–west direction such that at the Finnish side the bottom slope is gentle whereas at the Swedish side the coast is steep and the bottom topography is uneven. There are small fjords at the Swedish side in the northern Sea of Bothnia called the "High Coast" (*Höga Kusten*). The southern part of the basin has a variable topography. A large area with rather large depths extends from the southern Sea of Bothnia first to the northeast; the Finngrundet Shoal is located in the southwest corner close to Sweden. The depression extends furthermore to the north turning later towards the Swedish coast. The deepest place, Ulvö Deep (293 m), is located close to the High Coast. In the southern part of the Sea of Bothnia there is a Paleozoic limestone area, which was formed some 400 million years ago. In this area the bottom is flat.

The Northern Quark, which connects the Sea of Bothnia and the Bay of Bothnia, is a rather shallow area with a maximum depth of 65 m. There are two sills in the area with depths of 25 m, and there are large archipelago areas at Holmöarna at the Swedish coast and off the town of Vaasa at the Finnish side. The Northern Quark is often treated as a separate basin (southern boundary from Hörnefors to Vaasa, northern boundary from Käringskär to Stubben) whereas in oceanography it is divided between the Sea of Bothnia and the Bay of Bothnia. This is motivated because the Northern Quark is not a basin with its own hydrographic characteristics but a strait.

The northernmost basin of the Baltic Sea is the Bay of Bothnia. It reaches the latitude 65°50′, just short of the Artic Circle. It is asymmetric in its topography in the east–west direction like the Sea of Bothnia. There are two depressions in the basin, the northernmost being the smaller. There is a large archipelago area in the north with Hailuoto as the largest island. The Bay of Bothnia is a mostly rather shallow

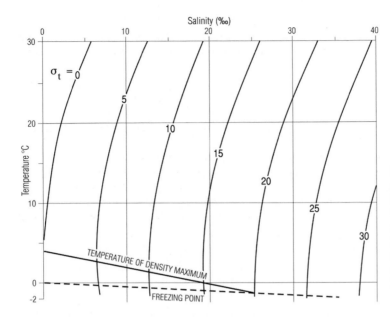

Figure 3.7. The density of seawater, freezing point, and the temperature of maximum density. Density (kg/m^3) equals $1{,}000 + \sigma_T$. (Redrawn from Pickard and Emery, 1990.)

area with a mean depth of 41 m. The deepest spot, Luleå Deep[3] (146 m) is located southeast of the town of Luleå.

3.2 BRACKISH WATER

3.2.1 Equation of state

The equation of state of seawater is $\rho = \rho(T, S, p)$, where ρ is density, T is temperature, S is salinity, and p is pressure[4] (Figure 3.7). The UNESCO (1981) standard is expressed in the form

$$\rho = \frac{\rho(T, S, 0)}{1 - p/K(T, S, p)} \tag{3.1}$$

where $K(T, S, p)$ is the bulk modulus, which is the reciprocal of compressibility. Empirical expressions are presented separately for $\rho(T, S, 0)$ and $K(T, S, p)$. This formula and the UNESCO (1981) equation of salinity[5] vs. conductivity are valid and highly accurate for all marine waters over the salinity range 0‰–40‰, which includes brackish marine waters. The accuracy is 0.04‰ for salinity and 5×10^{-3} kg m^{-3} for

---

[3] Again, an unofficial name.
[4] Taken as the gauge pressure (i.e., pressure in excess of atmospheric pressure at sea-level).
[5] In the UNESCO standard the unit is defined as the PSU (Practical Salinity Unit), used as, for example, $S = 34.75$ PSU or later just $S = 34.75$. Here ‰ is used for clarity for non-oceanographers, $S = 34.75$‰. They both represent the fractional mass of dissolved salts in seawater.

the density of Baltic Sea waters when the UNESCO formulas are used (Millero and Kremling, 1976). The full equation contains a large number of parameters, and for many purposes simplified expressions can be employed. A simplified form for Baltic Sea water is

$$\rho(T, S) = \rho_0 - \alpha(T - T_m)^2 + \beta S \qquad (3.2)$$

where ρ_0 is a fixed reference density; $T_m = T_m(S)$ is the temperature of maximum density; and α and β are parameters: $\alpha \sim 5 \times 10^{-3}$ kg m^{-3} °C^{-2} and $\beta \sim 0.8$ kg m^{-3}. This expression reflects the relatively rapid decrease in water density when temperature increases, proportional to the square of the temperature difference, and linear increase in density with the increase in salinity. The influence of pressure is neglected in Equation (3.2), which is usually a good approximation in the relatively shallow Baltic Sea where the pressure in the water column is limited by the shallowness of the basin to about 47 bar (see next paragraph). The temperature T_m is approximately a linear function of salinity

$$T_m[°C] = 3.98 - 0.216 \times 10^3 S \qquad (3.3)$$

where the brackets show the unit (salinity is dimensionless). Thus the fixed reference density equals $\rho_0 = \rho(3.98°C, 0, 0) \approx 1,000$ kg m^{-3}. The freezing point temperature is according to the UNESCO (1981) formula a linear function of the pressure but a more complex function of salinity

$$T_f[°C] = -57.5S + 54.0915S^{1.5} - 215.4996S^2 - 7.53 \times 10^{-3} p[\text{bar}] \qquad (3.4)$$

The validity range is 0‰–40‰ for salinity. A good linear approximation of Equation (3.4) is $T_f[°C] = -55S$. For the salinity of $S = 7.4$‰ and $p = 0$, Equation (3.4) gives $T_f = -0.403°C$, while the linear form gives $T_f = -0.407°C$.

The bulk modulus K of seawater is around[6] 2×10^4 bar. For example, at a 200 m depth the pressure is 20 bar and the resulting volume contraction is consequently 10^{-3}. In the deep basins of the Baltic Sea the density of a seawater parcel is therefore *in situ* larger by 1 kg/m^3–2 kg/m^3 than at atmospheric pressure. In Landsort Deep the thickness of the whole water column is thus contracted by 0.5 m due to the pressure. In Baltic Sea stratification, the changes caused by variation of temperature and salinity far exceed the compressibility effects. This feature leads to one of the key differences in convection and mixing between the Baltic Sea and large oceanic basins. In oceanic deepwaters, temperature and salinity are nearly homogeneous, and the nonlinear dependence of bulk modulus on temperature and salinity substantially influences the conditions for convection. Consequently, in the Baltic Sea, apart from very special investigations in or beyond margins of physical oceanography, the pressure effect on density needs not be considered.

---

[6] Bar is widely used in oceanography as a unit of pressure (1 bar = 10^5 Pa). One decibar (dbar) corresponds to a one-meter seawater layer with good accuracy.

3.2.2 Adiabatic changes and stability

In the structure of vertical stratification *adiabatic* temperature changes need to be considered. These are due to changes in the surrounding pressure field: with increasing pressure the temperature of a water parcel increases and *vice versa*, when ignoring external input or output of heat to and from the system. *Potential temperature* refers to the temperature that a water parcel would have if it were taken adiabatically to the surface, and *potential density* is the corresponding density. The adiabatic rate of change of temperature per depth is the adiabatic temperature gradient, denoted by Γ.

In exact form we have $\Gamma = \alpha T/(\rho c_p)$, where α is the temperature expansion coefficient and c_p is specific heat at constant pressure. The magnitude is about $0.003°C/\text{bar}$ in seawater (i.e., a water parcel cools adiabatically by $0.03°C$ as it is taken from a 100 m depth to the surface). The stability of the position of water particles in a water column (usually called the stability of stratification) is characterized by the following quantity

$$E = -\frac{1}{\rho}\left[\frac{\partial \rho}{\partial S}\frac{\partial S}{\partial z} + \frac{\partial \rho}{\partial T}\left(\frac{\partial T}{\partial z} - \Gamma\right)\right] \qquad (3.5)$$

where z is the vertical coordinate (positive upward). The watermass is stable if $E > 0$, neutral if $E = 0$, or unstable if $E < 0$. Equation (3.5) shows that if salinity is constant, water temperature must increase in depth by the adiabatic rate for neutral stratification.

The density of Baltic Sea waters is mostly in the range $1,005\,\text{kg/m}^3$–$1,010\,\text{kg/m}^3$. The density is sensitive to temperature, especially when the water is warm, see Equation (3.2). At $T = 20°C$ the density changes by $0.2\,\text{kg/m}^3$ for a temperature change of $1°C$, while at temperatures below $5°C$ the corresponding density change is below $0.1\,\text{kg/m}^3$. An increase in salinity of $1‰$ always increases density by about $0.8\,\text{kg/m}^3$. Thus, if the salinity difference between the upper layer and lower layer is more than $0.5‰$ and the temperature of the lower layer is below $5°C$, there is no way for temperature variations to force local deepwater convection.

In many applications, absolute density can be replaced by a fixed reference density. In ocean dynamics relative differences in density are more critical than the absolute value, and in this respect brackish seas are comparable with oceanic basins. That is, density variations due to salinity gradients are so large that the equation of state must include salinity in addition to temperature and pressure.

3.2.3 Properties of brackish water

Brackish water is loosely characterized as "saline water with salinity between freshwater and normal ocean water levels." The exact definition of brackish water is often taken as saline water that has *a freezing point temperature below the temperature of maximum density*. Thus, the distinction comes from the equality $T_f = T_m$; this gives

$S = 24.7$‰, $T = -1.3°C$ (Figure 3.7), which coincides with the average salinity of the lower layer at the entrance to the Baltic Sea. Sometimes 24.7‰ has been chosen to represent the salinity of water inflowing to the Baltic Sea (e.g., Brogmus, 1952). For a salinity of 5‰–10‰, the temperature of maximum density is 1.8°C–2.9°C and the freezing point is from $-0.27°C$ to $-0.56°C$.

The lower boundary of salinity for brackish waters has not been exactly defined, but 0.5‰ would be a good estimate. In physics, when the salinity level is more than this, its influence on stratification and consequently circulation becomes significant. The upper limit for usable household water is 1.5‰, but for freshwater biota the limit of feasible salinity in the environment is largely specific to each species.

Brackish seawater is rare in nature, and, apart from hydrographical surveys, there are few direct measurements of its physical properties. Rather, one must estimate them from the physical properties of freshwater and normal seawater; brackish water is somewhere between these extremes. For example, the equation of state says that the density of Baltic Sea water is $1,005 \text{ kg/m}^3$–$1,010 \text{ kg/m}^3$, while the densities of freshwater and seawater are about $1,000 \text{ kg/m}^3$ and $1,025 \text{ kg/m}^3$, respectively. Care must be taken when using empirical relations designed for seawater with normal oceanic salinities. Extrapolation to Baltic Sea salinity may bring about a severe bias, and the consistency of the extrapolation should be checked by means of a freshwater reference.

Table 3.2 compares the physical properties of Baltic Sea water represented by a mean salinity of 7.5‰ together with freshwater and oceanic water. As can be anticipated the differences are quite large regarding electromagnetic properties. This can be used for measurement techniques. The mechanical and thermal properties of brackish water differ quantitatively from those of freshwater and oceanic water, but the differences are within a few percent. In many applications they can be held constant across the salinity interval from 0‰ to 40‰. A very important exception is that in stratified waters even very small density differences are crucial to vertical and horizontal motion, and therefore the density field must be accurately mapped or modeled. Qualitatively, brackish water is like oceanic water in this respect.

The osmotic pressure[7] of seawater is sensitive to salinity, and this has important biological consequences. Freshwater and marine species, in particular vertebrates, have developed their strategy to survive with higher or lower osmotic pressure in their cells than the external medium, and therefore they have limited possibilities for adaptation to life in brackish water.

In marine biology, brackish water systems have specific characteristics. These systems exert pressures on marine species, and thus species composition is limited since only few freshwater and oceanic species are adapted to Baltic Sea conditions. Important factors are the dependencies of density of the water and osmotic pressure on salinity. There are major differences even inside the Baltic Sea (e.g., cod breed only in the southwestern Baltic Sea and in the Gotland Sea, while vendace is found only in the northernmost Gulf of Bothnia).

---

[7] Osmotic pressure is the pressure due to different solute concentrations produced in a solution in a space divided by a semipermeable membrane.

Table 3.2. Physical properties of Baltic Sea brackish water compared with freshwater and normal oceanic water at a temperature of 10°C and an atmospheric standard pressure of 1.013 bar.

	Freshwater $S = 0$	Baltic Sea $S = 7.5‰$	Oceanic $S = 35‰$	Reference
Mechanical properties				
Density (kg/m^3)	1,000	1,005	1,025	UNESCO (1981)
Dynamic viscosity (kg m^{-1} s^{-1})	1.31	1.33	1.40	Dietrich et al. (1963)
Compressibility (10^{-5} bar^{-1})	4.82	4.69	4.39	UNESCO (1981)
Speed of sound (m/s)	1,447	1,457	1,490	Chen and Millero (1977)
Thermal properties				
Specific heat (kJ kg^{-1} °C^{-1})	4.192	4.146	3.988	Cox and Smith (1959)
Thermal conductivity[a] (W m^{-1} °C^{-1})	0.59	0.58	0.56	Krümmel (1907)
Electromagnetic properties				
Static electric conductivity (S m^{-1})	<0.1[b]	0.92	3.77	Apel (1987)
Static dielectric constant	84.0	81.9	74.9	Meissner and Wentz (2004)
Osmotic pressure (bar)	0	5.2	22.5	Stenius (1904)

[a] At 17.5°C, temperature dependency is weak.
[b] For drinking water ~0.01, much less for distilled water.

3.3 BALTIC SEA WATERMASSES

3.3.1 Watermasses, stratification, halocline, and thermocline

The art of *hydrography* aims to characterize the physical and chemical properties of oceanic waters and their circulation by empirical methods.[8] Usually, this includes the fields of physical state variables—temperature, salinity, pressure, and density—and certain chemical properties such as oxygen concentration, nutrients, and turbidity. Here we mainly examine the physical properties. We use the abovementioned definition throughout this book.

In Section 3.2, the equation of state and the physical properties of seawater were presented. Now, a detailed look at the hydrography of the Baltic Sea will be made. The key factor is stratification and related gradients in hydrographic parameters. *Water type* refers to water with a given temperature and salinity, while *watermass* refers to a given set in the temperature–salinity (TS) space (e.g., Pickard and Emery,

---

[8] This definition is widely used among Baltic Sea oceanographers. In some other connections "hydrography" refers to the topography of the sea floor.

1990). For example, in the North Atlantic the warm and salty Mediterranean watermass can be clearly recognized. Water types and masses help to trace the general structure of circulation in the World Ocean. Temperature and salinity are so-called conservative quantities. The mean value or total amount of these quantities can only change at boundaries, whereas inside the water body only advection and mixing take place. Strictly speaking, solar radiation directly influences the temperature of a thin surface layer (about 10 m in the Baltic Sea). Watermasses are transported by currents, mixed by diffusion, and when watermasses with different densities meet the denser one sinks below the lighter one.

Due to convection and wind-induced mixing, homogeneous layers with constant density are formed in the water body. When the vertical density difference between two layers increases, mixing across the interface decreases. A thin boundary between two layers with a pronounced gradient is called a jump layer or a "cline". The term "cline" comes from the graphical form of vertical profiles: these are steep inhomogeneous layers but inclined at layer boundaries. The "cline" in temperature is called a *thermocline*, in salinity it is called a *halocline*, and in density it is called a *pycnocline*. Normally, a pycnocline is associated with a thermocline or halocline. The name "halocline" has its origin in the old method of determining salinity by titrating chlorinity giving the sum of halogens as the result.

A watermass can be bounded by coastline, bottom, or other watermasses with different TS characteristics. Due to the shape of bottom topography, like sills and channels, the movement of a watermass can be restricted to a certain area. At the horizontal boundary of two watermasses, hydrographic characteristics change abruptly across a narrow transition zone called a *front*. There the denser watermass sinks under the lighter one. Fronts may be found for climatological reasons, and in such areas convection can be continuous. In a frontal zone the pressure difference drives the horizontal flow parallel to the isopycnal lines of the front.

The signatures of watermasses can be studied graphically using a T–S diagram, where temperature and salinity are shown on a two-dimensional plot. This approach was first introduced in 1916 by Norwegian marine scientist Bjørn Helland-Hansen (1877–1957). With the help of a T–S diagram, different watermasses can be distinguished, and the origin and mixing of the watermasses can be examined. A T–S diagram forms a scatter plot, with clusters made up of the prevailing water types (see Figure 3.13).

3.3.2 Interaction between the Baltic Sea and the North Sea

Before discussing the detailed stratification conditions in the Baltic Sea, its interaction with the North Sea is presented due to the North Sea's major contribution to the overall stratification conditions in the Baltic Sea and beyond it in the Skagerrak and Kattegat (Figure 3.8).

The high net freshwater supply to the Baltic Sea (on average, 480 km^3/yr) leads to a strong outflow of low-density surface waters to the North Sea constituting about 60% of the total freshwater supply to the North Sea (Figure 3.9). This equates to a discharge of 15,000 m^3/s approaching that of the world's major rivers (it is next in the

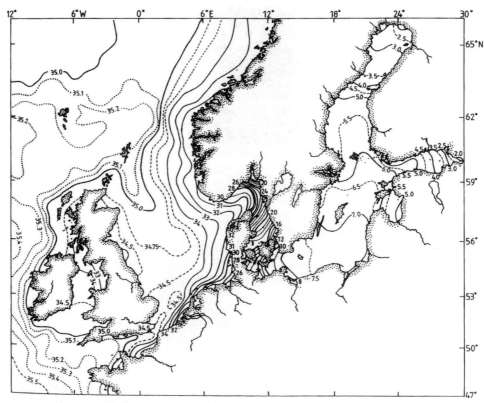

Figure 3.8. Climatological mean surface salinity (in ‰) in the Baltic Sea and North Sea in August. (Reproduced from Rodhe, 1999, with permission from Wiley.)

league table of discharge to the Yangtse River, 22,000 m³/s, which is the fourth largest river in the world). On the other hand, the more saline and dense North Sea waters flow into the Baltic Sea in the bottom layer. Actually, there is a permanent gradient in surface salinity from the eastern and northern extremes of the Baltic Sea toward the Danish Straits. The increase up to normal oceanic salinity in the North Sea takes place in the Belt Sea–Öresund region, the Kattegat, and the entrance to the Skagerrak. Here, the main fronts are the Belt Sea Front and the Skagerrak Front. In other parts of the Baltic Sea, temporal variations in salinity are mostly small compared with this dynamically active area. After leaving the Kattegat, the low-saline water of Baltic Sea origin follows the coast of Sweden and Norway, forming the Norwegian Coastal Current. Salinity fronts are common in this region.

The low-saline water of the Baltic Sea undergoes intense mixing with the high-saline bottom layer water in the Danish Straits. Due to the difference in length and topography between the Belts and Öresund, the water entering the Kattegat through the Öresund has a considerably lower salinity than that coming from the Belts. These two flows follow different paths through the Kattegat, along the Swedish and Danish

Baltic Sea watermasses

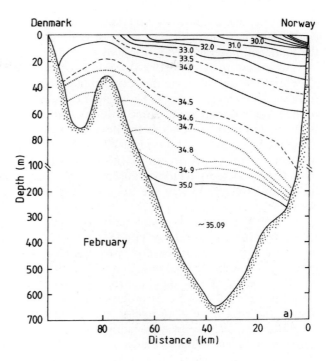

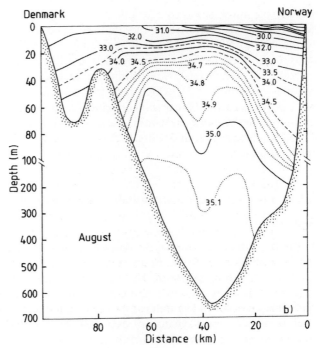

Figure 3.9. Cross-section of salinity in February and in August (in ‰) in Skagerrak. (Reproduced from Rodhe, 1999, with permission from Wiley.)

coasts, thus maintaining a salinity gradient. Furthermore, salinity increases northwards due to entrainment. The water of Baltic Sea origin covers the entire surface of the Kattegat, bounded from below by a pronounced halocline.

According to Rodhe (1999), the low-saline surface water forms a pool in the Kattegat. The outflow from the Kattegat to the Skagerrak compensates for the inflow from the Danish Straits to the Kattegat and the entrainment of the water below. This outflow generates an anticyclonic vorticity and thus maintains the westward intensification of surface currents in the northern Kattegat. This flow goes farther north along the Swedish and Norwegian coast in the Skagerrak. The outflow also feeds the low-salinity side of the Skagerrak front, running in a southeast–northwesterly direction. The compensating inflow of deepwater is strongly guided by the topography and follows the western side of the trench that penetrates from the Skagerrak into the Kattegat.

The flow through the Danish Straits and over the sills from the Belt Sea and Öresund to the interior of the Baltic Sea is by no means a steady two-layer flow. Here water exchange is predominantly barotropic (i.e., depends on the changing sea-level difference between the Baltic Sea and the North Sea). The instantaneous outflows and inflows are up to $25\,\mathrm{km}^3/\mathrm{day}$ (an order of magnitude larger than the mean) and have salinities within 8‰–28‰. It is important to mention that in the Danish Straits, outside the sills, there is an intense entrainment of surface water, originating from the Baltic Sea, into deepwater, leading to a considerably lower salinity of the inflowing water than that found in the deep layer of the Kattegat. The strength and duration of the inflows, as well as mixing in the straits, are important factors in determining the salinity of inflowing deepwater into the Baltic Sea. Since the halocline in the Baltic Sea proper is found far below the sill, there is no feedback from deepwater conditions to the inflow (Rodhe, 1999). The major part of the inflows that maintain Baltic Sea stratification are related to more or less continuously occurring events with a short period and with a moderate salinity in each case. On the other hand, the major inflows (see Chapter 5), which take place on average once in 10 years, determine not only the renewal of near-bottom layers with oxygen-rich inflowing waters but also strongly affect the overall stratification of the Baltic Sea. Figure 3.10 summarizes the Baltic Sea and North Sea water exchange together with Baltic Sea stratification in a schematic plot.

3.3.3 Vertical structure of the Baltic Sea water body

Salinity

In the Baltic Sea, salinity mostly determines the stratification of watermasses. Inflowing watermasses enter the Baltic Sea, move farther in, sink, and fill the deepwater pools. As a result the Baltic Sea water body has a permanent two-layer structure (Figure 3.11a): the *upper layer* and the *bottom layer* (or *lower layer*), separated by a halocline. The depth of the location of the halocline is usually 40 m–80 m, but in the shallow southwestern basins it is even less. The upper layer is homohaline, while the bottom layer is continuously stratified.

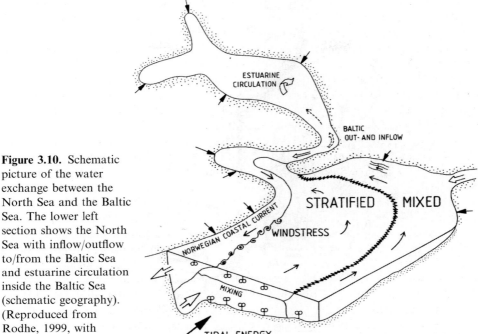

Figure 3.10. Schematic picture of the water exchange between the North Sea and the Baltic Sea. The lower left section shows the North Sea with inflow/outflow to/from the Baltic Sea and estuarine circulation inside the Baltic Sea (schematic geography). (Reproduced from Rodhe, 1999, with permission from Wiley.)

Salinity stratification is influenced by the formation of a seasonal *surface layer* in summer. This is warm, well-mixed, and isolated from the deeper water due to its lower density, added to which spring and summer runoff lowers its salinity. Figure 3.11b shows the mean vertical gradient of salinity and temperature in the Gotland Deep. A single, clear peak is shown at a 70 m depth.

The depth of the halocline is determined by advection, wind-induced and convective mixing, and sill depths. This depth changes very little in time except in the Danish Straits. Here, haline stratification is different from that in the Gotland Sea. It is wedge-shaped and this structure moves back and forth in relation with prevailing wind conditions. In areas where the vertical stratification of salinity is weak, the halocline can disappear in certain specific conditions. For example, in the western Gulf of Finland the dominating southwesterly winds work against standard estuarine circulation. Long-lasting and strong winds push a large amount of relatively fresh surface water into the Gulf of Finland increasing hydrostatic pressure, which may lead to gradual export of the salt wedge at the bottom layer of the basin (Elken *et al.*, 2003). The reversal may occur if southwesterly winds exceed the mean value of $4 \mathrm{m\,s^{-1}}$–$5.5 \mathrm{m\,s^{-1}}$.

A special case is the easternmost Gulf of Finland where freshwater from the River Neva has a strong effect on stratification. In this shallow section, salinity increases approximately linearly with depth. In the Gulf of Bothnia the halocline

66 Topography and hydrography of the Baltic Sea [Ch. 3

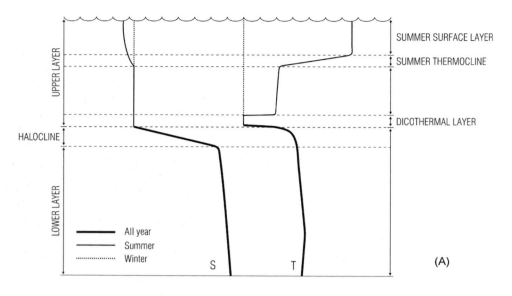

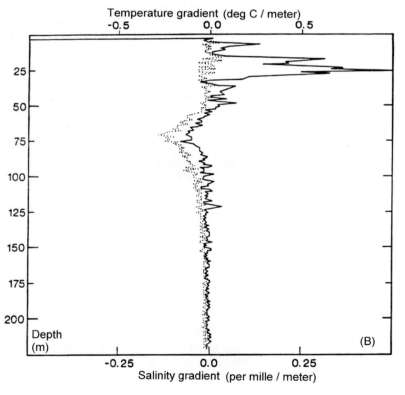

Figure 3.11. (top, A) A schematic diagram of vertical stratification of Baltic Sea watermasses. (bottom, B) Mean vertical gradient of salinity and temperature vs. depth (m). (Reproduced from Alenius and Leppäranta, 1982.)

is at its weakest, but still the salinity difference between the upper layer and bottom layer is more than 0.5‰ in the deeper areas, preventing vertical mixing of the entire water body.

The thickness of the halocline is 10 m–20 m. Watermasses below the halocline have no direct interaction with the wind-induced and convective mixing that is restricted to the homohaline upper layer. Stratification in the lower layer, maintained by advection, is almost linear with depth. A transient secondary halocline may be formed between the "old" water in the lower layer and recently advected saltier bottom water at a depth of some 125 m (Mälkki and Tamsalu, 1985). However, such a secondary halocline does not show up in statistical data (Figure 3.11b).

Salinity stratification also has seasonal variability, but these changes are much weaker than the corresponding variability in temperature. A salinity minimum is observed in spring in the surface layer due to the freshwater flux from rivers. The formation of a thermocline and weak vertical mixing in the upper layer through the thermocline keeps the freshwater in the surface layer and thus further reduces salinity there. This salinity minimum is usually 0.5‰ lower than the winter maximum. There is a lag between the salinity minimum and maximum river runoff. For example, Launiainen (1982) found that minimum is observed in Utö about 2.5 months after the runoff peak (Figure 3.12, see color section). This corresponds to an average current velocity of a few centimeters per second. The stable summer stratification with weak vertical mixing induces a salinity maximum in the lower layer. In the fall and winter, mixing is strong and the salinity difference between the surface layer and the lower layer is reduced again. This development is most pronounced near the coast where a halocline might not exist.

The T–S curves in Baltic Sea waters are hook-shaped and indicate a continuous stratification rather than discrete watermasses (Figure 3.13a), based on long-term cruise data (Figure 3.13b). The vertical lines show surface layer mixing and temperature stratification when approaching the thermocline, whereas the horizontal line represents lower layer and salinity stratification. The hook shape reflects temperature increasing with depth, since saltier water also carries heat from the North Sea. These curves illustrate well how the halocline separates the advected lower layer and vertically mixed upper layer from each other. If the hook falls to the horizontal and the T–S line is L-shaped, then the lower water is isothermal but stratified in salinity. The hooks glide along the S-axis in line with distance from the Danish Straits.

Considering the vertical temperature–salinity structure of the Baltic Sea, it seems likely that double diffusive convection does not play an important role in watermass formation there. No studies are known to us that suggest opposite conclusions.

Temperature

In the Baltic Sea, temperature first follows the two-layer structure determined by salinity (Figure 3.11a). The upper layer experiences a remarkable annual cycle forced by the radiation balance and air–sea interaction, while the lower layer is decoupled from atmospheric forcing and dictated by advection from the North Sea. A warm

68 Topography and hydrography of the Baltic Sea [Ch. 3

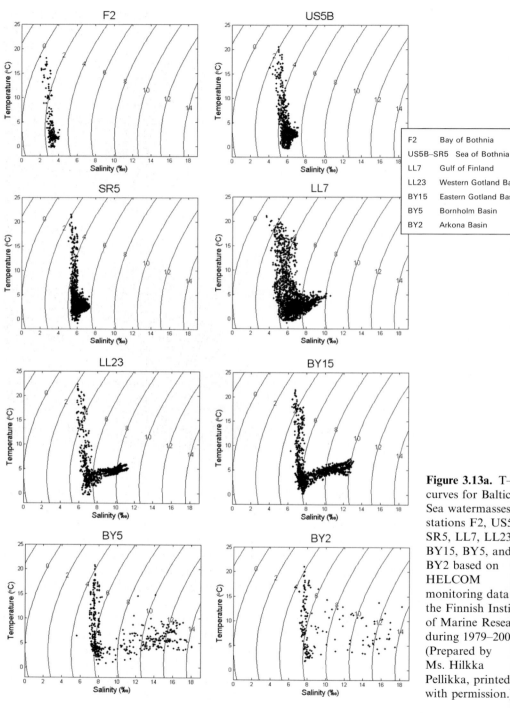

Figure 3.13a. T–S curves for Baltic Sea watermasses stations F2, US5B, SR5, LL7, LL23, BY15, BY5, and BY2 based on HELCOM monitoring data of the Finnish Institute of Marine Research during 1979–2007 (Prepared by Ms. Hilkka Pellikka, printed with permission.)

Figure 3.13b. Hydrographic surveying aboard R/V *Nautilus* using Nansen bottles. (Courtesy of the Finnish Institute of Marine Research.)

seasonal surface layer develops during summer with a seasonal thermocline in the upper layer. In the northern part of the Baltic Sea an inverse thermocline is formed in wintertime; temperature is at the freezing point at the surface and increases with depth to the temperature of maximum density (2°C–3°C). A dicothermal layer forms in summer at the bottom of the upper layer because the watermass is heated from the top and cold water remains farther down. This cold layer has also been called "winter water". The layer structure is clearly indicated in the mean temperature gradient (Figure 3.11b).

The seasonal evolution of temperature in the Baltic Sea is very different in the upper and lower layers. Due to the large seasonal variations in the energy balance at the sea surface, the surface water temperature reaches its maximum in summer and minimum in winter—and in winter at least a part of the sea freezes. The seasonal variability in temperature in the lower layer is weak. Largest variations are observed in the southern Baltic Sea where it mostly depends on advection from the North Sea through the Danish Straits. When the watermasses flow northwards they sink due to their higher salinity compared with the ambient water. This is why deep bottom waters are relatively warm: typically in the Gotland Basin the bottom temperature is 4°C–6°C, and in the Gulf of Finland, Gulf of Riga, and Gulf of Bothnia it is 2°C–4°C.

In the spring, after the ice melts, a thin upper layer is heated due to solar radiation (Figure 3.14, see color section). The surface waters quickly reach the temperature of maximum density T_m (1.5°C–3°C). In the southern Baltic the wintertime surface temperature may remain above T_m. When the temperature rises above T_m, the surface waters become lighter than the watermasses below. Thus, convection stops and a thermocline is formed separating the warm surface layer waters and the remarkably colder waters below it. Wind and solar radiation absorbed in the surface layer affect the depth of the thermocline. The thickness of the thermocline is typically 5 m–10 m and its shape varies a lot.

During summer the surface layer and thermocline are at the top of the upper layer. The surface layer is not always homogeneous but can possess a transient echelon structure consisting of minor thermoclines. The well-mixed surface layer is often defined to be such that the vertical change in temperature does not exceed some prescribed value, say 0.1°C/m (Table 3.3). A thermocline is usually easily determined from a single profile, but there are also cases where the interpretation is complicated. A statistical analysis of the distribution of thermocline depth can be made using an extensive dataset (Alenius and Leppäranta, 1982; see, e.g., Figure 3.11b).

The seasonal thermocline in summertime is located at a depth of 15 m–30 m in all basins of the Baltic Sea. The formation of the seasonal thermocline starts in the southern Baltic Sea at the beginning of May and in the Bay of Bothnia only one month later. The deepening of the thermocline due to autumn cooling starts in the north in late August whereas in the south it happens one month later (Figure 3.14, see color section). Sea surface temperature follows air temperature with a lag because the heat capacity of the sea is much larger than that of the atmosphere. In other words, the sea slowly reacts to the heat input from the atmosphere because of its large thermal inertia.

Table 3.3. The main characteristics of Baltic Sea stratification.

Layer	Thickness	Maintenance	Occurrence
Upper layer	40–80 m	Wind, convection	All year
Surface layer	10–20 m	Heating	Summer
Summer thermocline	5–10 m	Wind, Sun	Summer
Dicothermal layer	5–30 m	Cold winter	Summer
Halocline (HC)	10–20 m	Advection, water balance, wind, convection	All year
Lower layer	HC–bottom	Advection	All year
Secondary halocline	125 m–bottom	Advection	Transient

During the summer the thickness of the surface mixed layer deepens to some extent mainly due to mechanical mixing caused by wind forcing (Figure 3.15). The summer thermocline is strong, with a temperature drop of as much as 10°C over a distance of a few meters. It prevents to a large extent wind-induced mixing from affecting the layer below. The exchange of material and heat is also strongly weakened, which has important effects on biochemical processes (see the upwelling section in Chapter 8). The relatively warm and calm summertime weather restricts the deepening of the surface mixed layer to 10 m–20 m. The thermocline enables freshwater from rivers to stay in the surface layer due to the weak mixing across it, and this in turn strengthens the density difference across the thermocline. Thus, warming and the freshwater content of the surface layer have a positive effect on each other. In turn, bottom layer salinity increases in summer.

A specific feature of the Baltic Sea and many other ice-covered seas is a *dicothermal* layer at the bottom of the upper layer. The formation of such a layer takes place in the following way. During the spring the upper layer starts to warm from the surface while the temperature lower down is close to freezing. On the other hand, deep-layer water remains throughout the year at temperatures between 3°C and 6°C, but has a much higher density than upper layer water due to its higher salinity (Figure 3.15). The dicothermal layer remains throughout the summer and disappears only in autumn convection. The dicothermal layer is sometimes called "old winter water"

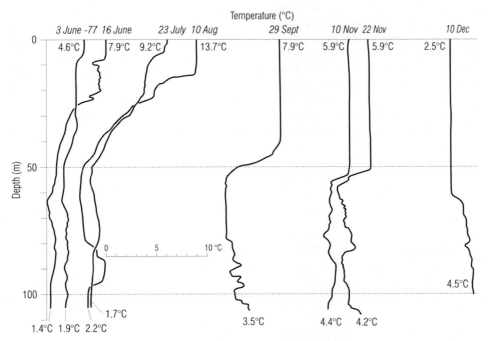

Figure 3.15. Time evolution of the different stages of thermocline development in the Bay of Bothnia in 1977. (Redrawn from Alenius, 1978 and Mälkki, 1978.)

due to its mechanism of formation. Consequently, temperature has a four-layer structure during the summer season: surface layer, thermocline, dicothermal layer, and lower layer.

Usually in late August, the energy balance of the sea surface becomes negative, the surface waters cool, become heavier, and sink due to convection. This leads to deepening of the surface mixed layer and to weakening of the thermocline (i.e., weakening of the vertical temperature gradient) (Figure 3.16). This process continues during the autumn. At the same time the lower parts of the upper layer are still warming due to thermal diffusion. For example, at a 30 m depth the temperature maximum is reached only in October while at the surface the highest values are measured usually during late July–early August (Figure 3.14, see color section). During late autumn or early winter the upper layer becomes isothermal as a consequence of thermohaline convection and wind-induced mixing. However, such mixing does not destroy the halocline, which remains throughout the year in the central basins.

As soon as the surface temperature has dropped below the temperature of maximum density, transient weak winter thermoclines may form close to the surface. But, in general, mixing continues due to forced mechanical convection. It is worth noting that as the temperature of maximum density decreases with increasing salinity, the strength of this maximum decreases at the same time. In other words, the difference between maximum density and the density at freezing point decreases with increasing salinity. The inverse winter thermocline is therefore more stable in freshwater than in brackish water.

In fact, the permanent winter thermocline, where temperature increases approximately from freezing point to that of maximum density, is located usually at the halocline. But, due to the effect of horizontal advection, water temperature beneath the winter thermocline can be much higher than the temperature of maximum density.

Ice forms in the Baltic Sea annually for 5–7 months. Landfast ice occurs in coastal and archipelago areas; significant salinity stratification may form even in these shallow waters in the neighborhood of river mouths. Farther out, drift ice fields exist, where mechanically forced mixing continues due to the motion of the ice whose strength is similar to that in the ice-free winter season.

3.3.4 Salinity and temperature in different basins

Salinity in the Baltic Sea decreases with distance from the Danish Straits (Table 3.4). The mean circulation pattern is anticlockwise (cyclonic) in the main basins. Thus, upper layer temperature and salinity are usually higher in the eastern parts of the basins than in the western parts (Figures 3.17–3.18; see also Chapter 5). In the Kattegat, salinity has a pronounced two-layer structure. Between these watermasses there is a permanent, strong halocline at a depth of about 15 m. Surface salinity varies from 18‰ to 26‰ and bottom salinity from 32‰ to 34‰ correspondingly.

Sec. 3.3] Baltic Sea watermasses 73

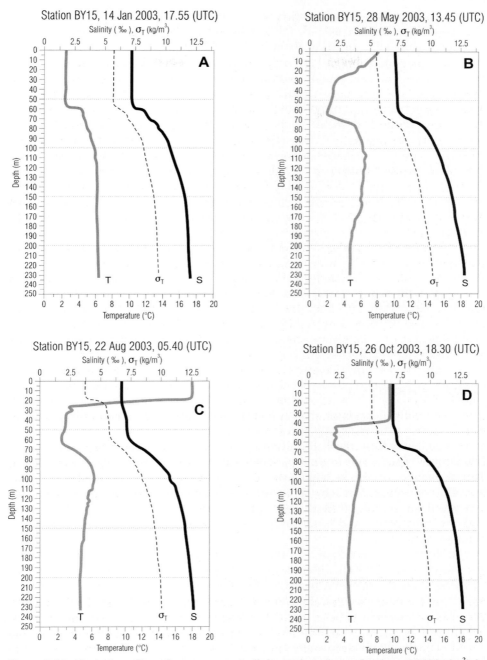

Figure 3.16. Typical profiles of temperature, salinity and σ_T (σ_T = density—$1{,}000\,\mathrm{kg/m^3}$) in the Gotland Deep in 2003 (source: Finnish Institute of Marine Research). (A) Winter; (B) spring; (C) summer; (D) autumn. (Prepared by M.Sc. Riikka Hietala, printed with permission.)

Table 3.4. Typical values of salinity in different basins of the Baltic Sea (according to Bock, 1971). In the southwestern Baltic Sea and in the Gulf of Finland, variations are pronounced.

Basin	Salinity (‰) Upper layer	Salinity (‰) Lower layer	Halocline depth (m)
Bay of Bothnia	2–4	4–4.5	50–60
Sea of Bothnia	5–6	6–7	60–80
Gulf of Finland	0–6	3–9	60–80
Gulf of Riga	4.5–6	6–7	20–30
Gotland Sea	6.5–8	9–13	60–80
Bornholm Basin	7.3–8.5	13–17	50–60
Arkona Basin	7.5–8.5	10–15	20–30
Belt Sea	8–24	10–30	—
Kattegat	18–26	32–34	15–20

Southwestern Baltic Sea

There is usually a two-layer structure in the Belt Sea just like that in the Kattegat. Surface layer water is usually Baltic Sea water (salinity 8‰–12‰) and lower-layer water originates from the Kattegat with salinities up to 32‰–34‰. Thus, salinity conditions vary in a wide range depending on meteorological conditions. In the Great Belt the watermass may be homogeneous when the mixing is strong. In the Öresund, stratification is influenced by three types of watermasses: surface water from the Arkona Basin (salinity 7.5‰–8.5‰), surface water from the Kattegat upper layer (salinity 18‰–26‰), and lower layer water from the lower layer of the Kattegat (salinity 32‰–34‰).

In the southwestern Baltic Sea the basins are shallow and the halocline is strong because the Danish Straits with inflowing salty water are close and there is an outflow of Baltic Sea brackish water. In the Arkona Basin the depth of the halocline is 20 m–30 m. Surface salinity is typically 7.5‰–8.5‰, and under conditions of strong inflow it can be as much as 11‰ in the central basin. In the Darsser and Drogden Sills, salinity can be more than 17‰. In the lower layer, salinity is typically 10‰–15‰, even though larger instantaneous values have been measured. In the Bornholm Basin the upper layer is 50 m–60 m deep with a salinity of 7.3‰–8.5‰, whereas salinity in the lower layer is somewhat larger, 13‰–17‰. This variability is not seasonal but controlled by irregular saltwater inflows (Figure 3.19, Table 3.4).

In January in the southwestern Baltic Sea the average surface water temperature is 2°C–5°C. Bottom temperature is 3°C–5°C in the south, being up to 8°C in the Bornholm Basin. By the middle of April, warming has already started with sea surface temperatures of 3°C–5°C (bottom 2°C–5°C). In July–August the surface temperature is 16°C–17°C (bottom 5°C–14°C) while in October the temperature is still about 10°C–12°C. The corresponding bottom layer temperature is 7°C–12°C in the southwestern Baltic Sea (Figure 3.20, Table 3.5, see p. 84).

Gotland Sea

The Gotland Sea makes up about half of the Baltic Sea watermass. In the Eastern Gotland Basin the permanent halocline locates at a depth of 60 m–80 m. The salinity of the upper layer is 6.5‰–8‰; below the halocline, salinity increases more or less linearly with depth to 9‰–12‰ at a 100 m depth and to 11.5‰–13‰ at a 200 m depth. However, in shallow coastal areas, salinity is much less, especially near the river mouths. In the Northern Gotland Basin, surface salinity is lower than in the Eastern Gotland Basin due to cyclonic circulation in the Gotland Sea. The waters of the Northern Gotland Basin mix with those from the Gulf of Finland and Gulf of Bothnia, resulting in a surface salinity of 6.5‰–7.3‰ and a lower layer salinity of 9.8‰–11.5‰. In the Western Gotland Basin, surface salinity is 6.3‰–7.7‰, while below the halocline at a 100 m depth, salinity is 8.7‰–10.3‰. In the near-bottom layer in Landsort Deep, salinity is about 10‰–11.5‰.

Salinity data show how incoming North Sea water makes a counterclockwise basin-wide loop in the Gotland Sea. Pronounced salinity stratification in the relatively deep Gotland Basins leads to oxygen depletion in bottom water. Ventilation of the deepest basins takes place only when irregular major saltwater inflows enter from the North Sea. In January in the Gotland Basin, surface temperature is 2°C–3°C (bottom 3°C–6°C); in the middle of April, surface temperature is 1°C–2°C (bottom 2°C–5°C). In July–August, surface temperature is 15°C–17°C (bottom 5°C–10°C), and in October it is still about 10°C–12°C, while bottom temperature in October is 4°C–9°C.

Large gulfs in the east and north

Salinity has a two-layer vertical structure in the Gulf of Riga. The upper layer is thin, and the halocline is found at a depth of 20 m–30 m. The watermass in the upper layer has a local origin, whereas warm, saline water flows in from the Eastern Gotland Basin into the lower layer of the Gulf of Riga. Salinity in the upper layer is between 4.5‰ and 6‰ and at most 7‰ in the lower layer.

In the Gulf of Finland a two-layer structure partly exists. This gulf is a direct continuation of the Northern Gotland Basin with no sill between. The most voluminous river in the whole Baltic Sea—the Neva—is located at the eastern extremity of the Gulf of Finland. As a result, a permanent east–west gradient in salinity prevails. Surface salinity increases from 0‰ at the mouth of the River Neva up to 6‰–6.5‰ in the west. A halocline is found mainly in the western and central parts at a depth of 60 m–80 m. In the bottom layer, salinity is 7‰–9‰ in the western

76 Topography and hydrography of the Baltic Sea [Ch. 3

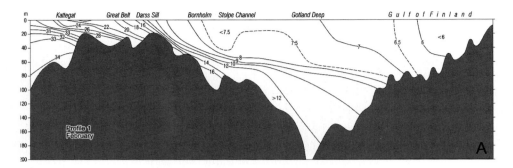

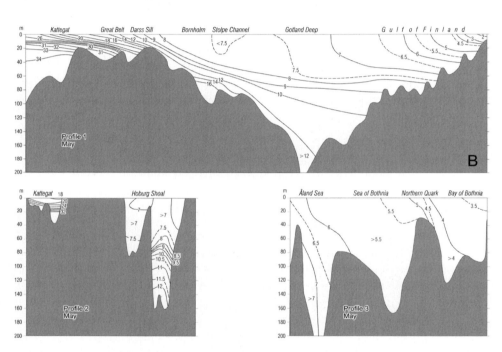

Figure 3.17. The salinity (‰) of the Baltic Sea water body in different seasons along different cross-sections: along the central axis Kattegat–Gulf of Finland, Kattegat–Eastern Gotland Basin in a west–east direction and along the central axis of the Gulf of Bothnia, Åland Sea—Bay of Bothnia. (A) February, (B) May, (C) August, (D) November. In February only the Kattegat–Gulf of Finland cross-section is available. (Redrawn from Bock, 1971.)

part, in the central part it is 5‰–8‰, and in the east it is 0‰–3‰. Near-bottom layers often suffer from oxygen depletion due to summer stratification and excess nutrient loading.

In January the eastern Gulf of Finland and the Gulf of Riga are frozen, and in the middle of April ice is still present. In July–August, surface temperature is 15°C–

Sec. 3.3]   Baltic Sea watermasses   77

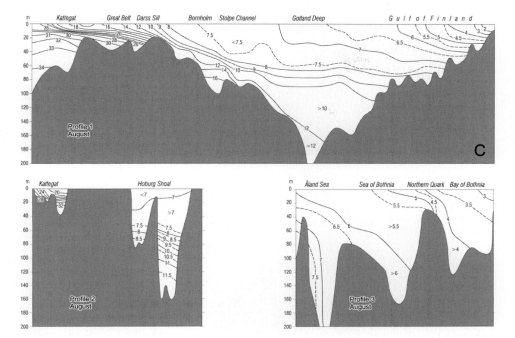

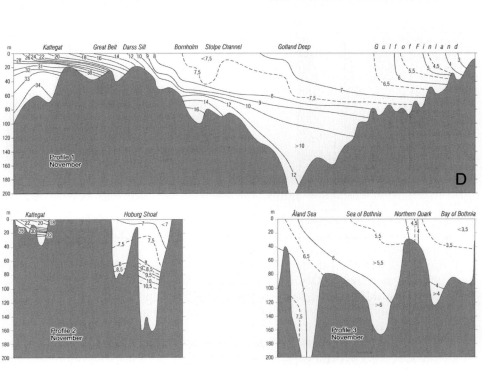

78   Topography and hydrography of the Baltic Sea   [Ch. 3

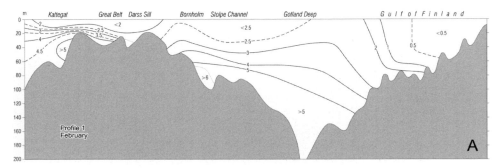

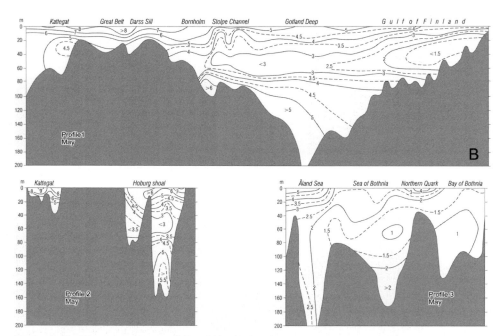

Figure 3.18. The temperature (°C) of the Baltic Sea water body in different seasons along different cross-sections: along the central axis Kattegat–Gulf of Finland, Kattegat–Eastern Gotland Basin in a west–east direction and along the central axis of the Gulf of Bothnia, Åland Sea—Bay of Bothnia. (A) February, (B) May, (C) August, (D) November. In February only the Kattegat–Gulf of Finland cross-section is available. (Redrawn from Lentz, 1971.)

17°C (bottom 2°C–5°C) in the Gulf of Finland and 17°C–18°C (bottom 7°C–15°C) in the Gulf of Riga, while the October surface temperature is 9°C–11°C in both basins. The temperature difference between the uppermost layer and the bottom layer clearly reduces from July, and in the Gulf of Finland the bottom temperature is 3°C–5°C whereas in the Gulf of Riga it is 5°C–9°C.

Stratification conditions in the Gulf of Bothnia greatly differ from those in the Gotland Basin. They depend on inflow and outflow between the Gulf of Bothnia and

Sec. 3.3] **Baltic Sea watermasses** 79

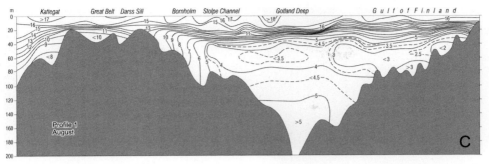

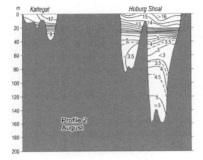

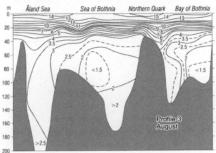

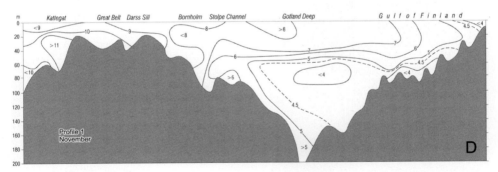

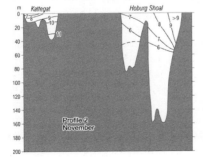

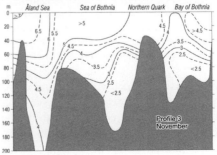

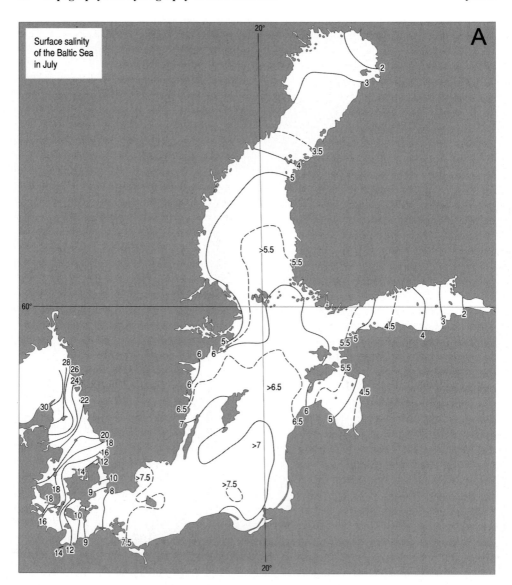

Figure 3.19. The mean distribution of salinity (in ‰) in the Baltic Sea in July. (A) At the surface. (B) At the bottom. (Redrawn from Bock, 1971.)

the Northern Gotland Basin and on river runoff. Overall stratification is rather weak and mixing extends to deep layers, and as the advection of Northern Gotland Basin upper layer water is fast enough there are normally no bottom areas without oxygen. However, in deep areas, stratification is strong enough to restrict autumn cooling from extending down to the bottom, and thus bottom waters are renewed only by

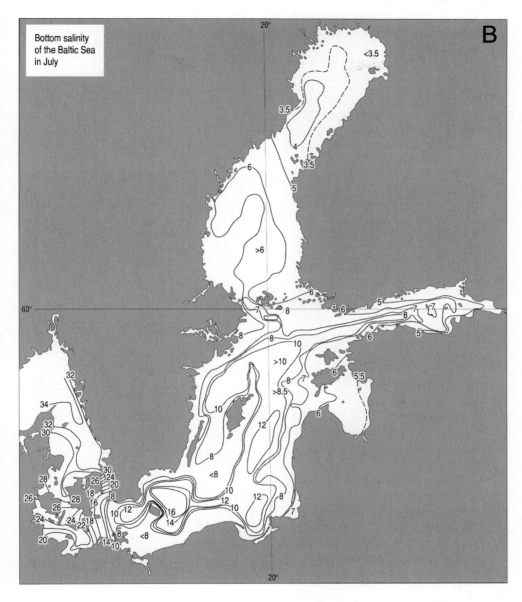

advection from the south as in other Baltic Sea basins. A weak halocline is found in the Sea of Bothnia at a depth of 60 m–80 m and in the Bay of Bothnia at 50 m–60 m.

In the Åland Sea, surface salinity is 5.25‰–6.25‰ whereas at a depth of 200 m salinity varies between 7‰ and 7.75‰. The lower layer watermass in the Åland Sea, as in the Sea of Bothnia, originates mostly from the upper homohaline layer of the Northern Gotland Basin; however, a small fraction of more saline deepwater flows in over sills. Sometimes in the near-bottom layer, saline water can flow in through the

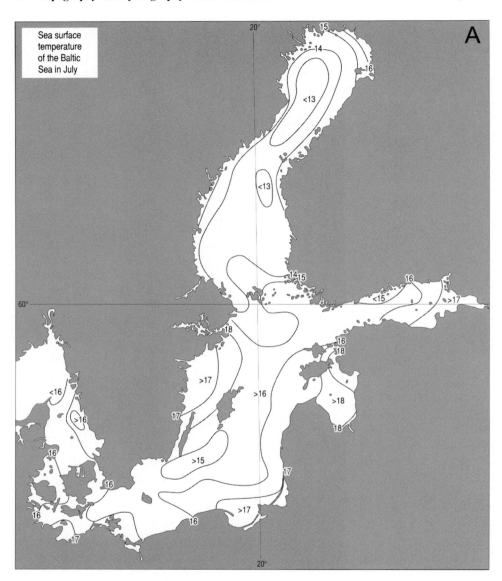

Figure 3.20. The mean distribution of temperature (in °C) in the Baltic Sea in July. (A) At the surface. (B) At the bottom. (Redrawn from Lentz, 1971.)

Åland Sea to the Sea of Bothnia, and a corresponding volume of fresher water flows out to the Gotland Basin. This strengthens stratification in the Sea of Bothnia.

In the Sea of Bothnia, surface salinity varies between 4.8‰ and 6.0‰; in the lower layer at a 150 m depth, salinity is 6.4‰–7.2‰. In the Bay of Bothnia, salinity is

Sec. 3.3] Baltic Sea watermasses 83

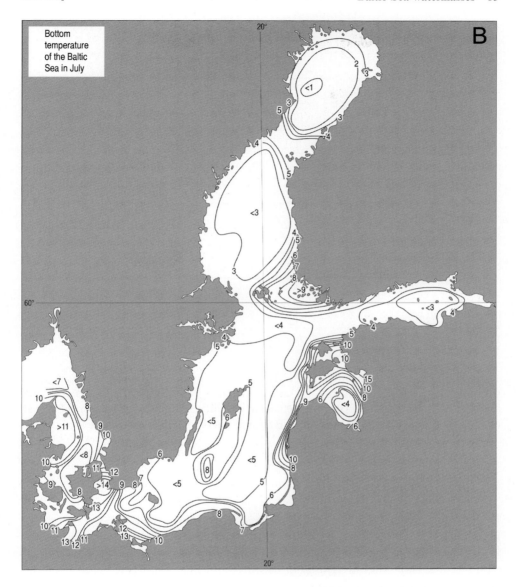

between 2‰ and 3.8‰ and at a 100 m depth near the bottom it varies between 4‰ and 4.5‰. Many rivers empty into the Gulf of Bothnia and near the river mouths salinity is close to zero (Figure 3.19).

In January the Bay of Bothnia is frozen, and in the middle of April there is still ice in the Gulf of Bothnia. In July–August, surface temperature is 13°C–15°C. The near-bottom water temperature is 1°C–4°C in the Bay of Bothnia whereas in the Sea of Bothnia the corresponding values are 3°C–7°C. In October, the cooling of surface

Table 3.5. Mean values of surface and bottom temperatures (in °C) in the Baltic Sea for different seasons (according to Lentz, 1971).

Basin	Temperature (°C): surface/bottom			
	January	April	July	October
Bay of Bothnia	F/×	F/×	13–15/1–4	4–6/3–5
Sea of Bothnia	F/×	F/×	13–15/3–7	5–8/3–8
Gulf of Finland	F/×	F/×	15–17/2–5	9–11/3–5
Gulf of Riga	F/×	F/×	17–18/7–15	9–11/5–9
Gotland Sea	2–3.5/3–6	1–2/2–5	15–17/5–10	10–12/4–9
Bornholm Basin	4/5–8	2–3/3–5	16–17/5–6	11–12/7–9
Arkona Basin	3–4/4–5	3–5/2–4	16–17/7–13	12/9–12
Belt Sea	2–3.5/3–5	5/4–5	16–17/8–14	12/12
Kattegat	3/5–6	5/5	16–17/8–11	10–11/10–12

F = the surface in some years is frozen; × = no data.

water has already started. In the Bay of Bothnia, sea surface temperature is only 4°C–6°C and in the Sea of Bothnia 5°C–8°C. The temperature difference between the uppermost layer and the bottom layer clearly reduces from July due to mechanical and thermal convection. In the Bay of Bothnia, near-bottom temperature is 3°C–5°C and in the Sea of Bothnia it is 3°C–8°C.

3.4 LONG-TERM VARIATIONS

In this section the focus is on examining the hydrography of the Baltic Sea during the instrumental period of the last 100 years. The collection of observations commenced in the late 1800s at fixed coastal and island stations and in standard oceanographic stations. The variability in hydrographic conditions has been large due to the large variability in regional climate, which hosts a continental climate on the eastern side and a polar front in the north. For example, the average January air temperature in Helsinki varied from −16.5°C to +1.4°C between 1900 and 2000. In the summertime, variations are smaller due to the regularity of the dominant forcing: solar radiation.

Regional climate variations can be characterized to some extent using the NAO index. A positive NAO implies the dominance of westerlies and therefore warmer winters and less ice. But the influence goes beyond the heat budget, since the strong westerlies and low atmospheric pressure produce higher sea levels in the Baltic Sea. In addition, the inflow of salty North Sea water is stronger during westerlies. Even specific ecological characteristics such as species composition and quantities of zooplankton have been noted to vary in line with the NAO index and with salinity variations (see, e.g., Vuorinen et al., 2003).

3.4.1 Temperature

Several investigations have indicated that the surface temperature of the Baltic Sea has risen by about 0.5°C during the last 100 years. Figure 3.21 shows results from the southern Baltic Sea with a warming trend. The reason is not yet exactly clear, but it is obviously connected to a similar rise in atmospheric surface layer temperature in the region. Fonselius and Valderrama (2003) made an extensive study of the climatology of deepwater temperature. Bottom layer temperature is connected to saltwater pulses, but it is difficult to find any clear trend here. However, some increase has been observed (Figure 3.22). In addition, in recent years warm summer inflows (see Section 5.4.4) have heated the water and consequently increased bottom layer temperature (see Chapter 5).

3.4.2 Salinity

Salinity variations on decadal timescales are dictated by the water balance (i.e., precipitation, evaporation, river runoff, and oceanic water exchange). The vertical distribution of salinity is also influenced by the stratification and mixing of watermasses. Salinity variations have been small though significant during the last 100 years.

In the Gotland Sea, salinity decreased at the beginning of the 1930s, but since World War II it has increased in the whole Baltic Sea by 0.5‰–0.9‰. The peak level was recorded at the beginning of the 1950s following the greatest known oceanic water inflow event in 1951. A local minimum

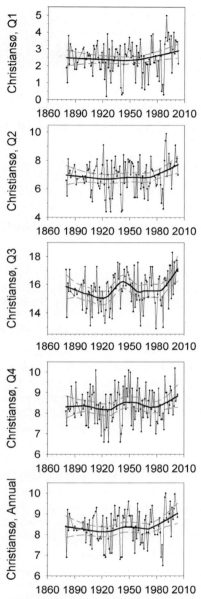

Figure 3.21. Temperature evolution in Christiansø, Denmark during 1880–1998 for different seasons (winter–spring–summer–fall) and for the annual mean. (MacKenzie and Schiedek, 2007, printed by permission from *Global Change Biology*.)

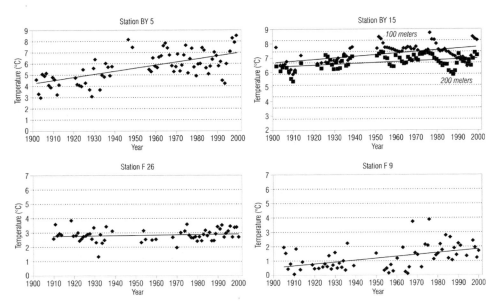

Figure 3.22. Long-term variability in deepwater temperature in Bornholm Basin (BY5), Eastern Gotland Basin (BY15), Sea of Bothnia (F26), and Bay of Bothnia (F9) in 1900–2000. (Redrawn from Fonselius and Valderrama, 2003.)

was recorded ten years later, and then salinity was first nearly constant but began to decline in 1978. In 1993 a new major oceanic inflow event took place stopping the decrease in salinity. Throughout the 100-year instrumental period, upper layer and lower layer salinities have been positively correlated, but the range in variability has been 3–4 times larger in the lower layer. The depth of the halocline has varied within 20 m, but it has always been strong in the Gotland Sea (Figure 3.23).

Variations in the salinity of the Baltic Sea are connected to regional climate conditions in the drainage basin. Air pressure and wind influence the occurrence of salt inflows, and precipitation influences river runoff. It has been hypothesized that increasing precipitation would lead to an increase in freshwater outflow from the Baltic Sea. Then the Kattegat watermass would become more fresh, new salt inflows would be less likely to occur, and the salinity level of the Baltic Sea would go down. According to Launiainen and Vihma (1990) variations in river runoff are only reflected in the salinity level after a certain delay (see Section 5.4.5).

3.4.3 Ice conditions

The Baltic Sea is in the seasonal sea ice zone of the World Ocean, at the climatological sea ice margin. Therefore, interannual variability in ice conditions is very large. In very cold winters the whole Baltic Sea freezes over, while in very mild winters ice occurs in only 12.5% of its surface area: the Bay of Bothnia and the Eastern Gulf of Finland (an all-time minimum was reached in winter 2007–2008). On average, the

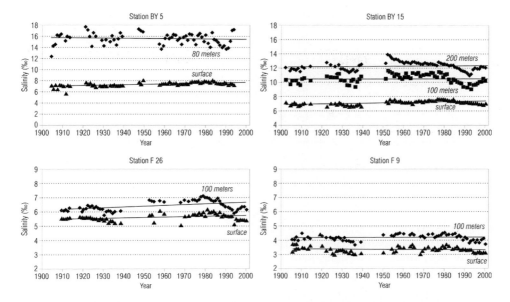

Figure 3.23. Long-term variability in salinity in Bornholm Basin (BY5), Eastern Gotland Basin (BY15), Sea of Bothnia (F26), and Bay of Bothnia (F9) in 1900–2000. (Redrawn from Fonselius and Valderrama, 2003.)

annual maximum ice extent is 45% of the surface area of the Baltic Sea. Because the thermal memory of the Baltic Sea is just 2–3 months, there is no correlation in the ice conditions between consecutive winters.

There are long time-series on ice conditions in the Baltic Sea, since such a phenomenon as ice break-up in spring is an important well-noted event and keen naturalists keep records of such dates. The longest time-series is about the ice break-up in River Daugava (from 1550) for the Baltic Sea drainage basin and ice break-up in Helsinki (from 1829) for the Baltic Sea itself. There are indirect long time-series concerning sea traffic conditions (i.e., the commencement and end of shipping at certain harbors each year). Probably the best known and most widely analyzed time-series is on the maximum annual ice extent in the Baltic Sea (begun in 1720) initially constructed by Risto Jurva (see Jurva, 1952; Palosuo, 1953) and updated thereafter by the FIMR Ice Service (Figure 3.24). These data are indirect for the early years since information from the central basins was quite limited from the days of sailboats right up to the times of aerial or spaceborne ice reconnaissance. For those years Jurva constructed numbers based on coastal ice conditions.

The maximum annual ice extent of the Baltic Sea shows very large variability with no temporal structure; it is white noise with large variance. The whole basin is completely ice-covered on average once every 30 years, more often during the 1800s, the most recent case being 1947. However, in 1987 the coverage was as high as 96%. Very mild winters occurred in the 1930s and in 1988–1993. In 1992 the Bay of Bothnia

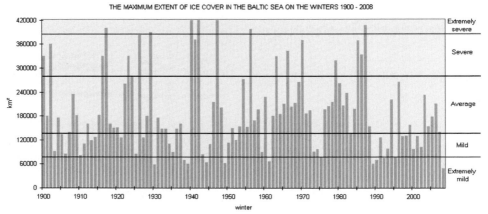

Figure 3.24. Maximum annual ice extent in the Baltic Sea during 1900–2008 in square kilometers; 420,000 km^2 corresponds to the Baltic Sea and Kattegat together. (Prepared by M.Sc. Jouni Vainio, printed with permission.)

was fully ice-covered only for about three weeks, in 2008 only about one week. Ice seasons have been very closely connected to the NAO, which governs wintertime temperature conditions. The correlation coefficient has been quite high (0.7–0.8) between ice coverage and mean air temperature (Palosuo, 1953) and between ice coverage and the NAO index (Tinz, 1996).

This chapter began with an introduction to Baltic Sea topography. It is a shallow sea with a mean depth of just 54 meters and with specific topographic characteristics in each basin. It is a brackish water basin with considerably lower salinities than in the oceans and with pronounced horizontal gradients due to opposite effects caused, on the one hand, by incoming saline waters via the Danish Straits from the North Sea and, on the other hand, by the large freshwater surplus from rivers. The watermasses of the Baltic Sea are characterized by strong salinity stratification throughout the year, whereas temperature stratification has a seasonal cycle due the annual course of solar radiation. Surface temperatures in the Baltic Sea have increased slightly during the last 100 years, whereas mean salinity is at present more or less the same as 100 years ago. Chapter 4 will give a detailed description of the water, salt, and heat budgets of the Baltic Sea.

4

Water, salt, and heat budgets

Martin Hans Christian Knudsen (1871–1949) was a Danish physicist who taught and conducted research at the Technical University of Denmark. He is primarily known for his study of molecular gas flow. He was also very active in physical oceanography, developing methods for defining the properties of seawater. He was the editor of *Hydrological Tables* (Copenhagen–London, 1901). Knudsen's formulae (Knudsen, 1900) are often applied to investigations of ocean water exchange in semi-enclosed basins. Knudsen also played an important and active role in the early years of the ICES. (Courtesy of Dr. Niels Kristian Højerslev.)

4.1 WATER BUDGET

The temperature and salinity fields of the Baltic Sea were presented in Chapter 3, and in this chapter we shall examine the water, salt, and heat budgets. These budgets guide the evolution of the physical conditions in the Baltic Sea. The water budget determines the flushing of Baltic Sea watermasses, the salinity level is externally forced by the exchange of water between the North Sea and the Baltic Sea, and the heat budget is coupled with atmospheric conditions. Budget questions are among the key objectives of the BALTEX research program (Omstedt *et al.*, 2004). Sections 4.1–4.3 discuss the budget based on observational material, and Sections 4.4–4.5 deal with modeling approaches using one-dimensional models and box models.

4.1.1 Physical problem

The volume of the Baltic Sea is 21,205 km^3. Annually about 480 km^3 of freshwater is added to the basin by river runoff and atmospheric net flux (precipitation minus evaporation), whereas a compensatory loss takes place through the Danish Straits as net water exchange with the North Sea (Figure 4.1). The conservation of Baltic Sea watermass requires that

$$\frac{d\xi}{dt} + \frac{H}{\bar{\rho}}\frac{d\bar{\rho}}{dt} = P - E + \frac{V_r + V_i - V_o}{A} \qquad (4.1)$$

where ξ is the average sea-level elevation; H is mean depth; $\bar{\rho}$ is mean density; P is precipitation; E is evaporation; V_r is river runoff; V_i and V_o are the inflows and outflows through the Danish Straits; and A is the surface area of the Baltic Sea (taken as constant here). Annual atmospheric net flux is one order of magnitude smaller than corresponding river inflow or open-boundary exchange $V_i - V_o$ (Ehlin, 1981). If the mean density of water was changed by 1 kg/m^3, the resulting sea-level change would be 50 mm—see Equation (4.1). Groundwater runoff is very small and can be thought of as included in river runoff. Ice formation and melting do not influence sea-level elevation except for grounded ice, but such ice, however, constitutes a very small portion of the whole watermass. Inflowing river ice and inflowing/outflowing drift ice in the Danish Straits must be added to liquid water for the total discharges.

The freshwater budget $P - E + V_r/A$ is strongly positive in the Baltic Sea throughout the year. This feature permanently supports the overall presence of a two-layer salinity structure with a strong halocline and gives rise to certain features of estuarine-type transport in which at the ocean boundary relatively fresh surface water flows out in the upper layer, while more salty oceanic water flows in the lower layer into the basin. Baltic Sea deepwaters are so salty and dense that they are renewed by horizontal advection rather than by convection (see Chapter 5 for details). A similar basin is the Black Sea with its anoxic deepwaters, whereas an opposite example is the Red Sea, where the incoming Indian Ocean water is less dense than the Red Sea water and comes in as a surface flow. Baltic Sea water balance is one of the principal research topics of the BALTEX program (Omstedt et al., 2004).

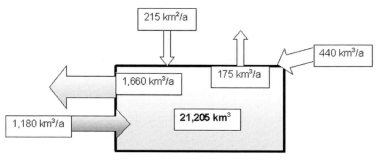

Figure 4.1. The annual Baltic Sea water balance. The Baltic Sea is shown as a box with water exchange with the North Sea at the left, atmospheric exchange of freshwater on the top, and river discharge on the right.

4.1.2 Precipitation and evaporation

Precipitation in the Baltic Sea region was discussed in detail in Section 2.4, which dealt with weather and climate characteristics. In this section, precipitation directly falling on to the Baltic Sea will be investigated. This is mainly connected to cyclone activity and thus occurs as frontal precipitation, with the level reaching 500 mm–600 mm per year (Table 4.1). More recent estimates seem to be on the higher side. Some of the precipitation comes down as convective showers in late summer and autumn caused by the conditional instability of the troposphere. However, in winter and spring the cold surface of the Baltic Sea prevents formation of convective precipitation to a large degree.

Observations have been taken using precipitation gauges in weather stations, but there are uncertainties in using these direct measurements. First, the weather stations are located on the coast and their readings may be affected by the proximity of the coast that, for example, considerably modifies wind properties. Only a few stations are located on small islands where mostly marine weather prevails. There are also some rain gauges mounted on ships (Smedman *et al.*, 2005). Second, much (in the north about half) of the precipitation comes down in the solid phase. This flux is more difficult to measure accurately than liquid precipitation. Expansion in the coverage of the Baltic Sea weather radar system has to some degree improved mapping of factual precipitation areas. Yet these data are basically qualitative information for real-time weather mapping and forecasting.

Figure 4.2 shows the mean precipitation chart over the Baltic Sea. Although the source is more than 50 years old, it still represents well present understanding (see Bergström *et al.*, 2001). The amount of precipitation is greater on the eastern coast of the Baltic Sea due to the governing westerly winds. The Scandinavian mountain chain produces major orographic lifting of air for these winds (which usually bring humid air from the Atlantic Ocean to the Baltic Sea basin) and consequent heavy precipitation over Norway and a dry descending airmass over the western side of the Baltic Sea. Additional influence on the eastern side is brought by windward conditions:

Table 4.1. Estimations of mean annual precipitation and evaporation in the Baltic Sea (in millimeters per year).

Reference	Period	Precipitation	Evaporation	Net
Simojoki (1949)	1886–1935	525	460	65
Brogmus (1952)	(40–50 years)	474	474	0
Dahlström (1977)	1931–1960	619	—	—
Henning (1988)	Climatological	639	493	146
Bergström *et al.* (2001)	Climatological	475	—	—
Rutgersson *et al.* (2001)	1981–1998	596	467	129

92 Water, salt, and heat budgets [Ch. 4

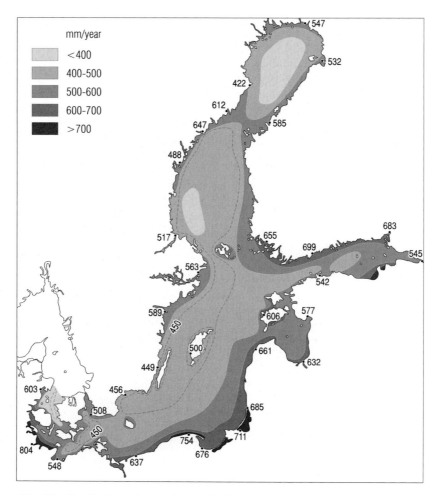

Figure 4.2. The distribution of annual precipitation in the Baltic Sea. (Redrawn from Brogmus, 1952).

friction, thermal convection, and orographic lifting over land areas near the sea. As a consequence, annual precipitation is more than 700 mm at the Polish coast and 450 mm–500 mm over southern Sweden. In the north a similar difference is observed, but the precipitation level is lower (e.g., in the Bay of Bothnia it is below 600 mm). In the open sea, precipitation is likely less than on the coast because of the lack of orographical variations and less convection.

Monthly precipitation in the Baltic Sea ranges between 15 mm and 95 mm showing a maximum (80 mm–95 mm) in July–August in Poland and a minimum (15 mm–30 mm) in February–March in the northwest (Figure 2.13). In the wintertime, precipitation accumulates on sea ice in the ice-covered region of the Baltic Sea to be released into seawater only during the ice-melting season. Storage time on the ice can be as long as seven months in the north, while melting takes place only during the last 1–2 months of the ice season. However, in terms of the sea-level, the influence of snowfall on sea ice is felt immediately.

Evaporation accounts annually for 450 mm–500 mm in the Baltic Sea (Table 4.1). Its estimation is quite difficult, because direct methods are not feasible in the open sea and for good results with indirect methods the sea surface temperature, humidity, wind, and waves would need to be known. Even so, the range of available estimates is quite narrow, all obtained with indirect methods. Monthly values are within 10 mm to 80 mm (Figure 4.3). On average, the minimum occurs in spring, when the sea surface temperature is low, stratification of the atmospheric surface layer is stable, and winds are weak. The maximum is found in late fall when the turbulent air–sea exchange is intensive. The level of evaporation is higher in the south due to the higher sea surface temperature and shorter ice season. In winter, evaporation takes place at the ice or snow surface but the level is low in such cold conditions.

There are three kinds of indirect methods to estimate evaporation over a sea surface. The *aerodynamic method* is based on the turbulent boundary layer theory:

$$\rho E = -\rho_a \overline{w'q'} \tag{4.2}$$

where ρ_a is air density; and w' and q' are turbulent fluctuations in vertical velocity and specific humidity. In practice, the bulk formula is often used to estimate the turbulent flux:

$$\overline{w'q'} = C_E(q_a - q_0)U_a \tag{4.3}$$

where $C_E \approx 1.1 \times 10^{-3}$ is the bulk exchange coefficient of moisture at a 10 m height (e.g., Launiainen, 1979; DeCosmo et al., 1996); q_a and q_0 are the specific humidities in air and at the sea-surface; and U_a is the wind speed. The exchange coefficient varies within a factor of 2 depending on the stability of stratification and surface roughness,

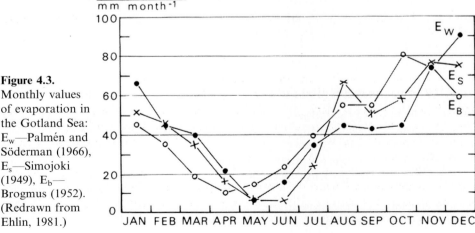

Figure 4.3. Monthly values of evaporation in the Gotland Sea: E_w—Palmén and Söderman (1966), E_s—Simojoki (1949), E_b—Brogmus (1952). (Redrawn from Ehlin, 1981.)

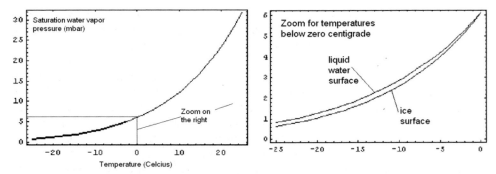

Figure 4.4. The saturation water vapor pressure of air (millibars) as a function of temperature (°C). The small plot shows the negative axis in more detail, upper curve for liquid water surface and lower curve for ice surface.

as will be discussed in Section 4.3.4. Sea surface humidity can be taken as the saturation value, shown in Figure 4.4 for water vapor pressure.[1]

With $q_a - q_0 = 10^{-3}$ and $U_a = 5$ m/s, we have $E \approx 1$ mm/day (the annual mean evaporation is 1 mm–1.5 mm/day in the Baltic Sea). In the cold ice season the air and sea surface temperature and the consequent saturation humidities become very small and therefore evaporation is also very small.

The *energy balance method* takes evaporation from the residual ΔQ of the surface heat balance after other terms have been determined: $\rho L_E E = -\Delta Q$, where L_E is the latent heat of evaporation. We have $E \approx 1$ mm/day for $\Delta Q = -50$ W/m². The *aerological method* considers the water balance of a certain large air volume and takes evaporation as the residual to balance advection q_w and precipitation (Palmén and Söderman, 1966): $\Delta M_w = E - P + q_w$.

In all, according to estimates based on observations, the atmospheric net freshwater flux into the Baltic Sea is about 100 mm per year (Table 4.1), summed over the whole Baltic Sea surface the value becomes 40 km³/yr. The range in net flux is fairly large mainly due to inaccuracies in the estimation of precipitation; however, all estimates are non-negative. In practice, studies based on observations indicate that this net flux is positive and that it would be highly unlikely to have it more than twice the above estimate.

4.1.3 Drainage basin and river inflow

The drainage basin of the Baltic Sea has a surface area of 1.633×10^6 km², 4.2 times as large as the sea area (see Figure 1.1). In addition to the coastal countries of the Baltic Sea, the drainage basin includes parts of Belarus, Ukraine, Slovakia, the Czech Republic, and Norway, but its boundary lies generally rather close to state boundaries. The drainage basin of the Kattegat is 86,980 km².

---

[1] Specific humidity is obtained from water vapor pressure as $q = 0.622 e/p$, where p is air pressure.

The latitudinal extent of the drainage basin is from 49°N to 69°N and the longitudinal extent is from 8°E to 37°E. A large range of climatic regions is encountered, from mild to cold and from maritime to continental. In the northern part stable snow conditions occur in winter for 6–7 months, the snow thickness reaching up to one meter. The Scandinavian Mountains and forests are in the northwest, even the tundra zone is reached in the north, and in the northeast the Finnish and Karelian taiga lake districts are found. Runoff to the Bay of Bothnia takes place in three large rivers in the north (the Kemijoki, Tornionjoki, and Luleå älv) and several medium-size rivers. The waters from the western Finnish Lake District flow into the Sea of Bothnia along several rivers, principally the Kokemäenjoki. Central Lake District waters flow into the Gulf of Finland via Kymijoki River, while the Neva delivers waters from the Finnish Saimaa Lake District and Karelian lakes at the end of the Gulf of Finland. The two largest lakes in Europe are located in Karelia: Ladoga and Onega.

The eastern and southern sectors of the drainage basin cover the wide Baltic and Polish plains, which are mostly under cultivation. Here, runoff is collected into several large rivers: the Narva to the Gulf of Finland, the Daugava to the Gulf of Riga, the Neman to the Eastern Gotland Basin, the Vistula to Gdansk Bay, and the Oder to the border of Arkona and Bornholm Basins. In addition, many small rivers exist in the region. There is one large lake (Peipsi) at the border between Estonia and Russia and one major lake district (Mazuria in northeast Poland). In the south the limit of the drainage basin reaches the Tatra Mountains at the border between Poland and Slovakia and Poland and the Czech Republic. The western side of the Baltic Sea drainage basin includes a system of large lakes across Sweden. Vänern is the largest lake, and much of its waters flow into the Kattegat via the Göta älv. This system once acted as the ocean channel between the Gotland Basin and the Kattegat for the saline Yoldia Sea. Farther south there are the cultivated plains of Skåne (southern Sweden) and Denmark.

Annual river inflow into the Baltic Sea (excluding the Danish Straits) is about 440 km^3 (Table 4.2). The Danish Straits and the Kattegat would add about 40 km^3 to this number, half of the water coming from the Göta älv into the Kattegat. The total discharge is comparable with the world's largest rivers (e.g., the Yangtse sums to about 700 km^3/yr), and therefore it is a highly important factor in the hydrography of the North Sea and Norwegian Sea. Total river inflow is about twice the precipitation, while the area where runoff is collected is 4.3 times the area of the Baltic Sea. Thus about half of the precipitation over the land areas in the drainage basin is lost by evaporation. Baltic Sea river discharge corresponds to a water layer of 1,170 mm when spread over the whole Baltic Sea.

The largest part (almost 200 km^3) comes from the Gulf of Bothnia, and this partly explains its low salinity. The Gulf of Finland receives 112 km^3, the largest inflow relative to the surface area of the basin, and the Gulf of Riga receives about 30 km^3. In all, 80% of the inflow comes into the northern and eastern basins or farthest from the ocean boundary. This feature is the major driver for the horizontal smoothly changing salinity field in the Baltic Sea. The largest river is the Neva, which accounts for 77.6 km^3/yr or 18% of total river inflow; its mouth is located in

Table 4.2. Mean annual river inflow (cubic kilometers) into the Baltic Sea after Mikulski (1970, 1972) during 1950–1970 and Bergström et al. (2001) during 1950–1990.

Basin	Mikulski	Bergström et al.
Bay of Bothnia	100	98
Sea of Bothnia	83	91
Gulf of Finland	112	112
Gulf of Riga	30	32
Baltic Sea proper	111	114
Altogether	436	447
Danish Straits and Kattegat	—	37
Baltic Sea and Kattegat	—	484

St. Petersburg in the eastern end of the Gulf of Finland (Figure 4.5). The ten largest rivers take care of 57% of total inflow (Table 4.3).

There are temporal variations in river runoff on timescales from years to decades due to changes in precipitation and evaporation in the drainage basin. Bergström et al. (2001) report decadal averages between 440 km^3/yr and 526 km^3/yr in 1921–1990 for inflow to the Baltic Sea and the Kattegat. The extremes in annual river inflow have been 350 km^3 in 1924 and 615 km^3 in 1976. Geographical and climatic variations cause major regional differences. The seasonal variations in river inflow are large in individual basins (Figure 4.6). The maximum occurs in spring after snowmelt. Due to the timing of snowmelt the spring peak occurs later in the north, in March in the southwest, April in the central regions, May–June in the Gulf of Finland, and July in the Bay of Bothnia. The minimum occurs in late summer in the south, but only in winter in north. Because the extremes are well-spread, total inflow into the Baltic Sea does not show much interannual variability. The monthly extremes have been 20 km^3 in December 1959 and 87 km^3 in May 1966 (Bergström et al., 2001).

The lakes of the drainage basin act as a strong smoothing filter for river inflow. In recent decades, man-made regulation of watercourses due to the needs of hydroelectric power production has also smoothed the annual course of river inflow. On the other hand, drying of the wetlands has lowered their water-buffering capacity, and this has strengthened runoff peaks in non-regulated rivers.

4.1.4 Freshwater budget

The total freshwater budget ($P - E + V_r/A$) of the Baltic Sea is continuously positive on a monthly level (Figure 4.7). The budget is dominated by river runoff, and the

Figure 4.5. The Neva in St. Petersburg is by far the largest river flowing into the Baltic Sea.

Table 4.3. The largest rivers in the drainage basin of the Baltic Sea and the Kattegat (Bergström et al., 2001).

River	Drainage area (km^2)	Inflow (km^3/yr)	Basin
Neva	281,000	77.6	Gulf of Finland
Vistula	194,400	33.6	Gdansk Bay
Daugava	87,900	20.8	Gulf of Riga
Neman	98,200	19.9	Eastern Gotland Basin
Oder/Odra	118,900	18.1	Bornholm Basin/Arkona Basin
Göta älv	50,200	18.1	Kattegat
Kemijoki	51,400	17.7	Bay of Bothnia
Ångermanälven	31,900	15.4	Sea of Bothnia
Luleåälven	25,200	15.3	Bay of Bothnia
Indalsälven	26,700	14.0	Sea of Bothnia
Altogether	*985,700*	*250.5*	*Baltic Sea and the Kattegat*

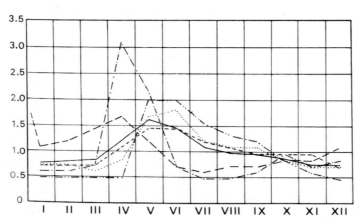

Figure 4.6. Monthly river inflows into Baltic Sea basins. The vertical axis shows flows relative to the annual average.
—— Baltic Sea;
– – – Baltic Sea proper; - - - - Gulf of Finland;
· · · · · Bothnian Sea;
· · — · · — Bothnian Bay;
— · — · — Gulf of Riga
(Mikulski, 1970, printed by permission from Nordic Hydrology.)

monthly mean is within 10,000 m^3–25,000 m^3/s, maximum in May and minimum in January. Only in October–December does evaporation exceed precipitation, but the net level is one order of magnitude lower than the level of river inflow. Thus, the sign of freshwater forcing does not change with seasons. In ice-covered areas in winter, precipitation is stored on the top of the ice cover and evaporation takes place at the same time at the snow or ice surface. Their net influence is felt in sea-level elevation by floating ice displacing liquid water. River inflow is still significant in winter and keeps the freshwater budget positive.

Taking 100 mm (37 km^3 over the whole Baltic Sea) to represent the net annual atmospheric exchange and 436 km^3–447 km^3 to represent annual river inflow into the Baltic Sea, total annual freshwater flux reaches about 480 km^3, which corresponds to a water layer of 1.23 m over the whole basin. This net flux is evidently compensated by the net outflow of Baltic Sea water into the North Sea.

Interannual variations show timescales from years to decades, with different behavior in different basins (Figure 4.8). Annual mean discharges from the Tornio–Kalix, Vistula, and Neva have ranged from 500–1,000 m^3/s, 800–1,700 m^3/s, and 2,000–3,000 m^3/s. Seasonal variations are also large, in particular in the north where the snow cover lasts for more than half year (Bergström et al., 2001). Therefore, in the Bay of Bothnia, peak runoff is 8,000 m^3/s in May, whereas the mean level is 3,200 m^3/s and the minimum below 2,000 m^3/s in winter. In the Danish Straits and the Kattegat, seasonal runoffs are 1,000 ± 500 m^3/s.

4.1.5 Water exchange through the Danish Straits

Oceanic water exchange between the Baltic Sea and the North Sea takes place through the Kattegat and the Danish Straits. The Kattegat is the transition zone or mixing basin between Baltic Sea and North Sea watermasses, and the section where water exchange is determined is here the boundary between the Kattegat and

Sec. 4.1] Water budget 99

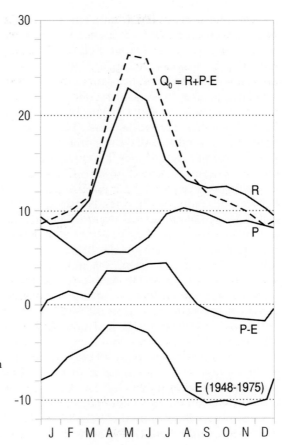

Figure 4.7. Mean freshwater budget in the Baltic Sea: precipitation P and river inflow R from 1931 to 1960 and evaporation E from 1948 to 1975 in 10^3 m^3 s^{-1}. (Redrawn from Jacobsen, 1980.)

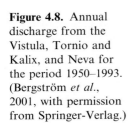
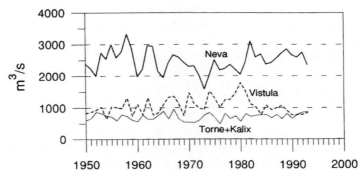

Figure 4.8. Annual discharge from the Vistula, Tornio and Kalix, and Neva for the period 1950–1993. (Bergström *et al.*, 2001, with permission from Springer-Verlag.)

the Danish Straits where salinity has already diluted to about 25‰ (see Figure 3.7). There are considerable differences in water exchange estimates depending on where the exchange is exactly placed. Following the present view (Kattegat–Danish Straits) the work by Soskin (1963) is the best reference, but for the boundary between the Danish Straits and Baltic Sea proper the values by Brogmus (1952) are representative.

The average annual inflow into the Baltic Sea from the Kattegat is generally taken as about 1,180 km^3 (Soskin, 1963), or 38,100 m^3/s in terms of mean discharge. The extremes have been 1,000 km^3 (1937) and 1,500 km^3 (1921). This is equal to three times the river inflow and corresponds to a 3,200 mm thick water layer over the whole Baltic Sea. On a monthly level the maximum is 120 km^3 in November–January and a minimum occurs in May of 70 km^3. Inter-annual and intra-annual variations are very large, dictated mainly by air pressure, wind distributions, and sea-level differences between the Baltic Sea and the North Sea. Very strong and intensive inflow events, called *Major Baltic Inflows*, occur irregularly, at present on average about once in a decade (see Section 5.4).

Outflow of water from the Baltic Sea is determined first of all by the sea-level difference between the Baltic Sea and the North Sea, strengthened by easterly winds and weakened by westerly winds. According to Soskin (1963), in the period 1898–1944 average annual outflow was about 1,660 km^3 (52,600 m^3/s), which corresponds to a 4,230 mm water layer over the whole Baltic Sea. Annual extremes were 1,300 km^3 (1942) and 2,100 km^3 (1927), and monthly extremes were 115 km^3 (June) and 150 km^3 (December–March).

The calculations by Soskin (1963) represent well our present understanding. Net outflow must of course be equal to net freshwater flux: 450 km^3–500 km^3/yr. The active water storage capacity of the Baltic Sea, taken as the difference between monthly maximum and minimum volumes, is 500 km^3. This is about the same as annual river inflow into the Baltic Sea. On short timescales, water exchange is mainly by means of back-and-forth movement controlled by atmospheric conditions. In calm weather the boundary of Baltic Sea and North Sea watermasses is at the actual Baltic Sea mouth, for easterly winds the boundary moves to Skagerrak, while in westerly winds it is at Darss Sill and Öresund (see Chapter 3).

Knudsen's formulas (Knudsen, 1900) are often applied to investigations of ocean water exchange in semi-enclosed basins. They are based on the conservation laws of mass and salts. Let us denote the freshwater inflow by F, ocean water inflow and salinity by (V_i, S_i), and outflow and salinity of the basin water by (V_o, S_o). These conservation laws are

$$V_o = F + V_i, \qquad S_i V_i = S_o V_o \qquad (4.4)$$

Here are five unknowns and two equations. If salinities and freshwater inflow are known, volume fluxes V_i and V_o can be determined (known as the Knudsen formula):

$$V_i = \frac{S_o}{S_i - S_o} F, \qquad V_o = \frac{S_i}{S_i - S_o} F \qquad (4.5)$$

Using Soskin's (1963) data for the entrance to the Baltic Sea ($V_i = 1,180$ km^3/yr, $V_o = 1,660$ km^3/yr), we have $S_i/S_o = 1.41$, and with $S_i = 25$‰, we have $S_o = 18$‰.

The input salinities in Equation (4.5) vary according to which section the formulas are applied. The approach represents the situation just at the section, since some inflowing water returns back soon and does not affect the internal dynamics of the sea. At the Darss Sill and Drodgen Sill salinity inflow is already below 20‰, and it is often assumed that the ratio of the salinities of inflowing and outflowing water is $S_i/S_o \approx 2$ and therefore the ratio of the corresponding water volume inflow and outflow would be about $\frac{1}{2}$ (e.g., Ehlin, 1981; Rodhe, 1999).

Example. Brogmus (1952) used Knudsen's formula at the Darss Sill. He fine-tuned his data for $S_o = 8.7‰$, $S_i = 17.4‰$, and $F = 472\,\text{km}^3/\text{yr}$, and then obtained $V_i = 472\,\text{km}^3/\text{yr}$ and $V_o = 944\,\text{km}^3/\text{yr}$. These numbers are smaller than the total exchange, since part of the exchange goes through the Öresund and does not pass the Darss Sill. ∎

4.2 SALT BUDGET

4.2.1 Salt exchange through the Danish Straits

The salinity of seawater is a conservative quantity. The total level can be affected only through boundaries, and inside the water body the salinity distribution is changed by advection and diffusion. In the Baltic Sea, salt inflow and outflow take place through the Kattegat (Figure 4.9). River inflow and precipitation cause dilution at river mouths and the sea surface, while evaporation acts in the opposite direction. Ice formation and melting act as evaporation and precipitation, respectively, but have no influence on an annual timescale.

Example. For a juice model, Baltic Sea water is a mixture of freshwater and normal seawater, with relative portions, respectively, of ν and $1 - \nu$. The mean salinity of Baltic Sea water is

$$S_I = (1 - \nu)S_O \qquad (4.6)$$

where $S_O = 35‰$ is the salinity of normal seawater. Since $S_I = 7.4‰$, we have $\nu = 0.80$ (i.e., the dilution is 1:4). The volume of the Baltic Sea is 21,205 km³ and mean freshwater inflow is 480 km³/yr, and consequently the renewal of freshwater takes 35 years. This value is truly representative, since freshwater is mainly fed far from the Danish Straits. ∎

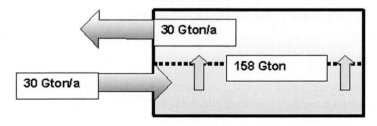

Figure 4.9. Baltic Sea salt storage and balance and idealization of the Baltic Sea as a two-layer box.

The fresh water renewal time in the Baltic Sea was found to be 35 years in the above example based on simple balance calculations. To determine the renewal time of oceanic water is a complicated question, since a large part of inflowing water soon returns to the North Sea via the Kattegat without taking much part in the renewal process of Baltic Sea watermasses. The volume of oceanic water is about 4,200 km^3 in the Baltic Sea, and in theory average ocean water inflow could change it in 5.3 years. But much of the inflow soon flows back out, and therefore the replacement of oceanic waters in distant basins takes a much longer time.

The following reasoning gives more insight into this. First, if we take deepwater velocity as $U \sim 1$ cm/s and the lengthscale along the deep part of the Baltic Sea as $L \sim 10^3$ km, the advective renewal time would be $T_A \sim 10^8$ s ≈ 3 years. Second, in random walk mixing[2] the timescale becomes $T_D \approx L^2/D$, where D is the diffusion coefficient; with $D \approx 10$ km^2/day we have 270 years. The average lifetime for an arbitrarily chosen water particle in the Baltic Sea is an intermediate value between these: 50–100 years fits into the middle of advection and diffusion timescales and can be taken as the renewal timescale of oceanic waters in the Baltic Sea. It may be an overestimate, since deepwater circulation is rather persistent. For a representative renewal time of the whole watermass we can choose 50 years, since freshwater renewal time must have more weight in the estimator. The water renewal time problem is discussed more thoroughly in Chapter 5 where numerical modeling tools are used to give more refined estimates.

Although the salinity level of the Baltic Sea water is low, its temporal and spatial variations are large and therefore it has a major role in maintaining stratification and circulation. Consequently, the conservation of salinity must be accounted for when researching the dynamics of the Baltic Sea. Let the x- and y-coordinates point toward the east and north as usual and the z-axis vertically upward, the level $z = 0$ is then beneath the sea, and the bottom and surface are given by $z = H$ and $z = \xi$. The salinity evolution equation presents how advection and diffusion modify salinity inside a water body and how net changes are produced by boundary conditions:

$$\frac{\partial S}{\partial t} + \boldsymbol{u} \cdot \nabla S = \nabla \cdot (\boldsymbol{K}_S \nabla S) \tag{4.7a}$$

$$z = \xi : K_{SV} \frac{\partial S}{\partial z} = -(P - E)S; \qquad z = H : \frac{\partial S}{\partial z} = 0 \tag{4.7b}$$

$$(x, y) \in \partial \Gamma : K_{SH} \nabla S = \Theta \tag{4.7c}$$

where $\boldsymbol{K}_S$ is the diffusion coefficient tensor[3]; $\partial \Gamma$ is the horizontal boundary of the sea; and Θ is the flux across horizontal boundaries. The tensor $\boldsymbol{K}_S$ is usually diagonal with $K_{S,xx} = K_{S,yy} = K_{SH}$ equal to the horizontal diffusion coefficient and $K_{S,zz} = K_{SV}$

---
[2] Water particles are assumed to move with random steps in a horizontal grid with equal probabilities of reaching the next grid point. This leads to normal diffusion when the grid size goes to zero.
[3] A tensor is a generalization of scalars and vectors and can here be taken as a matrix; this tensor $\boldsymbol{K}_S$ specifies how salinity diffuses into different directions in the pesence of a salinity gradient.

equal to the vertical diffusion coefficient, $K_{SH} \gg K_{SV}$. Such a formulation is needed because of the anisotropy between vertical and horizontal directions in geophysical fluid dynamics. Here surface and bottom boundary conditions are of the Neumann type.[4] Sometimes Dirichlet[5] boundary conditions are taken at one or both vertical boundaries. Lateral boundary conditions are defined by (i) zero flux area taken at solid boundaries and freshwater flux at river mouths to reflect the absence of salt exchange through them, and (ii) by describing salt exchange at the open ocean boundary (Danish Straits).

Representative diffusion coefficients for Baltic Sea conditions are $K_{SH} = 10^3$ m^2/s and $K_{SV} = 0.01$ m^2/s. Since $P - E > 0$ normally in the Baltic Sea, the watermass is slightly diluted from the top, while the bottom is a passive boundary. Local dilution is compensated by net advection for stationary conditions.

The influence of an absolute change in salinity on density is approximately the same whether the water is oceanic or brackish (i.e., a change of 1‰ in salinity changes the density by about 0.8 kg/m^3). On the other hand, the influence of freshwater input or evaporation on the value of salinity depends on existing salinity (e.g., 10 cm evaporation from a 10 m deep mixed layer takes ocean water salinity from 35.00‰ to 35.35‰ and brackish water salinity from 7‰ to 7.07‰. This relatively small change in water properties combined with existing strong stratification implies that it is very difficult to influence stratification by atmospheric water exchange processes in the Baltic Sea (in particular, the triggering of deeper convection) by evaporation at the surface. Similarly, brine rejection in sea ice formation does not result in significant convection in the Baltic Sea. Stratification is a stable feature forced primarily by the balance between the inflows of riverine and oceanic waters and advection.

4.2.2 Sea ice growth and melting

Ice formation and melting do not produce net changes in the salt content of the Baltic Sea, but they have an important role in the annual salinity cycle, mainly during the melting season. In winter the salinity of surface water increases, since some of the salt is rejected from the growing ice sheet. The salinity of ice becomes $\frac{1}{3}$ that of water, and thus growth of ice to thickness h corresponds to evaporation of $\frac{2}{3}h$. The growth rate may be 10 mm to 20 mm per day for new ice and 5 mm per day for normal midwinter ice. The influence may therefore be much stronger than that of evaporation, but is not strong enough to trigger convection to deepwater layers, because of existing pronounced stratification; thus, a process similar to that forming cold deep Atlantic ocean waters in the Labrador Sea and driving the so-called "Conveyor Belt" cannot happen in the Baltic Sea. The saltier water formed during ice formation is normally mixed with remaining upper layer water during the winter.

Sea ice is much less saline than seawater and therefore in spring the melting of ice lowers surface salinity and strengthens the vertical stability of water. Spring ice is almost fresh and therefore melting corresponds to precipitation, its rate being

---

[4] Fluxes are given—see Equation (4.7b).
[5] Absolute values are given at the boundary.

10 mm–30 mm/day, which is a considerable level since monthly precipitation in spring is 1 mm–2 mm/day. Meltwater flux is spread all over the ice-covered waters, and it precedes in time the river runoff peak. In addition, winter snowfall is stored on the ice and melting together with the ice it strengthens the ice's influence on the surface layer. The ice and snow result in a freshwater flux comparable with the spring runoff peak in the northern part of the Baltic Sea.

4.2.3 Analytic models for salinity

The simplest approach to describe the vertical distribution of hydrographic variables is by vertical one-dimensional models. They are also used for salinity, but in the Baltic Sea this should be done with great care due to the dominant role of horizontal advection. In a vertical diffusion model the steady state salinity gradient is

$$\frac{dS}{dz} = -\frac{S_0(P-E)}{K_{SV}} \tag{4.8}$$

where S_0 is surface salinity. Since $S_0 \approx 5‰$, $P - E \approx 100$ mm/yr and $K_s > 10^{-4}$ m^2/s, the salinity difference across the vertical water column in the Baltic Sea would be less than 0.1‰ units. The permanent strong two-layer stratification must therefore be maintained by horizontal advection.

A more adequate representation of salinity is offered by multi-dimensional models. A simple two-dimensional salinity model reads as

$$\frac{\partial S}{\partial t} = -u\frac{\partial S}{\partial x} + \frac{\partial}{\partial z}\left(K_{SV}\frac{\partial S}{\partial z}\right) \tag{4.9}$$

where the x-axis represents the horizontal direction. Consider a two-layer basin, indicating by subscripts 1 (upper layer) and 2 (lower layer) the quantities of these layers (Figure 4.9). We perform vertical integration across the layers (from top to bottom of each layer):

$$\frac{dS_k}{dt} = -\left\langle u\frac{\partial S}{\partial x}\right\rangle_k + \frac{1}{H_k}K_{SV}\frac{\partial S}{\partial z}\bigg|_{z_{1k}}^{z_{2k}} \tag{4.10}$$

where z_{1k} and z_{2k} stand for the lower and upper boundaries of the layers, $k = 1, 2$.

In general, advection can be estimated as $U_k\Delta_H S_k/L = a_k\Delta_H S_k$, where U_k and L are the horizontal velocity and lengthscale ($a_k = U_k/L$), and vertical exchange across the layer interface can be estimated as $H_k^{-1}K_{SV}\Delta_v S_k/\Delta z = b_k\Delta_v S_k$, where Δz is the vertical mixing lengthscale ($b_k = H_k^{-1}K_S/\Delta z$). We then have a pair of ordinary linear differential equations

$$\left.\begin{aligned}\frac{dS_1}{dt} &= -a_1 S_1 + b_1(S_2 - S_1) - S_1 F_a = -(a_1 + b_1 + F_a)S_1 + b_1 S_2 \\ \frac{dS_2}{dt} &= a_2(S_i - S_2) + b_2(S_1 - S_2) = b_2 S_1 - (a_2 + b_2)S_2 + a_2 S_i\end{aligned}\right\} \tag{4.11}$$

where $F_a = (P-E)/H_1$ is net atmospheric freshwater flux; and S_i is the salinity of inflowing water. This is a linear, interactive two-body system forced by freshwater and salt fluxes. The system has a simple steady state solution

$$\frac{S_1}{S_2} = \frac{b_1}{a_1 + b_1 + F_a}, \qquad S_2 = \frac{(a_1 + b_1 + F_a)a_2}{(a_1 + b_1 + F_a)(a_2 + b_2) - b_1 b_2} S_i \qquad (4.12)$$

It is easy to see that $F_a \ll b_1$, and therefore we have $a_1 \approx b_1$ in the Baltic Sea (i.e., in the upper layer, advection and vertical diffusion must be equally important to have $S_1 \approx \frac{1}{2} S_2$). Then if in the lower layer $a_2 \approx b_2$, we have $S_2 \approx \frac{2}{3} S_i$. Since in the Darss and Drodgen Sills $S_i \approx 18‰$, we would have $S_1 \approx 6‰$ and $S_2 \approx 12‰$. When mixing becomes small, it is clear that $S_1 \to 0$ and $S_2 \to S_i$; when advection becomes small $S_1, S_2 \to 0$.

Equations (4.11) can be analytically solved using standard techniques for the systems of linear ordinary differential equations. The time evolution of the system is characterized by two exponential adaptation timescales $\exp(r_k t)$, $k = 1, 2$, where the r_ks are inverse e-folding timescales

$$r_{1,2} = -\frac{a_1 + b_1 + a_2 + b_2}{2} \pm \sqrt{\left(\frac{a_1 + b_1 - [a_2 + b_2]}{2}\right)^2 + b_1 b_2} \qquad (4.13)$$

If the advective and diffusive timescales are close to each other and equal to $\tau \sim 100$ days, then the system has the local timescales of τ and 3τ. The shorter one is the immediate response of each layer and the longer one is the relaxation time of the whole system.

4.3 HEAT BUDGET

4.3.1 Heat transport in the water body

In the Baltic Sea the heat budget of the upper layer is driven mainly by solar and atmospheric forcing, while in the lower layer the horizontal advection of heat plays a very important role. The influence of horizontal boundaries seems to be very local.

In summer the upper layer warms up, and on top of it a surface layer forms with typical temperatures of 15°C–20°C. Beneath the surface layer the temperature decreases down to a minimum of 0°C–2°C (old winter water) in the halocline, and beneath the halocline the water is warmer due to horizontal advection. This is particularly clear in the Gotland Sea, where the advection of North Sea water is important and the lower layer temperature is around 5°C all year. In winter the upper layer cools down to freezing point or close to it. To create the annual cycle in the heat content of the upper layer necessitates heating and cooling rates of the order of 100 W m^{-2} at the air–sea interface.

The equation of heat conservation presents how advection, diffusion, and solar radiation modify the temperature inside a water body and how net changes are produced by boundary conditions. Heat capacity can be held constant here, $\rho c = 4.15\,\text{kJ}\,\text{m}^{-3}\,°\text{C}^{-1}$, and consequently we have the equation for changes of temperature as

$$\frac{\partial T}{\partial t} + \boldsymbol{u} \cdot \nabla T = \nabla \cdot (\boldsymbol{K}_T \nabla T) + q_s \quad (4.14a)$$

$$z = \xi : K_T \frac{\partial T}{\partial z} = \frac{Q_0}{\rho c}; \quad z = H : K_T \frac{\partial T}{\partial z} = \frac{Q_b}{\rho c} \quad (4.14b)$$

where $\boldsymbol{K}_T$ is the tensor diffusion coefficient (diagonal tensor with $K_{Txx} = K_{Tyy} = K_{TH}$ and $K_{Tzz} = K_{Tv}$ as the horizontal and vertical diffusion coefficients, respectively, and $K_{TH} \gg K_{Tv}$); q_s is the solar heat source; Q_0 is the surface absorption of heat; and Q_b is heat flux from the bottom. The penetration of solar radiation is significant from the surface down to the depth of visibility (5 m–10 m), and heat flux from the bottom may be important particularly in shallow regions. This equation is similar to the salinity conservation law—Equation (4.7)—the major difference being the term q_s reflecting the solar heat source. In addition, surface and bottom boundary conditions are usually of the Neumann type, although occasionally Dirichlet boundary conditions are taken instead for either or both boundaries.

Lateral boundaries are often represented by zero fluxes, which is equivalent to the assumption that river discharge and oceanic water inflow do not have the same significance for the heat budget as they do for the salt budget. However, in the eastern Gulf of Finland, for example, inflow from the Neva has an influence on temperature conditions and technically needs to be included in local studies. In practice, this is often difficult due to the absence of temperature measurements.

Pure vertical diffusion can be directly solved for certain simple cases. Consider the problem of a heat flux approximated by a *sine* function Dirichlet boundary condition at the surface and fixed temperature at infinite depths. Taking the $z = 0$ level at the surface and the z-axis positive downward, the equation and the solution then are

$$\frac{\partial T}{\partial t} = K_{TV} \frac{\partial^2 T}{\partial z^2}; \quad \begin{array}{l} z = 0 : T_0 + \Delta T \sin(\omega t), \\ z \to \infty : T(z,t) \to \text{constant} \end{array} \quad (4.15)$$

$$T(z,t) = T_0 + e^{-\lambda z} \Delta T \sin(\omega t - \lambda z), \quad \lambda = \sqrt{\frac{\omega}{2K_{TV}}} \quad (4.16)$$

where ω is the angular frequency. The solution is characterized by parameter λ. Wave-like temperature changes at the upper boundary (frequently called a temperature wave) are damped and lagged with depth. Damping is exponential with an e-folding scale of λ^{-1} and lagging is linear with a half-cycle delay at depth $\pi\lambda^{-1}$. At $z = 3\lambda^{-1}$, the amplitude is 5% and the phase lags by a half-cycle from the surface temperature cycle. Based on long-term temperature data at Finnish fixed oceanographic stations, a representative value is $\lambda^{-1} \approx 40$ m; consequently, the vertical diffusion coefficient is $K_{TV} \approx 10^{-4}$ m^2 s^{-1} for the annual cycle (see Figure 3.14). Thus, in deepwater ($H \approx 100$ m) the temperature cycle is about four months behind surface conditions and its amplitude is about 2°C.

It is convenient to take the zero reference of heat as the heat content of liquid water at a temperature of 0°C, and thus the heat content of liquid water per unit

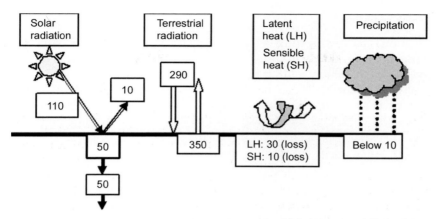

Figure 4.10. Surface heat balance in the Baltic Sea. The annual averages of solar radiation components, terrestrial radiation, turbulent fluxes (SH—sensible heat flux, LH—latent heat flux), and heat flux from precipitation are given for open sea conditions (W/m^2).

volume becomes $q = \rho c T$. In the case of ice, a significant amount of latent heat is stored in the solid phase: $q_i = -\rho_i(L - c_i T)h$, where $\rho_i = 910$ kg/m^3 is ice density; L is the latent heat of freezing; $c_i = 2.1$ kJ/(kg °C) is the specific heat of ice; and h is ice thickness. For example, for $T = -20°$C we have $q_i = -342$ MJ/m^3, of which -304 MJ/m^3 is due to phase change. In the Bay of Bothnia and in the Gulf of Finland the heat content becomes negative in normal winters (Jurva, 1937b). Ice formation not only serves as a heat sink but also protects heat storage due to its insulating capacity.

Surface heat exchange[6] consists of solar radiation (Q_s), incoming thermal radiation from the atmosphere, and outgoing thermal radiation from the surface (Q_{La} and Q_{Lo}), sensible and latent heat fluxes (Q_c and Q_e), and heat input from precipitation (Q_P):

$$Q_n = Q_s - Q_r + Q_{La} - Q_{Lo} + Q_c + Q_e + Q_P, \qquad (4.17)$$

where Q_n is net gain and Q_r is outgoing solar radiation (Figure 4.10). The sum $Q_R = Q_s - Q_r + Q_{La} - Q_{Lo}$ is the radiation balance and the sum $Q_c + Q_e$ is the turbulent heat exchange.

Solar radiation acts both as a surface heater and as an internal heat source for the surface layer, while the other factors act at the surface. Solar radiation can be written as $Q_s = Q_{so} + Q_{sw} + Q_r$, where Q_{so} is surface absorption and Q_{sw} is the surface layer source or radiation penetrating the surface (see Figure 4.10). Surface heating in boundary condition (4.17) then becomes

$$Q_0 = Q_n - Q_{sw} \qquad (4.18)$$

---

[6] Heat balance terms are taken to be positive when heat flows into the water.

4.3.2 Solar radiation

Solar radiation is the driving force behind the Earth system, and for the Baltic Sea it is also the primary heat source. In spring and summer, solar radiation melts the ice and snow and warms the water, while in fall and winter, thermal radiation and turbulent losses overcome solar radiation resulting in cooling of the water and freezing. The three parts of solar radiation—surface absorption (top millimeter), surface penetration, and outgoing—can be expressed as fractions of incoming radiation:

$$Q_{so} = (1 - \alpha)(1 - \gamma)Q_s, \qquad Q_{sw} = (1 - \alpha)\gamma Q_s, \qquad Q_r = \alpha Q_s \qquad (4.19)$$

where α is albedo; and γ is the fraction of solar radiation penetrating into the water. The proportions of surface absorption and penetration are approximately equal (i.e., $\gamma = \frac{1}{2}$). The penetrating part covers the optical band and narrow neighboring slices of the ultraviolet and near-infrared bands, and the magnitude of penetration depth is the depth of visibility or the Secchi depth. In some budget analyses, for simplicity, solar radiation is all taken at the surface ($\gamma = 0$). In exact terms, as used in marine optics, incoming and outgoing solar radiation at the surface are given by downwelling and upwelling planar irradiance, respectively.

Incoming solar radiation at the sea surface can be measured easily by various radiation sensors; however, only a few weather stations in the Baltic Sea region provide such data. It is often estimated by an astronomical–atmospheric formula of the type

$$Q_s = \cos Z T_{tr}(Z, e) F(N, Z)(r_0/r)^2 Q_{sc} \qquad (4.20a)$$

where Z is solar zenith angle; T_{tr} is atmospheric clear sky transmissivity; e is atmospheric water vapor pressure; F indicates the influence of cloudiness N; r and r_0 are actual and average Earth–Sun distances; and $Q_{sc} = 1.376\,\text{kW/m}^2$ is the solar constant. In clear sky conditions $F \approx 1$ and in overcast conditions $F \approx 0.1$. The astronomical formulas for Z and $(r_0/r)^2$ are given in the Appendix, and the transmissivity and cloudiness correction are provided, respectively, by the Zillman (1972) and Reed (1977) formulas

$$T_{tr} = \frac{\cos Z}{1.085 \cos Z + e(2.7 + \cos Z) \times 10^{-3} + 0.01} \qquad (4.20b)$$

$$F = 1 - 0.632N + 1.09 \times (\tfrac{1}{2}\pi - Z) \qquad (4.20c)$$

In Equation (4.20b) water vapor pressure is given in millibars. This formulation has been found suitable for the Baltic Sea (Saloranta, 2000; Ehn et al., 2002). Using system (4.20), corrections can be made to estimate solar radiation over the sea based on nearby land stations.

Albedo includes the influence of both surface reflection and backscattering from a water body. Typical values for different surface types are

Surface	Albedo (%)	Range (%)
Open water	7	5–10
Wet ice	30	20–40
Dry ice	50	30–60
Wet snow	60	40–70
Dry snow	85	80–95

In open water conditions, albedo is low and stable, while in the ice season it is high and varies over a wide range depending on the surface type (snow or ice), presence of liquid water, thickness of snow and ice, and the angular distribution of downwelling irradiance. Thin (10 cm) and bare ice has an albedo of 20%–30%, but when the ice thickness is more than 30 cm and the surface is dry, albedo is about 50%. Snow-covered ice has an albedo greater than 50%—even values as high as 95% have been recorded. In the melting season the surface becomes wet and albedo decreases. Albedo values are similar to those observed in polar seas (Perovich, 1998).

In the latitudes of the Baltic Sea there is a large annual cycle in radiation, in the north incoming radiation is close to zero in midwinter and on average 200 W/m²– 300 W/m² in summer. At latitude 60°N, the maximum level would be 630 W/m² according to formulas (4.20). The Jokioinen station in southern Finland gives a good overview of the monthly means of solar radiation (Table 4.4). The range is from 10 W/m² in December–January to 250 W/m² in June. Since Baltic Sea latitudes are within 6° of Jokioinen, relative differences in the radiation level due to the geographical location are within $\delta Z \tan Z$, which is about 10% in summer but much more in winter months. The fraction of diffuse radiation ranges from 40% in summer to 75% in winter.

4.3.3 Terrestrial radiation

Thermal radiation emitted by the sea surface is an energy loss term and given by the gray-body law[7]

$$Q_{Lo} = \varepsilon \sigma \underline{T}_0^4 \qquad (4.21)$$

where $\varepsilon = 0.96$–0.98 is the emissivity of the sea surface; $\sigma = 5.67 \times 10^{-8}$ W/(m² K⁴) is the Stefan–Boltzmann constant; and $\underline{T}_0$ is absolute surface temperature.[8] Outgoing terrestrial radiation is therefore between 265 W/m² (winter in north) and 380 W/m² (summer), following the annual cycle of surface temperature.

---

[7] A gray body radiates with a similar spectrum to that of a black body, but its total radiated energy is reduced by the emissivity factor, which is not the case with a black body.
[8] In parts of the theory it is more convenient to use Kelvin instead of Celsius, in which case the temperature symbol T is underlined ($\underline{T}$).

110 Water, salt, and heat budgets [Ch. 4]

Table 4.4. Mean monthly solar radiation in Jokioinen, southern Finland (60°49′N, 23°30′E) in 1971–1980 (FMl, 1982). The open sea, ice, and snow columns are based on albedos of 7%, 30%–50%, and 55%–85%, respectively (the ice and snow ranges are for wet and dry surfaces).

Month	Total (W/m²)	Diffuse (%)	Open sea (W/m²)	Ice (W/m²)	Snow (W/m²)	Cloudiness (1/8)
January	10	74	9	5–7	1–5	6.4
February	36	63	33	18–25	5–16	6.0
March	91	50	85	45–64	11–41	5.5
April	150	46	140	75–105	19–68	5.6
May	218	39	203	109–153	27–98	5.1
June	249	38	232	—	—	4.9
July	212	44	197	—	—	5.3
August	170	44	158	—	—	5.6
September	94	51	87	—	—	5.9
October	44	56	41	—	—	6.2
November	14	66	13	7–10	2–6	6.7
December	6	75	6	3–4	1–3	6.5

Thermal radiation by the atmosphere is a difficult task to solve since it does not come from a single surface but from gas molecules, water droplets, and aerosols at different altitudes and at different temperatures. In practice, one normally has to work with standard weather data, and therefore specific types of formulas have been established based on the gray-body analogy:

$$Q_{La} = \varepsilon_a \sigma \underline{T}_a^4 \qquad (4.22a)$$

$$\varepsilon_a = \varepsilon_a(N, e) = (a + be^{1/2})(1 + cN^2) \qquad (4.22b)$$

where $\underline{T}_a$ is the absolute temperature of air (at a standard 2 m altitude); ε_a is effective atmospheric emissivity ($0 \leq \varepsilon_a \leq 1$); and a, b, and c are emission law constants. In Baltic Sea modeling work these constants have often been taken as $a = 0.68$, $b = 0.036\,\text{mbar}^{-1/2}$, and $c = 0.18$ (Omstedt, 1990). It can be seen that $a < \varepsilon_a < 1.01$ in Baltic Sea conditions, and a normal value would be $\varepsilon_a \sim 0.8$. At the monthly average level, atmospheric radiation is between 215 W/m² (winter in the north) and 330 W/m² (summer).

Consequently, the terrestrial radiation balance $Q_{La} - Q_{Lo}$ is negative except for very rare extreme situations. When $T_0 = T_a$, the balance equals

$(\varepsilon_a - \varepsilon)\sigma T_a^4 \sim -50\,\text{W/m}^2$. Examination of Table 4.4 shows that the total radiation balance (terrestrial + solar) turns positive at some time in March–April, when albedo has become low enough (later in ice-covered areas). In September–October the balance turns negative for fall and winter.

4.3.4 Turbulent heat exchange

The turbulent exchange of momentum, heat, and moisture between the atmosphere and sea surface is examined using the turbulent boundary layer theory. The fluxes of these quantities are interdependent and therefore all need to be discussed together. They are given by

$$\tau_a = -\rho_a \overline{w' U_a'} \tag{4.23a}$$

$$Q_s = -\rho_a c_p \overline{w' \theta'} \tag{4.23b}$$

$$Q_e = -\rho_a L_E \overline{w' q'} \tag{4.23c}$$

where τ_a is the atmospheric shear stress at the sea surface; ρ_a is air density; $U_a = u\mathbf{i} + v\mathbf{j}$ and w are the horizontal and vertical wind velocity, respectively; c_p is the specific heat of air at constant pressure; θ is potential temperature; L_E is the latent heat of vaporization; q is specific humidity; and the primes stand for turbulent fluctuations of the quantity in concern. Moisture flux has already been considered in Section 4.1 for evaporation in the water budget.

For the estimation of turbulent fluxes, so-called *bulk formulas* are often used

$$\tau_a = \rho_a C_D U_a U_a \tag{4.24a}$$

$$Q_H = \rho_a c_p C_H (T_a - T_0) U_a \tag{4.24b}$$

$$Q_e = \rho_a L_E C_E (q_a - q_0) U_a \tag{4.24c}$$

where C_D is the drag coefficient or the momentum exchange coefficient; and C_H and C_E are wind-independent sensible and latent heat exchange coefficients in neutral conditions, $C_H \approx C_E \approx 1.1 \times 10^{-3}$ (DeCosmo, 1996). The drag coefficient C_D depends on wind speed—for example, $C_D = 0.0012(0.066 U_a + 0.63)$ according to Bunker, 1976)—for $U_a = 10\,\text{m/s}$, $C_D = 1.5 \times 10^{-3}$. These proportionalities $\tau_a \propto \rho_a U_a^2$, etc. result from dimensional analysis, and transfer coefficients are then obtained from empirical data. Transfer coefficients also depend on the stability of stratification in the atmospheric surface layer. The standard reference levels are 2 m for temperature and humidity and 10 m for wind velocity.

If $T_a - T_0 = 2°\text{C}$ and $U_a = 5\,\text{m/s}$, sensible heat flux is $Q_H \approx 20\,\text{W/m}^2$; if the wind speed reaches 15 m/s, sensible heat flux becomes $60\,\text{W/m}^2$. The temperature difference between the sea surface and air may have any sign, and sensible heat flux is usually within about $\pm 50\,\text{W/m}^2$. Latent heat flux is mostly negative and in magnitude close to sensible heat flux in the Baltic Sea, and therefore the turbulent fluxes together are usually below terrestrial radiation losses. When the air–sea temperature or humidity difference is large and wind speed is high, the turbulent flux can be a governing term. This is the case in late fall in the Baltic Sea.

A more accurate picture can be obtained from the structure of the velocity, temperature, and humidity profiles in the surface layer (see, e.g., Tennekes and Lumley, 1972). Turbulent exchange of a property Q is proportional to $K \, \partial Q/\partial z$, where eddy diffusivity is $K \sim u_* l$, with l and u_* being characteristic turbulent length and fluctuation scales. In the surface layer, fluxes are assumed constant. Therefore, the flow can be simply aligned along the x-axis, and according to Prandtl's *mixing length hypothesis*,[9] the size of turbulent eddies scales with distance from the boundary, $l \propto z$. In neutral stratification, logarithmic profiles result:

$$u(z) = \frac{u_*}{\kappa} \log\left(\frac{z}{z_0}\right) \qquad (4.25a)$$

$$\theta(z) = \frac{\theta_*}{\kappa} \log\left(\frac{z}{z_0}\right) \qquad (4.25b)$$

$$q(z) = \frac{q_*}{\kappa} \log\left(\frac{z}{z_0}\right) \qquad (4.25c)$$

where u_* is friction velocity defined by $u_* = (\tau_a/\rho_a)^{1/2}$, representing characteristic fluctuation velocity; θ_* and q_* are the characteristic friction layer temperature and humidity; $\kappa \cong 0.4$ is the von Kármán constant; and z_0 is *roughness length*. The bulk formulas also follow from this theory, and the roughness length and drag coefficient are related by $C_D = [\kappa/\log(z/z_0)]^2$.

In non-neutral conditions, buoyant lengthscales influence turbulent transfer as well as the mechanical mixing length. Consequently, surface layer profiles and exchange coefficients also depend on the stability of stratification, expressed using the Richardson number as $C_D = C_D(Ri)$,

$$Ri = \frac{g}{T_v} \frac{\partial \theta/\partial z}{(\partial u/\partial z)^2} \qquad (4.26)$$

where T_v is virtual temperature[10]; for $Ri > 0$ ($Ri < 0$), stratification is stable (unstable). There are empirical formulas to obtain exchange coefficients as a function of the Richardson number (e.g., Curry and Webster, 1999). A physically more explicit approach is to employ the Monin–Obukhov theory (e.g., Tennekes and Lumley, 1972; Vihma, 1995), which states that the logarithms in Equations (4.25) be replaced by so-called universal functions $\phi_m(\zeta), \phi_h(\zeta), \phi_s(\zeta)$, where $\zeta = z/L$ is the dimensionless height and the scaling length L is the Monin–Obukhov length

$$L = -\frac{\Theta_0 u_*^3}{\kappa g \overline{\theta' w'}} \qquad (4.27)$$

---

[9] According to Prandtl's mixing length hypothesis, turbulent transfer processes are described by a characteristic lengthscale over which turbulent eddies mix with fluid properties.

[10] An adjustment applied to the real air temperature to account for a reduction in air density due to the presence of water vapor.

where Θ_0 is the mean temperature of the adiabatic atmosphere (see Paulson, 1970 for unstable stratification and Webb, 1970 for stable stratification). Common forms of the universal functions are shown, e.g., in Vihma (1995) and Andreas (1998).

4.3.5 Heat input from precipitation

Heat transfer from precipitation takes place as sensible heat exchange and as phase changes. This can be expressed as

$$Q_P = [\rho c(T_P - T_0) + \rho L \chi_p]P \qquad (4.28)$$

where $\chi_p = -1, 0$, or 1 for solid liquid phase change, no phase change, or liquid solid phase change, respectively. For example, if $P = 10\,\text{mm/day}$ of liquid water and $T_P - T_0 = 5°C$, we have $Q_P \approx 2\,\text{W/m}^2$ on a liquid water surface. This is not much. But were the precipitation solid, the phase change would take $30\,\text{W/m}^2$ heat from the surface layer (i.e., solid precipitation would melt and the surface would cool down). Consequently, snowfall on open water or rain on a snow or ice surface are important heat transfer mechanisms.

4.3.6 Annual course

Net heat flux through the sea surface is of the order of $Q_n \sim 100\,\text{W/m}^2$, in summer positive and in winter negative (Table 4.5). This corresponds to heating or cooling at the rate $Q_n/(\rho c H)$; that is, about $0.04°C/\text{day}$ for the whole upper layer ($H \sim 50\,\text{m}$) or $0.2°C/\text{day}$ for the surface layer ($H \sim 10\,\text{m}$).

Upper layer thermodynamics takes the following annual course in the Baltic Sea (Figure 3.10):

Winter Ice exists in part of the basin, and the mixed layer extends to the halocline. Radiation balance is negative; and turbulent losses are strong as long as the sea surface is open, but ice formation and snowfall with the resulting low surface temperature strongly weaken turbulent transfer. The level of solar radiation is low.

Spring Due to increasing solar radiation, the radiation balance turns positive, ice and snow melt, and the surface temperature increases. Turbulent transfer is weak due to light winds and stable stratification of the atmospheric surface layer. The thermocline develops at a depth of around 10 m.

Summer Solar heating is strong and the maximum surface temperature in early August is 15°C–25°C. The thermocline deepens to 15 m–20 m. The radiation balance turns negative in September–October.

Fall The Baltic Sea cools toward the winter, and terrestrial radiation losses, sensible heat flux, and latent heat flux are all of equal magnitude. The thermocline extends down to the halocline.

Horizontal advective and diffusive heat transfer can be comparable with upper layer atmospheric heating. If the horizontal current velocity and temperature

Table 4.5. Annual course of magnitudes of heat budget components in the Baltic Sea, assuming that the sea surface is ice-free all year. These numbers are based on the regional climate and first-order heat flux formula presented in this chapter.

	Solar radiation	Terrestrial radiation		Turbulent fluxes	
		Incoming	Outgoing	Sensible heat	Latent heat
January	10	240	−300	−50	−30
February	40	240	−300	−50	−50
March	90	240	−300	−20	−50
April	140	270	−330	10	−50
May	210	300	−360	10	0
June	240	330	−390	10	0
July	200	340	−400	0	0
August	160	340	−400	0	−20
September	90	330	−390	0	−50
October	40	300	−360	0	−50
November	10	270	−330	−20	−50
December	10	240	−300	−40	−50
Mean	*100*	*290*	*−350*	*−10*	*−30*

gradient are $U \sim 5\,\text{cm/s}$ and $\Delta T/L \sim 1°\text{C}/100\,\text{km}$, advective change would be $U\Delta T/L \sim 0.04°\text{C/day}$. Taking the horizontal diffusion coefficient as $D \sim 10^2\,\text{m}^2/\text{s}$ and temperature variations as $1°\text{C}$ over $10\,\text{km}$ distances, the diffusive smoothing rate would be $D\Delta T/L^2 \sim 0.01°\text{C/day}$.

The lower layer is largely decoupled from the upper layer by the strong halocline, and its heating is the result of advection from neighboring basins. The heat content of inflowing North Sea waters therefore has an important role in maintaining the relatively high temperature of the lower layer in the Baltic Sea.

Few heat budget investigations have been carried out for the Baltic Sea based on observed data. A well-known case is the work of Hankimo (1964), in which he investigated data from one year in the southwestern Sea of Bothnia (Table 4.6). In this period the sea was ice-free at the site. Solar radiation clearly controls the situation, exceeding $100\,\text{W/m}^2$ between the spring and autumn equinoxes and peaking at $300\,\text{W/m}^2$ at midsummer. Net terrestrial radiation was between $-70\,\text{W}/$

Sec. 4.3]　　　　　　　　　　　　　　　　　　　　　　　　　　　　　Heat budget　115

Table 4.6. Sea surface heat balance at Finngrundet, southwestern Sea of Bothnia in March 1961–February 1962 based on observations (in Watts per square meter) (Hankimo, 1964).

	Net solar radiation	Net terrestrial radiation	Sensible heat flux	Latent heat flux	Net heat flux
March	91	−56	14	−30	19
April	175	−67	12	−13	107
May	230	−50	17	−1	196
June	295	−52	18	−3	257
July	259	−47	5	−17	200
August	184	−46	5	−46	97
September	113	−53	4	−53	11
October	48	−44	2	−49	−43
November	18	−49	−22	−67	−120
December	8	−54	−58	−81	−186
January	11	−49	−41	−53	−132
February	33	−44	−41	−56	−108
Mean	*122*	*−51*	*−7*	*−39*	*25*

m^2 and −40 W/m^2, and the total radiation balance was negative from November to February. Sensible heat flux was negative for the same period, dropping down to −60 W/m^2 in December and comparable with net terrestrial radiation only during December–February. Latent heat flux was always negative and comparable with net terrestrial radiation during August–February dropping down to −80 W/m^2 in December. The heat gain peaked at 250 W/m^2 in June and heat loss to 190 W/m^2 in December—the amplitude of heating was thus some 440 W/m^2.

Hankimo (1964) did not consider precipitation, but as far as monthly means are concerned that is a small factor. Should snowfall accumulate to 100 mm in a winter month, the resulting mean heat flux would be 5 W/m^2—see Equation (4.28).

In Hankimo's case the sea was open at the site all year. In the presence of ice, all heat fluxes would have been less. Solar radiation would decrease strongly due to the change in albedo, and turbulent heat fluxes would be much reduced, because the air–sea temperature and humidity differences would be expected to go down. Terrestrial radiation is more difficult to estimate, but likely any change would not be large.

4.4 SURFACE LAYER MODELS

4.4.1 Analytic models of temperature

Consider a vertical water layer with depth H extending down from the sea surface. The heat content per unit surface area is $\rho c H \tilde{T}$, where $\tilde{T}$ is the depth-averaged temperature. It is assumed that H is larger than the depth of visibility (i.e., solar radiation is all absorbed in this layer). The evolution of the mean temperature can be examined using the vertically integrated heat conservation law obtained from Equation (4.14):

$$\frac{\partial \tilde{T}}{\partial t} = -\frac{1}{H}\int_0^H (\boldsymbol{u}\cdot\nabla T)\,dz + K_T \nabla_H^2 \tilde{T} + \frac{Q_n + Q_b}{H\rho c} \qquad (4.29)$$

In vertical modeling, advection and horizontal diffusion are not explicitly solved but their influences can be analysed from parameterized forms. The last term contains forcing of the heat content through the top and bottom surfaces. A feasible approach in analytical modeling is to take a linear top-surface heat flux formulation:

$$Q_n = k_0 + k_1(T_a - T_0) \qquad (4.30)$$

where the coefficients k_0 and k_1 do not depend explicitly on T_0. This form has a simple interpretation: for $k_0 \to 0$ the surface layer becomes a low-pass filter for atmospheric temperature, and for $k_1 \to 0$ the surface temperature becomes decoupled from air temperature.

The approximation—Equation (4.30)—is obtained as follows. Solar radiation, incoming thermal radiation from the atmosphere, and heat flux from precipitation are purely external factors and are absorbed in k_0, while sensible heat flux is as such equal to constant $\times (T_a - T_0)$ and thus simply the "constant" goes to k_1. Outgoing terrestrial radiation is split into two parts[11] by $\underline{T}_0^4 \approx \underline{T}_a^4 - 4\underline{T}_a^3(T_a - T_0)$, where the underlined temperatures are expressed in Kelvin (as earlier). Then, net terrestrial radiation is $Q_{La} + Q_{Lo} \approx -\varepsilon(1-\varepsilon_a)\sigma \underline{T}_a^4 + \varepsilon 4\sigma \underline{T}_a^3(T_a - T_0)$, which is accurate to about 1% when $|T_a - T_0| < 10°C$ and $T_a \sim 0°C$. Latent heat exchange is split using the Clausius–Clapeyron equation (e.g., Gill, 1982) for the saturation water vapor pressure e_s. Assuming $e_0 = e_s$ and writing $e_a = Re_s$, where R is the relative humidity, latent heat flux can be expressed as

$$Q_e = -\rho_a L_E C_E \frac{0.622 e_s(T_a)}{p_a}\left[(1-R) - \frac{L_E}{R_v T_a^2}(T_a - T_0)\right]U_a \qquad (4.31)$$

where R_v is the gas constant of dry air. Finally, coefficients k_0 and k_1 become

$$\left.\begin{aligned} k_0 &= (1-\alpha)Q_s - \varepsilon(1-\varepsilon_a)\sigma\underline{T}_a^4 - \rho_a L_E C_E \frac{0.622 e_s(T_a)}{p_a}(1-R)U_a \\ k_1 &= \varepsilon 4\sigma \underline{T}_a^3 + \rho_a c_p C_H U_a + \rho_a L_E C_E \frac{0.622 e_s(T_a)}{p_a}\cdot\frac{L_E}{R_v T_a^2}U_a \end{aligned}\right\} \qquad (4.32)$$

---

[11] Use the binomial formula for $[\underline{T}_a + (\underline{T}_0 - \underline{T}_a)]^4$ and take the two leading terms.

These expressions look somewhat complicated but the resulting climatological values make up for it by being smooth, with seasonal variation coming mainly from solar radiation. In the equilibrium of $Q_n = 0$ we have $T_0 = T_a + k_0/k_1$.

In early autumn, typical cooling values of $(1-\alpha)Q_s = 100\,\text{W/m}^2$, $\varepsilon = 0.97$, $\varepsilon_a = 0.85$, $T_a = 10°\text{C}$, $R = 85\%$, and $U_a = 5\,\text{m/s}$ give $k_0 = 100 - 53 - 23 = 24\,\text{W/m}^2$ and $k_1 = 5 + 8 + 16 = 29\,\text{W/(m}^2\,°\text{C)}$. Thus, under these conditions the equilibrium for $Q_n = 0$ is $T_0 \approx T_a + 1°\text{C}$. In winter, solar radiation is weak and k_0 becomes negative.

Assuming the surface layer temperature to be uniform and equal to the vertically integrated mean temperature, we have the one-layer equation or the slab model (the tilde is omitted from the water temperature symbol for simplicity):

$$\frac{dT}{dt} = \lambda(T_a - T) + F + F_w \qquad (4.33)$$

where $\lambda = k_1/(H\rho c)$; $F = k_0/(H\rho c)$; and F_w represents bottom and horizontal heating. In principle the bottom and horizontal heating terms can be parameterized as in Equation (4.30), and the solution with $F_w = 0$ can be easily changed into solutions with $F_w \neq 0$ by modifying F and λ. The solution with $F_w = 0$ is

$$T(t) = T(0)e^{-\lambda t} + \int_0^t e^{-\lambda(t-\tau)}(F + \lambda T_a)\,d\tau \qquad (4.34)$$

The memory timescale of the system is λ^{-1} and its magnitude is $\lambda^{-1} \sim H\,\text{m}^{-1}$ day. The thermal memory of the Baltic Sea upper layer is thus about 50 days, defined as the e-folding timescale the upper layer takes to adjust to atmospheric forcing. For $T_a =$ constant, the solution is $T(t) = T(0)e^{-\lambda t} + (T_a + \lambda^{-1}F)(1 - e^{-\lambda t})$. Thus, after a time of around $3\lambda^{-1}$ the equilibrium $T \approx T_a + \lambda^{-1}F$ is obtained ($e^{-3} = 5\%$). The transient part is important in short-term forecasting, but when $t \gg \lambda^{-1}$ it can be removed.

Example 1. Autumn cooling model (Rodhe, 1952). The slab model was used in the past to forecast autumn cooling in the Baltic Sea. Air temperature is taken piecewise as daily means. Knowing actual/predicted air temperatures at days $1,\ldots,n$, prediction for the surface temperature on day n, $t = n\Delta t$, is

$$T(t) = T(0)e^{-\lambda t} + (1 - e^{-\lambda t})\left[\lambda^{-1}F + \sum_{i=1}^n T_a(i\Delta t)e^{-\lambda(n-i)\Delta t}\right]$$

This is known as the method of the weighted air temperature sum. It is clear that using $F = 0$ the solution is a weighted average of the initial temperature and predicted daily mean air temperatures with exponentially decaying weights. In other words, the air temperature is low-pass filtered for the upper layer temperature, and the filter weights depend on the depth of the upper layer. ∎

Example 2. Freezing date. Taking a linear atmospheric cooling rate, $T_a = -\alpha t$ and constant F, Equation (4.34) can be directly integrated. Leaving the transient terms out, the solution is $T_a(t) + \lambda^{-1}(F + \alpha)$. The freezing date t_F is then obtained from the solution as $T(t_F) = T_f(t)$, where T_f is the freezing temperature; since $T_F \approx 0°C$ we have $t_F \approx (1 + \alpha^{-1}F)\lambda^{-1}$. Thus, if $F \approx 0$ the freezing date is delayed by λ^{-1} days from the air temperature zero downcrossing. This is a first-order approximation in the Baltic Sea, with depth H representing the depth of the halocline in deeper areas (note that $\lambda^{-1} \sim H\,\text{m}^{-1}$ day). Coastal areas ($H = 5\,\text{m}-10\,\text{m}$) freeze after a 1- to 2-week cold period but central basins need more than two months. This freezing delay was first examined for Finnish lakes by Simojoki (1940).

Taking a parabolic form of air temperature evaluation, the conditions under which the sea surface does not freeze can be obtained. The result is that the length of the period that air temperature is below zero must be less than $2\lambda^{-1}$. ∎

Example 3. Periodic air temperature forcing $T_a = T_{ao} + \Delta T_a \sin(\omega t)$ gives the solution

$$T = T_{ao} + \Delta T_a \frac{\lambda}{\sqrt{\lambda^2 + \omega^2}} \sin(\omega t - \varphi), \quad \varphi = \arctan(\omega/\lambda)$$

For $\omega \gg \lambda$ the forcing cycles are damped proportional to ω^{-1} and the phase shift is asymptotically $\pi/2$, while for $\omega \ll \lambda$, surface water temperature follows air temperature with an asymptotically vanishing lag. ∎

4.4.2 Mixed layer models

The analytical slab model approach is expanded by letting the lower boundary of the surface layer be a free variable, $H = H(t)$, with $w_e = dH/dt$ equal to the entrainment[12] rate. These models are in general called mixed layer models, as it is assumed that the surface layer to depth H is homogeneous or well-mixed (see Niiler and Kraus, 1977). These models do not include horizontal effects. Having mixed layer depth as a dependent model variable brings major complications since the mechanical energy budget then needs to be solved for the entrainment rate. The model equations are

$$\frac{dT}{dt} = \frac{Q_n - Q_s e^{-\kappa H}}{\rho c H} - w_e \frac{\Delta T}{H} \quad (4.35a)$$

$$\frac{dS}{dt} = -\frac{P-E}{H}S - w_e \frac{\Delta S}{H} \quad (4.35b)$$

$$\frac{d\mathbf{U}}{dt} = -f\mathbf{k} \times \mathbf{U} + \frac{\tau_a - \tau_H}{\rho H} \quad (4.35c)$$

$$w_e(c_i^2 - s\mathbf{U}^2) = 2mu_*^3 + \frac{H}{2}[(1+n)B_0 - (1-n)|B_0|] + \left(H - \frac{2}{\kappa}\right)Q_s \quad (4.35d)$$

[12] Erosion of the underlying water layer by mixed layer turbulence.

Sec. 4.4] Surface layer models 119

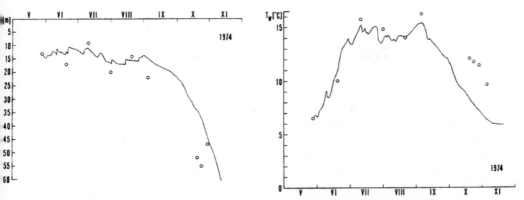

Figure 4.11. Comparison of the mixed layer model output (solid line) with observed data (circles) from June to November, 1974; h = thermocline depth, and T_w = mixed layer temperature. (Tyrväinen, 1978, with permission from *Nordic Hydrology*).

The first three equations are straightforward. Here ΔT and ΔS are the temperature and salinity differences across the pycnocline; f is the Coriolis parameter; and τ_a and τ_H are the surface and mixed layer bottom stresses. The fourth equation presents an equilibrium condition for the entrainment rate: the left side has energy consumption consisting of radiation by internal waves and the entraining of deepwater, and the right side has energy production due to wind stress, buoyancy, and solar radiation. Here c_i is internal wave speed; $U = u\mathbf{i} + v\mathbf{j}$ is horizontal velocity; B_0 is buoyancy flux at the surface; and the factors s, m, and n represent corresponding fractions of the kinetic energy of the mixed layer, wind stress, and buoyancy flux taking part in the entrainment (see Niiler and Kraus, 1977 for more details).

Tyrväinen (1978) applied the mixed layer model to the summertime surface layer in the Gulf of Finland and also calibrated the model against real data. Agreement was found to be good (Figure 4.11). In summer the thermocline deepened from 10 m to 20 m, while the temperature first increased until mid-June and then remained even with fluctuations of 2°C–3°C around the mean. In mid-September a rapid deepening of the thermocline commenced; the rate of deepening was somewhat overestimated in the model, which is likely due to ignoring advection. According to Tyrväinen (1978) the accuracy of the model was 2°C–4°C for temperature and 5 m for the depth of the mixed layer.

Not only is the mixed layer concept to some degree oversimplified since in some cases the mixed layer was not well defined on the validation data, but echelon thermocline structures and continuous stratifications were also present (Tyrväinen, 1978). The lack of horizontal processes is a limitation, as was the case with analytical models. On the other hand, mixed layer models can reproduce the seasonal evolution of the thermocline on the basis of the energy budget, and this helps in understanding the physics of thermocline depth. The case for the Gulf of Finland indicated that the temperature and depth of the mixed layer are to a large degree forced locally. There are a few further studies of mixed layer models in the Baltic Sea (e.g., Stigebrandt, 1985; Eilola, 1997).

4.4.3 Turbulence models

For a more detailed description of the vertical temperature–salinity–velocity structure, advanced turbulence models are needed. In the Baltic Sea a $k-\varepsilon$ type second-order closure model has been widely examined (Svensson, 1979; Omstedt et al., 1983; Omstedt, 1990), known better as the general equations solver "PROBE" (Svensson, 1998). Diffusion coefficients are obtained from turbulent kinetic energy (k) and the dissipation rate of turbulent kinetic energy (ε), which are obtained independently from second-order turbulence characteristics. The system of equations for average (averaged over turbulent fluctuations) temperature, salinity, and velocity is

$$\frac{\partial T}{\partial t} = \frac{\partial}{\partial z}\left(K_T \frac{\partial T}{\partial z}\right) + \kappa Q_s e^{-\kappa z} \tag{4.36a}$$

$$\frac{\partial S}{\partial t} = \frac{\partial}{\partial z}\left(K_S \frac{\partial S}{\partial z}\right) \tag{4.36b}$$

$$\frac{\partial \mathbf{U}}{\partial t} = f\mathbf{k} \times \mathbf{U} + \frac{\partial}{\partial z}\left(K \frac{\partial \mathbf{U}}{\partial z}\right) \tag{4.36c}$$

and for turbulence quantities we have the equations

$$\frac{\partial k}{\partial t} = \frac{\partial}{\partial z}\left(\frac{K}{\sigma_k}\frac{\partial k}{\partial z}\right) + P_s + P_b - \varepsilon \tag{4.36d}$$

$$\frac{\partial \varepsilon}{\partial t} = \frac{\partial}{\partial z}\left(\frac{K}{\sigma_\varepsilon}\frac{\partial k}{\partial z}\right) + \frac{\varepsilon}{k}(c_{1\varepsilon}P_s + c_{3\varepsilon}P_b - c_{2\varepsilon}\varepsilon) \tag{4.36e}$$

where $\sigma_k = K/K_T$ and $\sigma_\varepsilon = K/K_S$ are the Prandtl and Schmidt numbers;

$$P_s = K\left[\left(\frac{\partial u}{\partial z}\right)^2 + \left(\frac{\partial v}{\partial z}\right)^2\right] \quad \text{and} \quad P_b = \frac{K}{\sigma}\frac{g}{\rho_0}\frac{\partial \rho}{\partial z} \tag{4.36f}$$

are turbulence production due to shear and buoyancy; and $c_{1\varepsilon}$, $c_{2\varepsilon}$, and $c_{3\varepsilon}$ are model constants. When k and ε are known, eddy diffusivity can be obtained from $K = C_\mu k^{3/2}\varepsilon^{-1}$. The $k-\varepsilon$ model thus has six parameters, and using their standard values they are $C_\mu = 0.09$, $c_{1\varepsilon} = 1.44$, $c_{2\varepsilon} = 1.92$, $c_{3\varepsilon} = 0.8$, $\sigma_k = 1.0$, and $\sigma_\varepsilon = 1.3$. The most uncertain part of this system is the equation for the dissipation of turbulent kinetic energy—Equation (4.36e)—which involves several parameters. All in all, upper layer physics acquits itself fairly well in the model for central basins, but deeper horizontal processes become more significant, relatively speaking, and hence so-called "deep mixing" functions have been added to the model.

In the Baltic Sea environment the $k-\varepsilon$ model was first applied for the vertical structure of current velocity and turbulent mixing (Svensson, 1979), and later on its application extended to cover a wide range of problems, up to the climatology of the Baltic Sea.

An early application of the $k-\varepsilon$ model was autumn cooling in the Bay of Bothnia, with the aim of predicting the freezing date (Omstedt et al., 1983). Figure 4.12 shows the simulated time evolution of the vertical temperature profile together with observation from October 23 to December 13, 1979. In general, the agreement is reasonable, but there are a few significant deviations: for October 30–November 3, surface cooling is off in the model; in November there are several cases when the thermocline depth is far too large; and in the last week of simulation, as surface temperature reaches the temperature of maximum density, surface layer cooling is too strong in the model. The discrepancies were reported by Omstedt et al. (1983) and mainly concern horizontal advection.

PROBE has been applied to a large number of oceanographical problems in the Baltic Sea including supercooling and ice formation in turbulent water (Omstedt and Svensson, 1984; Omstedt, 1985), dispersion of particles in the turbulent Ekman layer (Rahm and Svensson, 1986), and atmosphere–Baltic Sea water exchange (Omstedt et al., 1997).

One-dimensional approaches have limitations in that advection cannot be properly accounted for and interactions between the coastal zone and central basins are not solved. However, for several central basin problems good results have been obtained with this approach, and furthermore a group of one-dimensional models have been used as the basis for a connected box model network.

4.4.4 Self-similarity approach

Marine dynamics is characterized by a wide range of spatial and temporal scales from microturbulence to global variations. Nihoul and Djenidi (1987) argued that the marine system can be described by fairly well-defined "spectral windows" (i.e., domains of lengthscales and timescales associated with identifiable phenomena). Marine weather (diurnal and synoptic variations) and long-term variability form a two-layer vertical structure for buoyancy (i.e., temperature or salinity). Vertical structure on the timescale of marine weather and long-term variability can be described by the so-called self-similarity concept.

Barenblatt (1996) proposed the following for a self-similarity definition: "a time-developing phenomenon is called self-similar if the spatial distributions of its properties at various different moments of time can be obtained from one another by a similarity transformation" (the fact that we identify one of the independent variables with time is of no significance). Establishing self-similarity has always represented progress for a researcher: self-similarity has simplified computations and the representation of the properties of the phenomena under investigation. In handling experimental data, self-similarity has reduced what would seem to be a random cloud of empirical points such that they lie on a single curve or surface, constructed using self-similar variables chosen in some special way. This self-similarity approach has been applied to the investigation of various problems, such as turbulent flows in atmospheric surface layers, wall layers of turbulent shear flows, the dynamics of turbulent spots in fluids with strongly stable stratification, etc.

122   Water, salt, and heat budgets   [Ch. 4

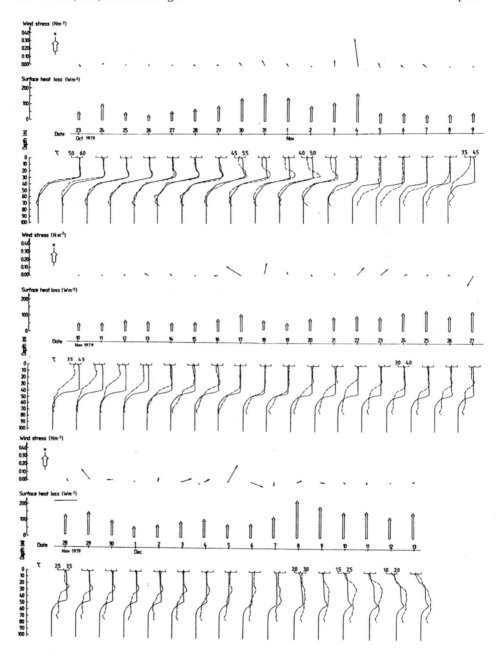

Figure 4.12. Simulation of surface layer cooling by the $k-\varepsilon$ model for October–November 1979 in the Bay of Bothnia. Surface boundary conditions (wind stress and heat loss) are shown above the profiles; measured profiles are shown by dashed lines and calculations by solid lines. (Omstedt *et al.*, 1983, with permission from *Tellus*.)

The self-similarity concept was introduced in marine sciences for the first time by Kitaigorodskii and Miropolski (1970). The self-similarity of a marine system variable (e.g., temperature θ and vertical co-ordinate ξ) is described in non-dimensional form as

$$\theta = \frac{T_1(t) - T(z,t)}{T_1(t) - T_H}, \qquad \xi = \frac{z - h(t)}{H - h(t)} \qquad (4.37)$$

where $T_1(t)$ is temperature in the mixed layer; $T(z,t)$ is the vertical temperature profile beneath the mixed layer; H is the depth of the ocean's active layer; T_H is the temperature at the lower boundary of the ocean's active layer, which is taken as a constant; z is the vertical co-ordinate; and $h(t)$ is the thickness of the upper mixed layer. The self-similarity approach assumes that $\theta = f(\xi)$. The main requirements for the self-similarity model to be valid are that the model be restricted to the main thermocline layer, that large-scale stratification be stable, and time integration be performed over the inertial period. This is applicable under summer stratification conditions in the Baltic Sea.

An approximate analytical expression for $\theta(\xi)$ can be found by using a fourth-order polynomial approximation of the classical type of von Kármán and Polhausen in boundary layer theory (Polhausen, 1921). So, this function together with its boundary conditions can be given as

$$\theta(\xi) = a_0 + a_1\xi + a_2\xi^2 + a_3\xi^3 + a_4\xi^4 \qquad (4.38a)$$

$$\theta(0) = 0, \qquad \theta(1) = 1, \qquad \theta'(1) = 0 \qquad (4.38b)$$

$$\int_0^1 \theta\, d\xi = \kappa_T, \qquad \int_0^1 \int_0^\xi \theta\, d\xi\, d\xi' = \bar{\kappa}_T \qquad (4.38c)$$

where θ' is a total derivative. The functions $\kappa_T, \bar{\kappa}_T$ can be determined from empirical data using Equations (4.37) and (4.38) (Mälkki and Tamsalu, 1985):

$$\kappa_T = \frac{T_1 - \int_0^1 T\, d\xi}{T_1 - T_H}, \qquad \bar{\kappa}_T = \frac{T_1 - \int_0^1 \int_0^\xi T\, d\xi T\, d\xi'}{T_1 - T_H} \qquad (4.39)$$

In considering low-frequency variations, the small-scale variability of these functions has to be filtered out by averaging over time. The function κ varies between 0.5 and 0.9, but after one inertial period (about 13 h), $t^{-1}\int \kappa_T\, dt$ and $t^{-1}\int \bar{\kappa}_T\, dt$ remain fairly stable. Mälkki and Tamsalu (1985) also found that the values of κ_T and $\bar{\kappa}_T$ depend on mixed layer development. In the case of entrainment (mixed layer depth increasing) it was found that $\kappa_T = 0.75$, $\bar{\kappa}_T = 0.3$, while in the case of detrainment (mixed layer depth decreasing) $\kappa_T = 0.6$, $\bar{\kappa}_T = 0.2$.

Assuming that θ is a function only of ξ and using Equation (4.38) and the abovementioned values for $\kappa_T, \bar{\kappa}_T$, we finally obtain for entrainment (Case A) and

124 Water, salt, and heat budgets [Ch. 4

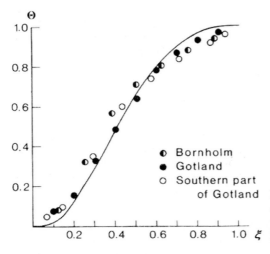

Figure 4.13. Self-similarity profile of salinity. (Mälkki and Tamsalu, 1985, with permission from the Finnish Institute of Marine Research.)

detrainment (Case B), respectively

$$\theta(\xi) = 1 - (1-\xi)^3 \qquad (4.40a)$$

$$\theta(\xi) = 1 - 4(1-\xi)^3 + 3(1-\xi)^4 \qquad (4.40b)$$

The self-similarity studies of Mälkki and Tamsalu (1985) described above are the most comprehensive ever undertaken for the Baltic Sea (Figure 4.13). The studies were carried out in the Gulf of Finland, the Gulf of Bothnia, the Gulf of Riga, and in the Baltic Sea proper over several years in the 1970s (D.Sc. R. Tamsalu, pers. commun.). Separating the curves into entrainment—Equation (4.40a)—and detrainment—Equation (4.40b)—classes, made the self-similarity approach more feasible. A single self-similar profile, derived from a vertical CTD sounding, is often located between Curves A and B (see Tamsalu et al., 1997). If the self-similar concept is applied to these profiles, then according to Equations (4.40a) and (4.40b) all the calculated (self-similar) profiles join the same non-dimensional curve in the thermocline layer when time integration over the inertial period is carried out.

4.5 BOX MODELS

4.5.1 Basin systems

The Baltic Sea consists of several basins that have straits and channels allowing interactions between the basins. The morphology of the Baltic Sea therefore suggests that the box model approach could be feasible. In this approach the different basins are taken as individual boxes: for each box a zero- or one-dimensional model is employed, and laws of communication between the boxes are established. The number of boxes is normally of the order of ten. This method provides a simple and easily understandable solution, but it is somewhat cruder than conventional numerical circulation models.

Communication between the boxes can be made using climatology, stochastic steps, or physical laws. The stochastic method means setting a random walk model into the system of boxes. The physical approach employs flow laws based on sea-level elevation and stratification.

4.5.2 Stochastic models

The Markov chain approach provides elegance to a stochastic model for the transport of properties and matter in a sea (Leppäranta and Peltola, 1986). The transport of properties is taken care of by model particles. In unit time each particle makes a displacement

$$\boldsymbol{d} = \langle \boldsymbol{d} \rangle + \boldsymbol{d}' \qquad (4.41)$$

where $\langle \boldsymbol{d} \rangle$ is the long-term mean displacement field; and $\boldsymbol{d}'$ is a random step. The sea is divided into N regions, and transport probabilities between regions can be estimated by Monte Carlo simulations. As a result, a transition probability matrix $P = \{p_{ij}\}$ is obtained, and this matrix contains all the information necessary for the time behavior of the system (see, e.g., Feller, 1968). The quantity p_{ij} indicates the probability of moving from box i to box j in one time step, and the Markov assumption states that the next step depends on the present location, not on the past location. Power n of the transition matrix gives the transition probability for n steps, and its eigenvalues indicate the stationary conditions and timescales of the system.

Leppäranta and Peltola (1986) used this approach in the Baltic Sea. The basin was divided into 19 parts, made up of the usual basins (Figure 2.3) with the Sea of Bothnia and Gulf of Finland divided into two parts, and Northern Quark and Skagerrak were added. Figure 4.14 shows, as an example, the outcome of a random walk after five years for particles leaving Helsinki with zero circulation and another walk with known mean surface layer circulation (see Section 5.2) as the deterministic component. In the former case the probability of location is still highest in the Gulf of Finland, while in the latter case the distribution has several local peaks guided by the circulation and morphology of the Baltic Sea.

4.5.3 Physics-based box models

The structure of the Baltic Sea can be considered as a system of basins separated by straits and channels (see Figure 3.2), which serve as a net of forcing and water exchange links. Therefore, the dynamics of these communication links is of key importance. In general, oceanic water travels north and east in the lower layer and the upper layer experiences weak cyclonic mean circulation disturbed by variable winds. At straits between basins sharp fronts form often, while within basins horizontal variations are small. An exception is the Gulf of Finland where inflowing Gotland Sea water and river discharge water form a continuously decreasing salinity distribution from west to east and south to north across the whole basin.

126 Water, salt, and heat budgets [Ch. 4

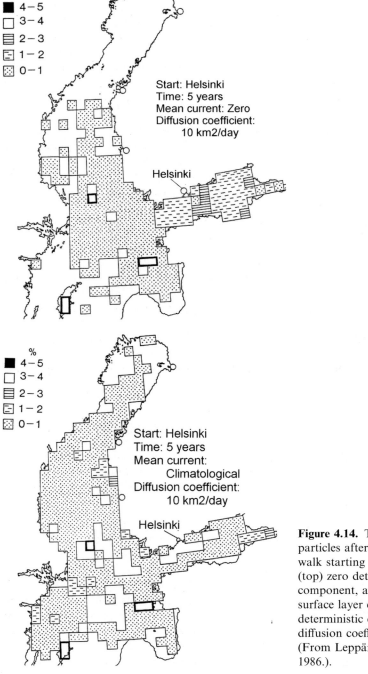

Figure 4.14. The distribution of particles after a 5-year random walk starting from Helsinki: (top) zero deterministic component, and (bottom) mean surface layer circulation as the deterministic component. The diffusion coefficient is $10 \, \text{km}^2/\text{day}$. (From Leppäranta and Peltola, 1986.).

Strait dynamics

Knudsen equations (see Section 4.2) have been applied to estimate the long-term exchange of water and salt through straits (Knudsen, 1900). However, they do not explain the dynamics of the water flow with its time–space variations. These dynamics include two basic types of flows: barotropic and baroclinic. In the former case the flow is forced by sea-level tilt along the strait, while in the latter case the pressure gradient due to density stratification provides the forcing (see Section 5.1). The flow is further influenced by friction, Coriolis acceleration, and topography, but local winds have less direct significance.

There are two different vertical circulation systems in straits. First, there is estuarine circulation where surface layer freshwater inflow is mixed with deeper saline water by the wind. Second, the bottom water below the sill depth level may flow out by mixing mechanisms. In many cases these circulation systems exist simultaneously but their interaction is weak. In some narrow straits these systems may come together (the so-called "overmixed estuary").

In a uniform channel with constant slope β, flow velocity can be estimated by the Manning formula (e.g., Li and Lam, 1964), famous from the study of river hydraulics, as

$$U = MR^{2/3}\sqrt{\beta} \tag{4.42}$$

where M is the Manning coefficient; and R is the hydraulic radius.[13] A representative value of the Manning coefficient is $50\,\mathrm{m}^{1/3}\,\mathrm{s}^{-1}$. This formula is valid for homogeneous water where tilt determines the velocity; it is based on a friction-corrected Bernoulli equation.

Stigebrandt (1999, 2001) came up with a similar law for the barotropic flow between neighboring basins,

$$Q = \sqrt{\phi^{-1}\Delta\xi} \tag{4.43}$$

where Q is barotropic discharge; ϕ represents friction; and ξ is sea-level elevation. Friction is due to bottom friction and topographic resistance at the end of the straits.

According to Stigebrandt (2001), two-layer baroclinic flow in a strait is hydraulically controlled by the equation

$$\frac{u_1^2}{g'H_1} + \frac{u_0^2}{g'H_0} = 1, \qquad g' = g\frac{\Delta\rho}{\rho_0} \tag{4.44}$$

where u_1 and H_1 are upper layer velocity and thickness; u_0 and H_0 are lower layer velocity and thickness; g' is reduced gravity; $\Delta\rho$ is the density difference between layers; and ρ_0 is a fixed reference density. In steady state situations Knudsen-type conservation laws for both water and salts are needed.

---

[13] $R = A/L$, where A is the cross-sectional area and L is the length of the wet perimeter.

Assuming that density depends on salinity only, the Stigebrandt equation is obtained:

$$P^3\left(1+\frac{\eta^3}{(1-\eta)^3}\right) - 2P^2 + P = \frac{\eta^3}{F_e^2} \qquad (4.45)$$

where $P = Q_1/Q_F$ is a mixing parameter; Q_1 is baroclinic transport; Q_F is net freshwater flux; $\eta = H_0/(H_0 + H_1)$; and F_e is a Froude number. This is defined as

$$F_e^2 = \frac{Q_F^2}{g\beta S_0 H^3 W^2} \qquad (4.46)$$

where $\beta = \rho^{-1}\partial\rho/\partial S$; $H = H_0 + H_1$; and W is the width of the strait. Equation (4.45) has two real roots only if $F_e \leq 1$. In estuaries the typical value of the Froude number is about 1. Baroclinic flow reaches a maximum when these two roots form a double root. If a maximum occurs, the estuary is said to be "overmixed".

Stigebrandt (2001) used his strait model for the Baltic Sea system. The basins are connected by narrow and shallow straits to the Kattegat, and external forcing is made by the sea-level elevation in the Kattegat. The mean sea-level of the Baltic Sea ξ_B is obtained from the known Kattegat level ξ_k ($\Delta\xi = \xi_k - \xi_B$) and the storage equation

$$A\frac{d\xi_B}{dt} = Q + Q_F \qquad (4.47)$$

where A is the Baltic Sea surface area. According to model simulations, higher-than-average monthly frequencies are strongly damped by the Danish Straits but at lower frequencies the amplitude is conserved. Thus, the Baltic Sea can be taken as a closed system for internally forced sea-level variations on a timescale less than one month.

This model has been further applied to Northern Quark, for two deep straits on both sides of Holmö Island. In both straits the sill depth is 20 m and their width is less than 2 km. Let $H = 20$ m, salinity of the Sea of Bothnia $S_0 = 5.5$ ‰, freshwater flux $Q_F = 3,425$ m^3 s^{-1}, and the sum of the strait widths $W = 3.3$ km. Then we have $F_e = 0.055$ and the condition of overmixing is $P = 3.1$, which implies that $Q_1 = 10,600$ m^3 s^{-1}, $Q = 7,125$ m^3 s^{-1}, and $S_1 = 3.7$‰. The resulting salinity difference between the Bay of Bothnia and Sea of Bothnia becomes 2‰, close to the observed value.

If the width of a strait is more than the Rossby radius $R_1 = c_i/f$, where c_i is the speed of internal waves, rotation may stop the two-layer system from extending across the whole strait. This situation occurs in Irbe Strait, where outflowing low-salinity water on the northern side forms a narrow coastal current. In broad straits, water exchange is estimated using the geostrophic method in a two-layer baroclinic flow. In the Baltic Sea the Rossby radius is 3 km–10 km, and sometimes it is even less than 1 km (Fennel et al., 1991; Alenius et al. 2003).

Density-driven currents have a significant impact on the deep-water characteristics of the Baltic Sea since they account for the water exchange between the deeper parts of neighboring basins. The essential quantitative problem is to determine flow rates that relate to a set of external parameters such as strait topography, stratification, and internal circulation of the upstream basin. Deepwater inflow along narrow channels between basins in the Baltic Sea has been examined by Lundberg (1983). The hydraulic theory for rotating channel flows has been applied to determine deepwater flows. The Bornholm Channel and Irbe Strait were studied by Laanearu et al. (2000) and Laanearu and Lundberg (2003), while the Stolpe Channel from the Bornholm Basin to the Gotland Sea was studied by Borenäs et al. (2007). Hietala et al. (2007) examined the flow through the Southern Quark.

Box system model

A 13-box model using strait dynamics for water exchange between the boxes was presented by Omstedt (1990) for the Baltic Sea. In each box a vertical turbulence model is employed to ascertain the vertical diffusion of momentum, heat, and salts (see Section 4.4.3), both barotropic and baroclinic horizontal flows are allowed between the boxes.

Figure 4.15 shows model results for the Arkona Basin and for the Åland Sea in winter 1986–1987. This winter was very severe and both basins froze over as reflected by temperature evolution. The model matched observed data quite well. Short-term fluctuations in sea-level in the Arkona Basin were not captured by the model, but the accuracy in surface temperature was better than 1°C.

The water, salt, and heat budgets of the Baltic Sea have been examined in this chapter. Observation material has been presented to help understand the level and variations of fluxes on timescales up to 100 years. Knowledge of these fluxes forms the foundation stone of understanding how the evolution of the Baltic Sea as a

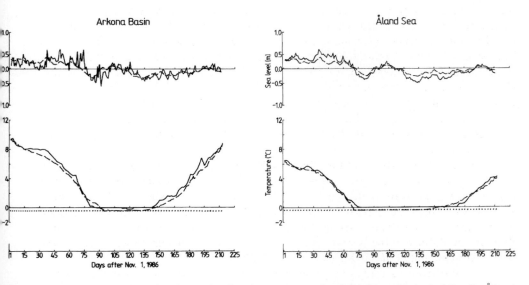

Figure 4.15. Model simulations for the sea-level and temperature for the Arkona Basin and for the Åland Sea from the 13-box Baltic Sea model in winter 1986–1987. Model values are shown by dashed lines and observations by solid lines. Sea-level observations are from Klagshamn in the Arkona Basin and Stockholm for the Åland Sea. (Omstedt, 1990, with permission from *Tellus*.)

physical system is forced externally. In the last two sections one-dimensional vertical models and box models used in the budget studies have been presented to give insight into the physical mechanisms in the problem. In Chapter 5 the dynamics of the Baltic Sea will be examined, starting from observed circulation characteristics and ending with three-dimensional models, which also provide more accurate answers to budget problems.

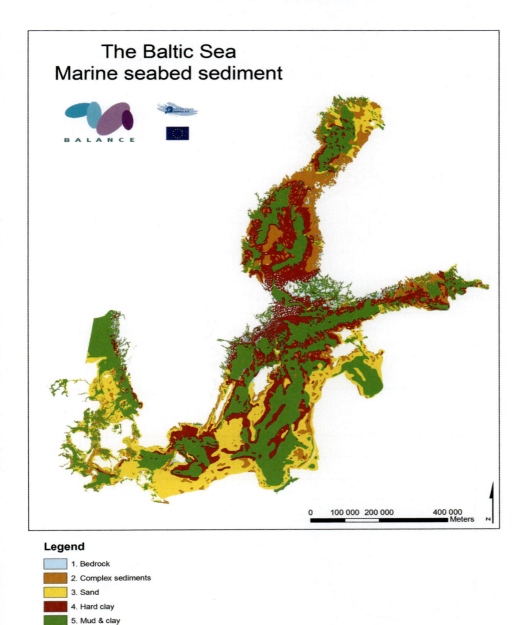

Figure 3.5. Distribution of Quaternary deposits in the Baltic Sea (Al-Hamdani *et al.*, 2007). © BALANCE-project. (Printed with permission from Professor Aarno Kotilainen.)

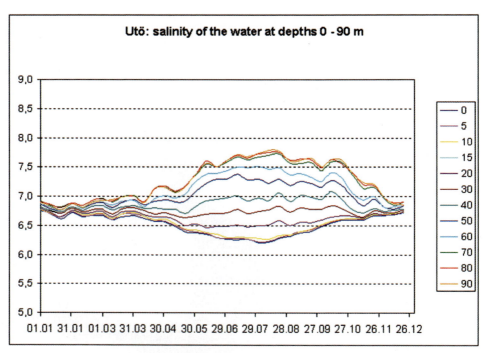

Figure 3.12. Long-term mean annual course of salinity (in ‰) at different depths in Utö during 1911–2005. Depths are shown on the right in meters. (Prepared by M.Sc. Riikka Hietala, printed with permission.)

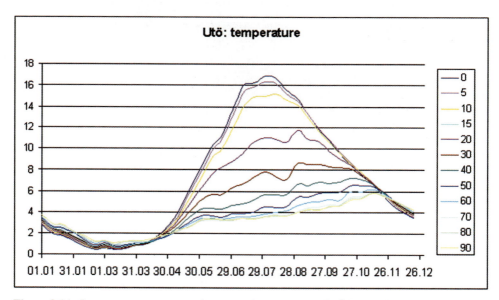

Figure 3.14. Long-term mean annual course of temperature (in °C) at different depths in Utö during 1911–2005. Depths are shown on the right in meters. (Prepared by M.Sc. Riikka Hietala, printed with permission.)

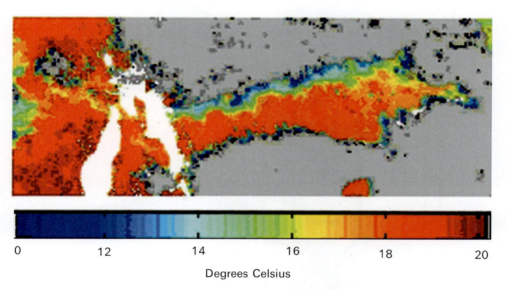

Figure 5.5. Upwelling off the northern coast of the Gulf of Finland on September 2, 2002 (the figure was processed at the Finnish Environment Institute (SYKE) from NOAA/AVHRR satellite data received by the Finnish Meteorological Institute).

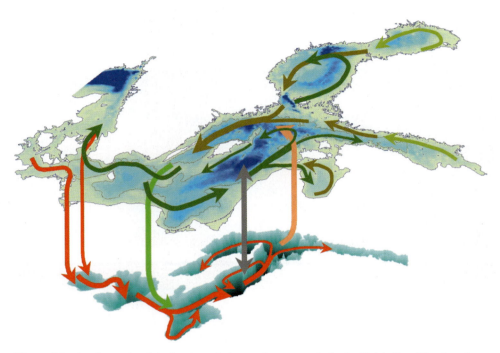

Figure 5.9. A schematic of the large-scale internal water cycle in the Baltic Sea. The deep layer below the halocline is given in the lower part of the figure. Green and red arrows denote surface and bottom layer circulation, respectively. The light green and beige arrows show entrainment, the gray arrow denotes diffusion. (Elken and Matthäus, 2008, printed with permission from Springer-Verlag.)

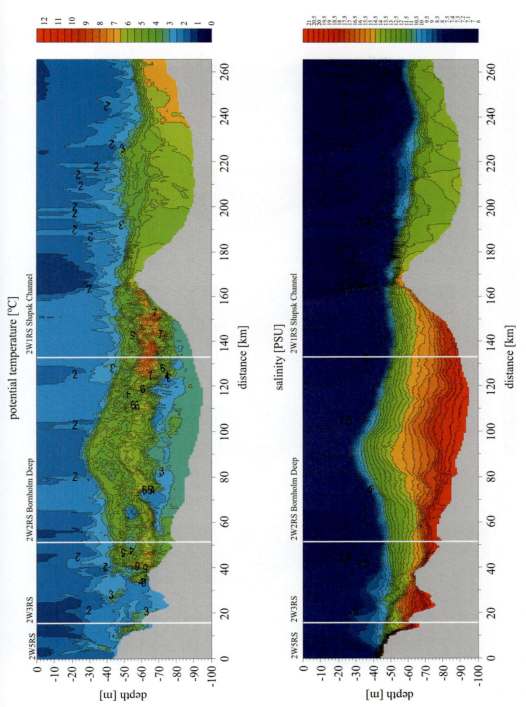

Figure 5.23. Potential temperature and salinity on February 16–18, 2003 along the axis Arkona Basin–Bornholm Gate–Bornholm Deep–Stolpe Channel–Gdańsk Deep. (Piechura and Beszczyńska-Möller, 2004, printed by permission from *Oceanologia*.)

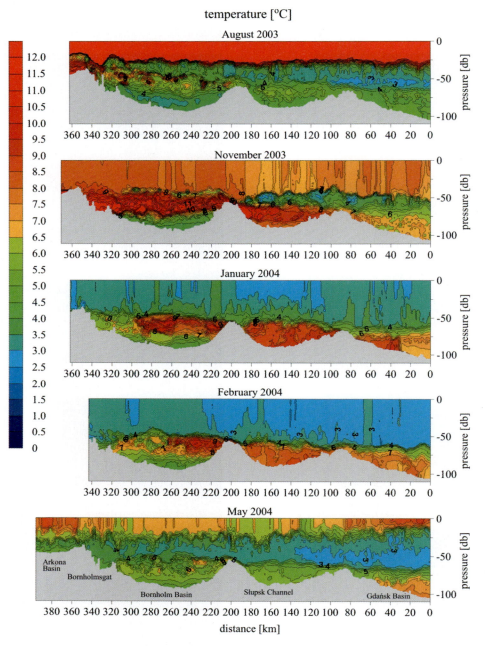

Figure 5.24. Bornholm Deep–Stolpe Channel–Gdańsk Deep temperature transects measured by R/V *Oceania* (from top down) in August 2003, November 2003, January 2004, February 2004, and May 2004 showing the eastward propagation of warm inflow water over Stolpe Sill into the southeastern Gotland Basin. (Feistel *et al.*, 2004, printed by permission from *Oceanologia*.)

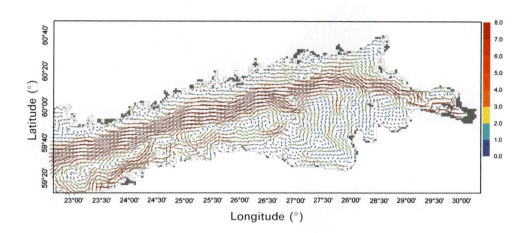

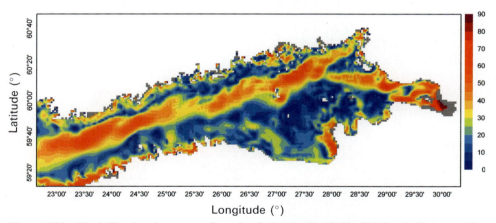

Figure 5.28. (Top) Simulated mean surface circulation in the Gulf of Finland. (Bottom) The stability of currents. Calculations are for the years 1987–1992. (Andrejev *et al.*, 2004a.).

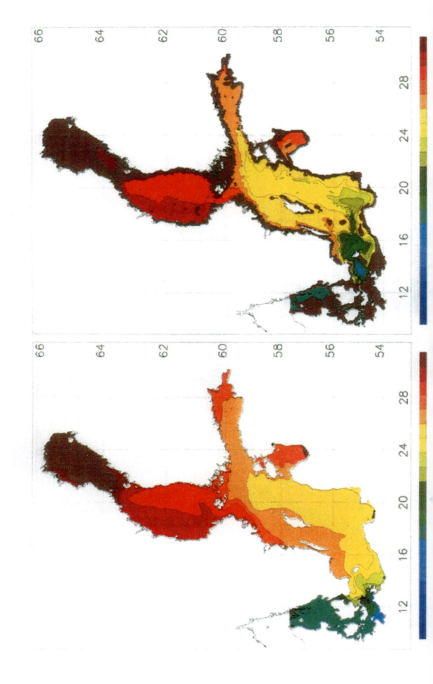

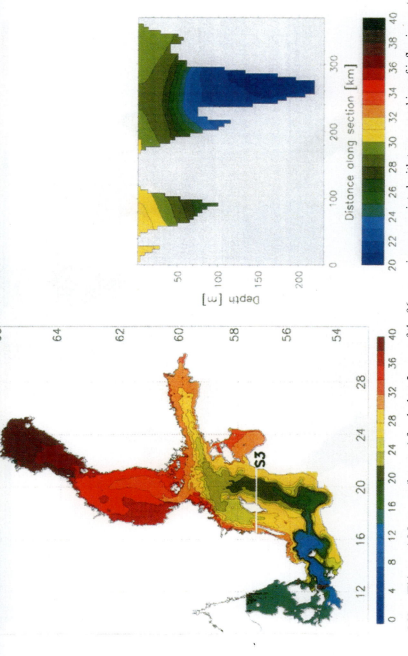

Figure 5.30a. (This page) Mean age (in years) for the last 5 years of the 96-year spin-up associated with tracer-marking of inflowing water at the Darss and Drogden Sills. The figures depict ages at the sea-surface (upper left panel) and at halocline depth (upper right panel), at the surface with a salinity of 17 (lower left panel), and at the longitudinal section S3 across the Gotland Basin (lower right panel). The location of the section is shown as a white line in the lower left panel. Note the different color bars. In the upper right panel, mean ages are not depicted in shallow areas without a permanent halocline, where the conceptual two-layer model does not apply. These shallow areas have been taken as those with depth less than 22.5 m. If salinity in the water column is below 17% the associated surface is set equal to the water depth. Hence, the lower panel shows bottom ages in most of the Baltic Sea interior. (Meier, 2007, printed by permission from *Estuarine, Coastal and Shelf Science*.)

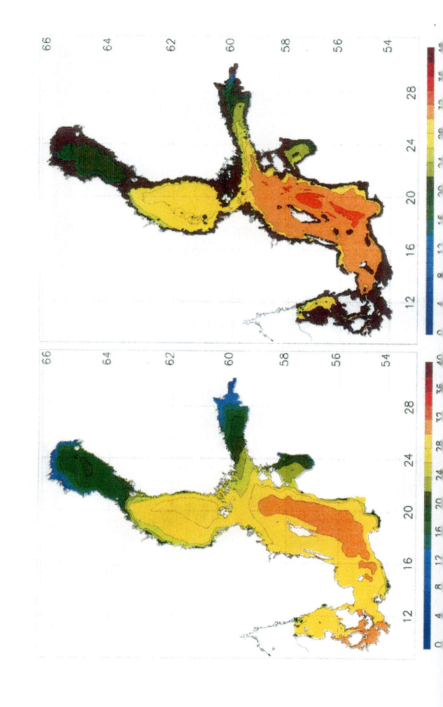

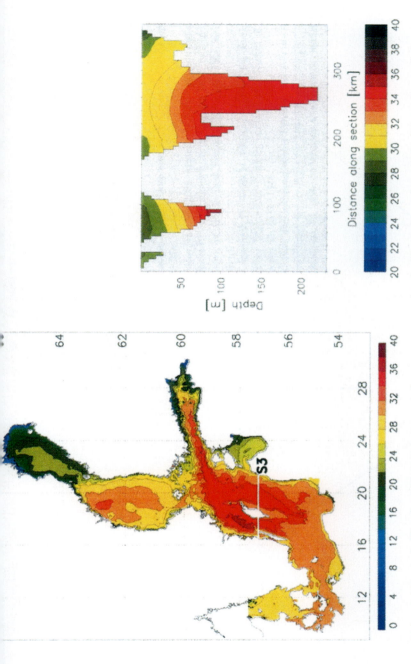

Figure 5.30b. As in Figure 5.30a (previous page), but ages are associated with tracer-marking of freshwater from all rivers.

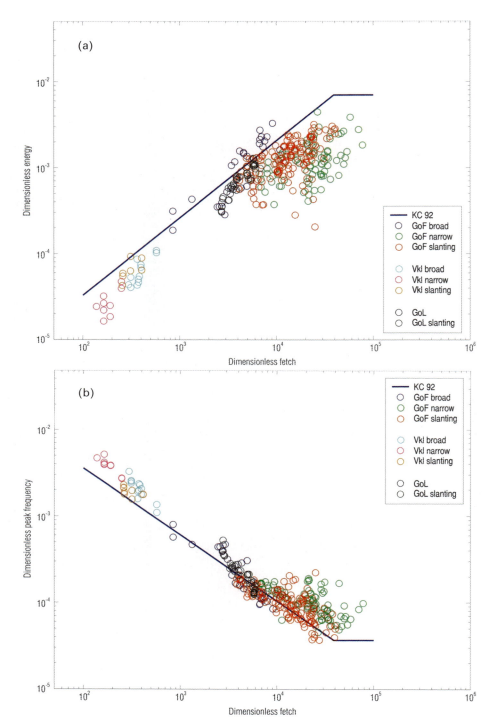

Figure 6.13. Wave growth as a non-dimensional diagram: (a) Energy as a function of fetch; (b) peak frequency as a function of fetch. Solid line shows the theoretical form, the symbols denote data from the Gulf of Finland (GoF), Vanhankaupunginlahti Bay at Helsinki (Vkl), and the Gulf of Lyon (GoL), Mediterranean Sea. (Redrawn from Pettersson, 2004.)

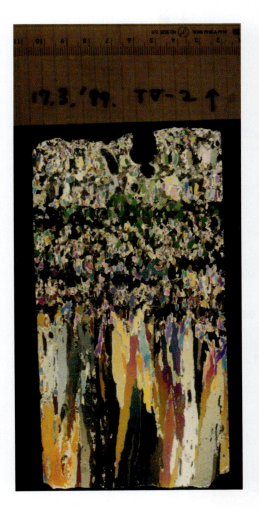 

Figure 7.4. Ice crystal structure in Baltic Sea ice (left) and lake ice (right). The top layer is fine-grained snow ice, and the lower layer is columnar-grained congelation ice. (Prepared by Dr. Toshiyuki Kawamura, printed by permission.)

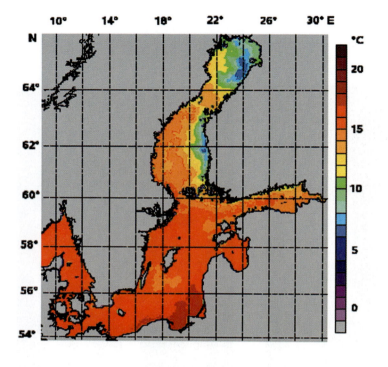
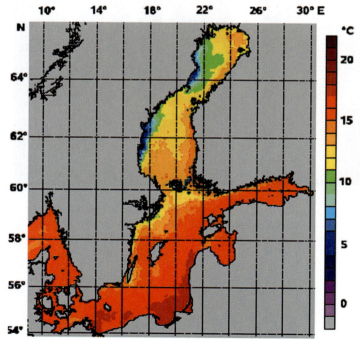

Figure 8.11. Monthly mean composites of sea surface temperature in °C for September 2003 (top) and September 2005 (bottom). Sea-surface temperatures are constructed by combining (averaging) available NOAA satellite overpasses for one month. The scale for sea surface temperatures is from −1.0°C to 22.0°C, in increments of 1°C. (Prepared by Dr. Andreas Lehmann, printed with permission.)

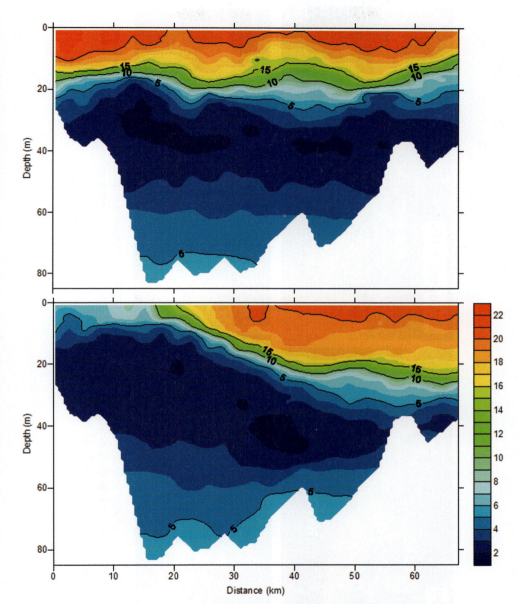

Figure 8.12. Measured temperature (°C) cross-section in the Gulf of Finland in summer 2006 (Estonia on the left side, Finland on the right side). (Top) Stratification is normal off the Estonian coast on July 11; (bottom) pronounced upwelling in this region on August 8. (Prepared by Inga and Urmas Lips, Marine Systems Institute, Tallinn Uniiversity of Technology, printed with permission.)

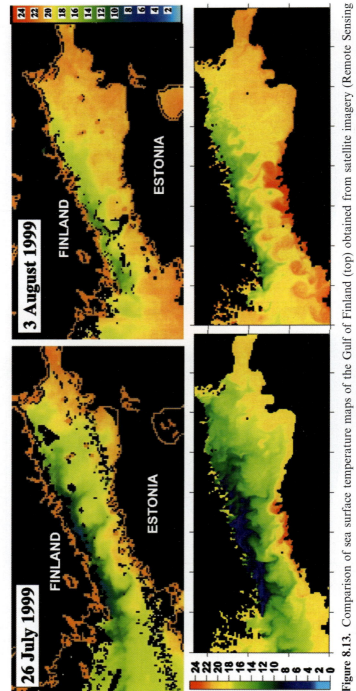

Figure 8.13. Comparison of sea surface temperature maps of the Gulf of Finland (top) obtained from satellite imagery (Remote Sensing Laboratory, Stockholm University) and (bottom) simulated by Zhurbas et al. (2007). The numbers beside the color scale refer to temperatures (°C). (Prepared by Dr. Victor Zhurbas, printed with permission.)

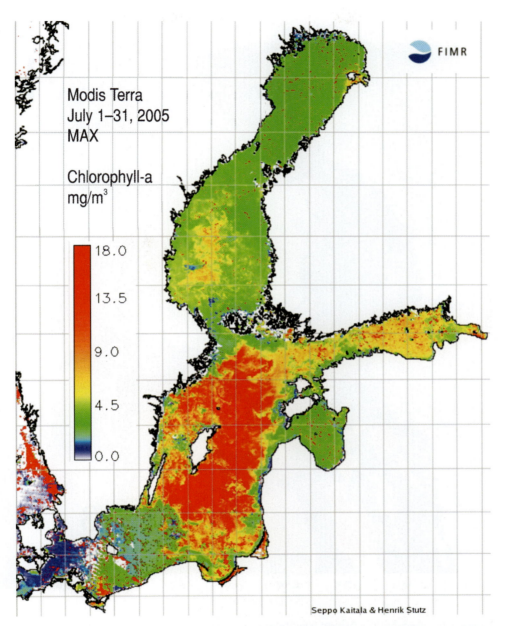

Figure 9.5. Chlorophyll *a* maximum in July 2005 in the Baltic Sea based on MODIS data from the Terra/Aqua satellite. This was produced by an empirical algorithm fitting *in situ* chlorophyll *a* data in 13 reflectance bands with PLS regression analysis. (Prepared by Dr. Seppo Kaitala and Mr. Henrik Stutz, Finnish Institute of Marine Research, printed with permission.)

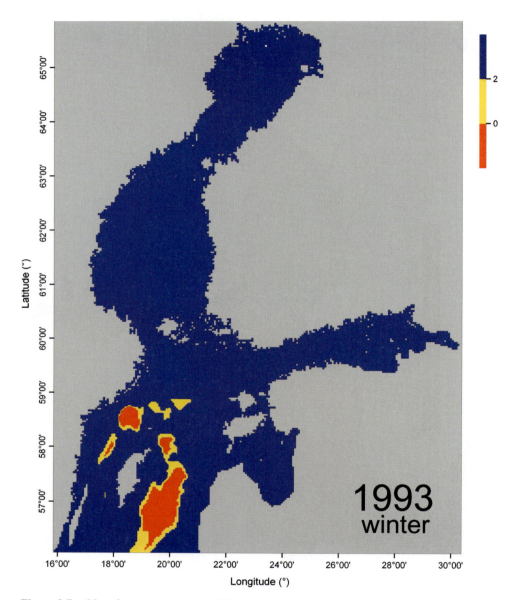

Figure 9.7a. Near-bottom oxygen conditions (mL/L) in the Baltic Sea in 1993. Red represents anoxic conditions and yellow critically low oxygen concentrations of less than 2 mL/L. (Courtesy of M.Sc. Jan-Erik Bruun, Finnish Institute of Marine Research.)

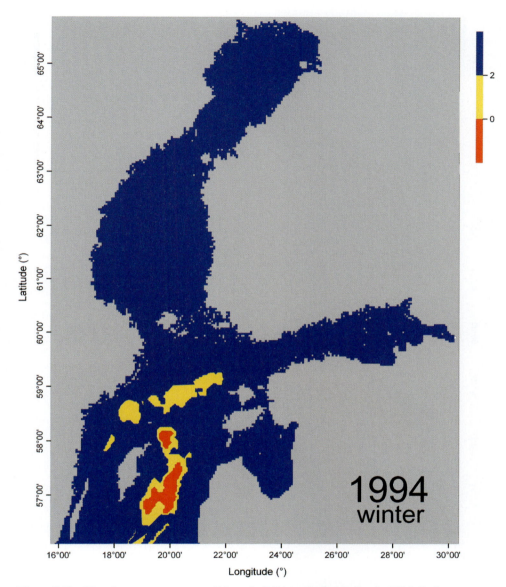

Figure 9.7b. Near-bottom oxygen conditions (mL/L) in the Baltic Sea in 1994. Red represents anoxic conditions and yellow critically low oxygen concentrations of less than 2 mL/L. (Courtesy of M.Sc. Jan-Erik Bruun, Finnish Institute of Marine Research.)

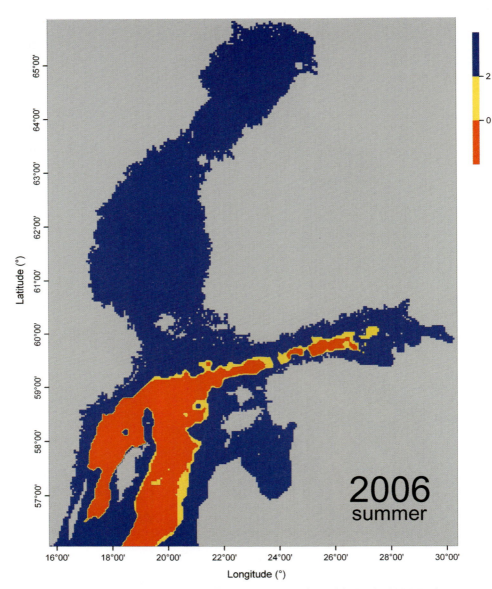

Figure 9.7c. Near-bottom oxygen conditions (mL/L) in the Baltic Sea in 2006. Red represents anoxic conditions and yellow critically low oxygen concentrations of less than 2 mL/L. (Courtesy of M.Sc. Jan-Erik Bruun, Finnish Institute of Marine Research.)

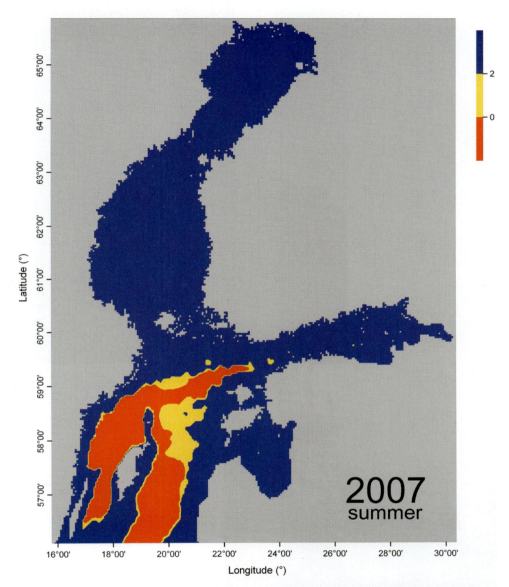

Figure 9.7d. Near-bottom oxygen conditions (mL/L) in the Baltic Sea in 2007. Red represents anoxic conditions and yellow critically low oxygen concentrations of less than 2 mL/L. (Courtesy of M.Sc. Jan-Erik Bruun, Finnish Institute of Marine Research.)

Figure 9.9. The shore of the island of Aegna in 2000 (left) and in 2002 (right) from the same location. The dramatic change is due to coastal erosion caused by ship-induced waves. (Courtesy of Indrek and Andres Kask.)

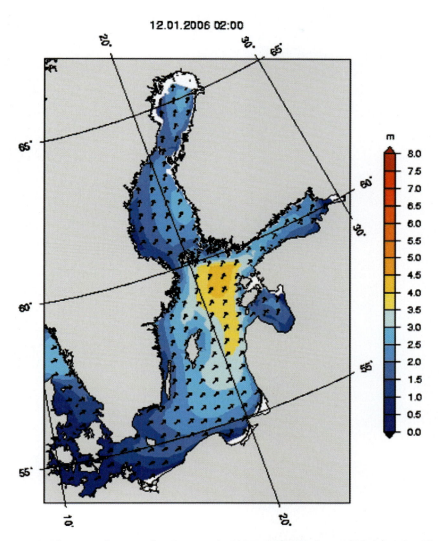

Figure 9.14. The wave forecast for January 9, 2005 at 08:00 GMT provided by the Finnish Institute of Marine Research. On that day a significant wave height of 7.2 meters was observed in the Northern Gotland Basin, which is close to the record value. In the forecast map the wave height (in meters) is shown by colors and wave direction is given by vectors. The areas with white color in the Bay of Bothnia are ice-covered. (Prepared by M.Sc. Laura Tuomi, Finnish Institute of Marine Research, printed with permission.)

5

Circulation

Professor Otto Krümmel (1854–1912) was a German oceanographer. He was educated principally at the University of Göttingen. He first approached the subject of geography through the study of classics and history. In 1883 he acceded to the chair of geography at Kiel, and there in the seaport he found a connexion between his subject and marine investigations, which directed his subsequent career. He retained the chair at Kiel until 1911, and during his tenure he introduced the science of oceanography to the world at large. He completed Boguslawsky's work on oceanography in Ratzel's series of geographical handbooks in 1887, published an account of the "Plankton Expedition" onboard the *National* in the North Atlantic Ocean in 1889, served on the International Council for the Study of the Sea (1900–1909), and finally produced the great work of his life, *Handbuch der Ozeanographie*, between 1907 and 1911 (Krümmel, 1907). In 1911 Krümmel left Kiel to take up the professorship of geography at Marburg. He died at Cologne on October 12, 1912. (Courtesy of Leibniz-Institut für Meereswissenschaften an der Universität Kiel.)

5.1 THEORETICAL BACKGROUND

In this chapter a detailed description of circulation in the Baltic Sea is given. In Section 5.1 the general theory of ocean dynamics in the Baltic Sea is discussed along with some important special cases. The conservation laws for heat and salt were treated in Chapter 4, and here the focus is on purely dynamic questions. In Section 5.2, general circulation is introduced using different examples based on observations. Section 5.3, "Deepwater circulation and ventilation", gives an overview of the internal water cycle and internal mesoscale dynamics and mixing. Section 5.4 gives a detailed view of the Major Baltic Inflow—one of the key features in Baltic Sea physics. Section 5.5 gives a brief summary on how three-dimensional numerical models can be used as tools to investigate circulation physics. The tidal phenomenon is not discussed here but will follow in Chapter 6.

5.1.1 Basic system of equations; scale analysis

The system of ocean currents is three-dimensional, and their dynamics is based on the Navier–Stokes equation and the continuity equation of incompressible fluids. These equations stand for the conservation of momentum and mass. In ocean dynamics the Boussinesq approximation[1] is used, which states that density differences are sufficiently small to be neglected, except where they appear in terms multiplied by acceleration due to gravity. The basic equations of ocean dynamics are

$$\frac{\partial \boldsymbol{u}}{\partial t} + \boldsymbol{u} \cdot \nabla \boldsymbol{u} + 2\boldsymbol{\Omega} \times \boldsymbol{u} = -\frac{1}{\rho}\nabla p + \nabla \cdot \boldsymbol{\tau} + \nu \nabla^2 \boldsymbol{u} \qquad (5.1a)$$

$$\nabla \cdot \boldsymbol{u} = 0 \qquad (5.1b)$$

where $\boldsymbol{u} = (u, v, w)$ is the current velocity; $\boldsymbol{\Omega}$ is the Earth's rotation rate ($\Omega = 0.7292 \times 10^{-4}\,\text{s}^{-1}$); $\boldsymbol{\tau}$ is the Reynolds stress tensor; and ν is the molecular kinematic viscosity. The coordinate system is chosen in such a way that the x-axis is directed eastwards, the y-axis is directed northwards, and the z-axis is directed upwards. Equations (5.1) are purely dynamic. For the complete circulation system the equation of state of seawater and the conservation laws of heat and salt must be added.

Molecular viscosity is generally neglected in circulation theory and only the horizontal part of Coriolis acceleration is included. There is a strong distinction between horizontal and vertical directions due to the large difference in lengthscale and due to the fact that gravity strongly limits vertical motions. Horizontal currents in the Baltic Sea have a typical magnitude of 10 cm/s whereas vertical velocities are typically less than 0.1 mm/s. This feature, a much more prominent property of the velocity field in the shallow Baltic Sea than that in the open ocean, implies that the equation for the vertical component of motion can be normally simplified using the hydrostatic approximation.

Let us denote horizontal and vertical velocity by $U = (u, v)$ and w and the horizontal gradient operator by ∇_H. The Reynolds stress is taken in the first-order approximation as $\boldsymbol{\tau} = 2\boldsymbol{A} \cdot \boldsymbol{\dot{\varepsilon}}$, where A is the eddy diffusion tensor and $\boldsymbol{\dot{\varepsilon}}$ is the strain rate tensor. Due to the anisotropy of ocean dynamics the mixing length is much larger

---

[1] According to Joseph Valentin Boussinesq (1842–1929).

in the horizontal than in the vertical direction, and therefore eddy diffusion tensor components are taken as $A_{xx} = A_{yy} = A_H$, $A_{zz} = A_v$, and $A_{pq} = 0$ when $p \neq q$. Then the equations of ocean dynamics can be written in the following form:

$$\frac{\partial U}{\partial t} + U \cdot \nabla_H U + w \frac{\partial U}{\partial z} + f\mathbf{k} \times U = -\frac{1}{\rho}\nabla_H p + A_H \nabla_H^2 U + \frac{\partial}{\partial z}\left(A_v \frac{\partial U}{\partial z}\right) \quad (5.2\text{a})$$

$$\frac{\partial p}{\partial z} = -\rho g \quad (5.2\text{b})$$

$$\frac{\partial w}{\partial z} + \nabla_H \cdot U = 0 \quad (5.2\text{c})$$

where $f = 2\Omega \sin \phi$ is the Coriolis parameter; and ϕ is latitude.

In Baltic Sea dynamics, typical scales are $U = 10$ cm/s, $T = 5$ days (synoptic scale), $L = 50$ km, and $H = 25$ m. Representative turbulent (eddy) viscosities are $A_H = 10^5$ m^2/s and $A_v = 0.05$ m^2/s. Sea-level measurements in the Baltic Sea show that the inclination of the sea surface is typically $\beta \sim 10$ cm/100 km, and then the horizontal pressure gradient is in the surface layer $\rho^{-1}\nabla p = -g\beta \sim 10^{-5}$ m/s^2. With the help of these levels the characteristic magnitudes of the horizontal equation of motion (5.2a) can be evaluated (Table 5.1). This process is frequently called "scaling of dynamical equations".

The vertical velocity scale W is obtained from the equation of continuity, $W/H = U/L$, and thus all advection terms have equal magnitudes. The dominating terms are Coriolis acceleration, pressure gradient, and vertical friction. The last term includes the transfer of wind stress down from the sea surface and the damping of motion by bottom friction. In the case of a vanishing pressure gradient, Coriolis acceleration and vertical friction contribute most to the balance. Advection and horizontal friction are smaller than Coriolis acceleration by almost one order of magnitude. However, advection plays an important role in intensive dynamics (i.e., when U becomes large), and horizontal friction becomes important near the coasts when L decreases. The inertia term is significant in relatively fast processes when the timescale T is just a few hours and in so-called inertial oscillations.

Due to the small size and the geographical location of the Baltic Sea the Coriolis parameter can normally be taken as a constant ($f = 1.26 \times 10^{-4}$ s^{-1}, when $\phi = 60°$N). Investigations can be made in a local plane projection, which rotates around the local vertical axis (so-called "f-plane approximation"). The β-plane approximation, in which the Coriolis parameter varies linearly in the north–south

Table 5.1. Typical values for different terms of the equations of motion (unit 10^{-6} m/s^2).

Inertia	Advection	Coriolis acceleration	Pressure gradient	Internal friction	
				Horizontal	Vertical
U/T	U^2/L	fU	$\rho^{-1}\nabla_H p/L$	$A_H U/L^2$	$A_v U/H^2$
0.2	2	10	10	4	10

direction is not always used in the Baltic Sea because of its small size, and the equations would be much more complicated in spherical coordinates. Boundary conditions are simple here. There is nearly everywhere a passive solid boundary. The inflow of freshwater takes place in the river mouths, and a dynamic inflow–outflow system exists in the Danish Straits. Water exchange in the Danish Straits plays a dominant role in deepwater circulation and indirectly in the general circulation of the Baltic Sea by regulating general sea-level elevation.

The main dimensionless numbers in Baltic Sea dynamics are the Rossby number $Ro = U/fL$, the Froude number $Fr = U/(Lg)^{1/2}$, and the Ekman numbers $Ek_v = A_v/(UH)$ and $Ek_H = A_H/(UL)$. They describe the significance of, respectively, Coriolis acceleration, gravity, vertical friction, and horizontal friction. The main characteristic lengthscales are the basin dimensions and the Rossby radius of deformation $R = c/f$ where c is the characteristic wave speed. The Rossby radius is one of the fundamental lengthscales in geophysical fluid dynamics.

Barotropic and baroclinic flows

There are two basic types of flow fields: barotropic and baroclinic circulation. In the barotropic case the density of seawater is assumed constant or the isopycnals (isolines of density) are parallel to isobars. In the baroclinic case the isopycnals and isobars are inclined with respect to each other (the inclination angle usually varies with depth). Consequently, the pressure gradient and the resulting currents also vary in depth. The baroclinic and barotropic components are usually called "baroclinic and barotropic modes". In practice, in homogeneous waters the circulation is barotropic, and in stratified waters, baroclinic circulation needs to be considered.

The Rossby radius is different in barotropic and baroclinic flows. For the barotropic Rossby radius the wave speed is the shallow water wave speed, and for the baroclinic Rossby radius the internal wave mode represents characteristic waves. A number of case studies based on current measurements have shown that the response of the Baltic Sea is baroclinic to wind events with a duration more than 50 hours, whereas in short-term wind events, duration 10–40 hours, the response is barotropic. The response to wind events shorter than 10 hours quickly vanishes. There is one order of magnitude more energy in the baroclinic mode than in the barotropic mode. Therefore, the Baltic Sea cannot be treated as a homogeneous water body.

Meteorological conditions govern to a large extent changes in the flow field, and, not surprisingly, there are similarities in the spectra of wind and current velocities. Especially in the surface layer the variability of the currents is strongly determined by wind forcing; however, inertial oscillations (see Section 5.1.2) and shallow water waves (see Chapter 6) grow from the internal dynamics of the sea. The wind transport of surface waters builds up pressure gradients, which provide forcing to deeper waters. The resulting motion is called "wind-generated secondary circulation".

5.1.2 The dynamics of surface currents

There are four mechanisms to induce currents in the Baltic Sea: wind stress at the sea surface, sea surface tilt, thermohaline horizontal gradient of density, and tidal forces. Currents are steered furthermore by Coriolis acceleration, topography, and friction. Voluminous river runoffs can produce local changes in sea-level height and consequently also in currents. Due to the small size of Baltic Sea basins, friction caused by the bottom and shores markedly damps the currents. The general circulation is typical for a stratified system. Inflowing waters into a basin settle at a depth where ambient water has an equal density. So, fresher water goes into the upper layer and more salty watermasses go into a certain lower layer.

On the longest timescale—from several months to years—a baroclinic, wind-independent basic circulation appears. This is due to a positive freshwater balance and the resulting large horizontal gradient of salinity. Freshwaters leave the Baltic Sea in the near-surface layers whereas the inflow of saline watermasses takes place in the lower layer.

On short time-scales (1–5 days) currents are caused by wind stress. Due to the large variability in winds, the resulting long-term wind-driven mean circulation is weak, and transient currents are one order of magnitude larger than average ones. Drift currents produce in coastal areas upwelling and downwelling features that are affected by Kelvin-type waves. The water body is laterally mixed by mesoscale eddies and deepwater circulation (see, e.g., Fennel and Sturm, 1992; Lass and Talpsepp, 1993; Raudsepp, 1998; Stigebrandt et al. 2002; Elken and Matthäus, 2008). On a timescale of 1 hour to 1 day, there are several periodic dynamical processes. The most important are inertial oscillations (13.2–14.5 hours; discussed in Section 5.1.3) and seiches (less than 40 hours; discussed in Chapter 6).

In conclusion, of the processes affecting long-term mean surface circulation, the observed outcome in the Baltic Sea is based on a non-linear combination of wind-independent baroclinic mean circulation and mean wind-driven circulation. Which is the more important is difficult to answer in such a non-linear system; it depends on the case studied and on the timescale under investigation.

5.1.3 Important special cases in sea dynamics

Inertial oscillations

In physical oceanography, inertial oscillations refer to the circular motion of water when inertia and Coriolis acceleration balance each other. This situation initiates, for example, as wind ceases to drive a surface current. Gustafsson and Kullenberg (1936) observed inertial oscillations in the Gotland Sea (Figure 5.1); this was the first time this phenomenon was documented in physical oceanography. Their study was based on current measurements collected during the summer period between 1931 and 1933. Later, it became clear that inertial motion is one of the dominating features in the physics of the Baltic Sea (Kullenberg, 1981).

The energy spectra of Baltic Sea current velocities show that the inertial period (13.2–14.5 hours) is the strongest single period (Figure 5.2). For periods between 10

136 Circulation [Ch. 5

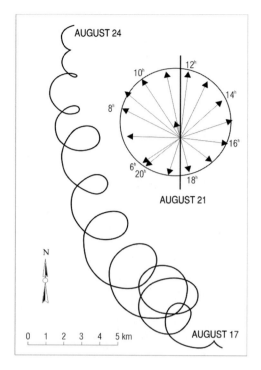

Figure 5.1. Inertial oscillations in the Baltic Sea between August 17 and 24, 1933. The circular diagram shows current velocity on August 21. (Gustafsson and Kullenberg, 1936.)

and 12 hours there are indications of an interaction between tidal waves and inertial waves. Also relatively small periods have been found (8, 7.5, 6.5, and 4 hours).

When external forcing ceases, rotational effects take over and the inertial oscillation becomes visible. Mathematical analysis of this situation is simple to perform using complex variables. Denote the velocity as $q = u + iv$, $i = \sqrt{-1}$. The basic equation for inertial motion is obtained from the horizontal equation of motion. External forcing and friction terms are zero and the flow field is assumed to be spatially homogeneous. Thus, the equation of motion is reduced to

$$\frac{dq}{dt} = -ifq \qquad (5.3)$$

Multiplication of q by the imaginary unit $-i$ (operation $-iq$) turns the velocity vector q by 90° to the right, thus pointing in the direction of Coriolis acceleration. This first-order equation is directly integrated into:

$$q = q_0 \, e^{-ift} \qquad (5.4)$$

The absolute value of velocity is constant ($|q| = |q_0|$) and the direction of velocity turns clockwise by the angular velocity of $-f$ in the northern hemisphere ($f > 0$). One circle makes the so-called inertial circle with radius R_I and period T_I:

$$R_I = \frac{|q_0|}{f}, \qquad T_I = \frac{2\pi}{f} \qquad (5.5)$$

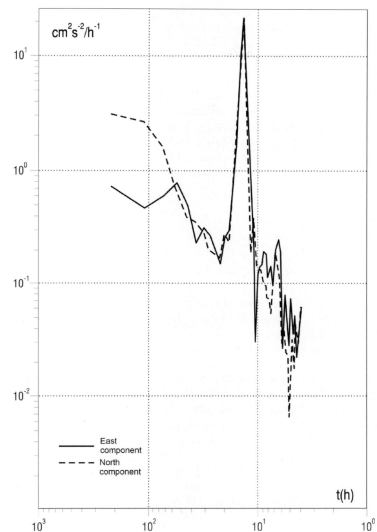

Figure 5.2. The spectral density of current velocity at a depth of 13 m in the Gulf of Bothnia. The continuous line represents the east component of currents (perpendicular to the coast) and the dashed line represents the north component of currents (parallel to the coast). (Redrawn from Alenius, 1980.)

In the Baltic Sea the inertial period is 13.2–14.5 hours (at latitude 60° it is 13.8 hours). If we assume that initial velocity is 10 cm/s, the radius of the circle is about 800 meters. However, in reality the circle reduces in time due to frictional forces. This can be modeled in a simple way by adding a damping term $-rq$ on the right-hand side of Equation (5.3), where r is a friction coefficient. The solution is then $q = q_0 \, e^{-rt} \, e^{-ift}$. The inverse number $1/r$ is the relaxation time[2] of the system; data from Gustafsson

---

[2] Relaxation time is the time in spin-down when velocity has decreased to a fraction e^{-1} of the original, also called an e-folding timescale.

and Kullenberg (1936) suggest that $1/r \approx 1$ week in summer conditions. Inertial motion is in different phases above and below the thermocline, and near the coasts it is quickly damped due to friction.

Ekman drift

Swedish oceanographer V. W. Ekman theoretically explained more than 100 years ago (Ekman, 1905) how wind-driven surface layer currents are formed. Wind induces a shear stress on the sea surface. Vertical turbulent friction transfers the momentum of the wind downwards while Coriolis acceleration turns the flow direction to the right in the northern hemisphere. The Ekman equations are based on the steady state in a horizontally homogeneous ocean. They can be written in the following form:

$$A_v \frac{d^2 q}{dz^2} - ifq = 0 \tag{5.6}$$

The general solution is

$$q = C_1 \exp(\lambda z) + C_2 \exp(-\lambda z), \qquad \lambda = \frac{(1+i)}{2}\sqrt{\frac{f}{2A_v}} \tag{5.7}$$

In classical Ekman layer theory the length $D = \pi\sqrt{2A_v/f}$ is defined as the *Ekman depth*; when the depth of the sea (H) satisfies $H \gg D$, the velocity profile takes the shape of the famous Ekman spiral, where at $z = D$ the speed drops to a fraction $e^{-\pi} \approx 4\%$ of surface speed and the spiral has rotated clockwise by angle π. If $H < D$, above the sea bottom there is another Ekman layer where the velocity decreases to zero and has a spiral profile. For a representative turbulent viscosity of the whole Ekman layer, $A_v \sim 10^{-3}$–10^{-2} m^2 s^{-1}, we have $D \sim 10$ m–40 m. In the Baltic Sea, when eddy viscosity is large, $H \sim D$, which means that speed decay is faster and rotation less than in deep-ocean cases, and the surface and bottom Ekman layers merge. The boundary conditions are taken as

$$z = 0 : q = 0; \qquad z = H : A_v \frac{\partial q}{\partial z} = \frac{\tau_a}{\rho} \tag{5.8}$$

where $z = 0$ is the sea bottom; $z = H$ is the sea surface; and τ_a is the wind stress. The solution is

$$q = q_s \frac{\sinh(\lambda z)}{\sinh(\lambda H)}, \qquad q_s = \frac{\tau_a}{\rho A_v \lambda} \tanh(\lambda H) \tag{5.9}$$

where q_s is surface velocity. This profile is a truncated spiral, which approaches the Ekman spiral as $\lambda^{-1} \gg H$ (Figure 5.3). The vertical velocity integral of Ekman flow or Ekman transport becomes

$$Q = \int_0^H q\, dz = -i\frac{\tau_a - \tau_0}{\rho f} \tag{5.10}$$

If $\tau_0 = 0$, then the classical result of Ekman transport perpendicular to wind stress is obtained. In the Baltic Sea this is true for the upper layer in deep areas where there is a

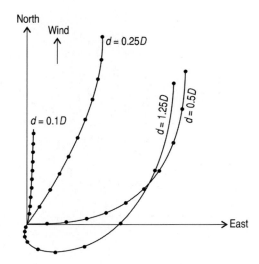

Figure 5.3. The Ekman profile of current velocity in shallow seas. d is sea depth and D is Ekman depth.

halocline. But in shallow areas, bottom friction reduces the transport and the spiral is less pronounced. Using the velocity solution directly, we have

$$Q = -i\frac{T_a}{\rho f}\left(1 - \frac{1}{\cosh(\lambda H)}\right) \qquad (5.11)$$

Thus, Ekman transport at a finite sea depth and bottom friction depend on eddy viscosity via parameter λ.

The Ekman solution qualitatively holds true. Current speed reduces and the direction turns to the right with increasing depth. However, the approximation that vertical viscosity is constant is not very accurate, hence the reason the profile is not observed in detail in nature. For large sea depths, observed surface currents are about 2%–3% of wind speed, but the turning angle is 20°–30°. This simple surface drift model is used in many practical applications, such as forecasting the drift of floating objects. Figure 5.4 presents observations of freely moving sea ice and surface layer water velocity. In a qualitative sense the Ekman spiral is observed beneath the ice.[3] Observations showed that ice drifted 15° to the right of the wind direction and the speed of ice drift was 2% of wind speed.

Ekman theory has a direct application in understanding wind-induced upwelling. If the wind blows parallel to the coast and the coast is on the left, then Ekman transport is directed offshore. This water transport is compensated due to continuity by the vertical movement of water towards the sea surface. As an example, Figure 5.5 (see color section) shows upwelling taking place off the northern coast of the Gulf of Finland due to southwesterly winds. Upwelling theory, its measurement, and modeling are described in Chapter 8.

---

[3] The Ekman spiral was first observed under the ice by Frithiof Nansen in the 1890s. The spiral was documented much later for the open ocean.

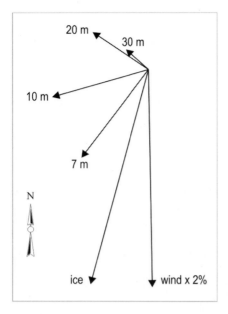

Figure 5.4. The wind-driven motion of ice and currents at different depths in the Ekman layer in the Bay of Bothnia, April 1975. (Leppäranta, 1990.)

Geostrophic flow

Geostrophic flow—the current driven by the pressure gradient on a rotating planet—is a steady, frictionless current in which Coriolis acceleration is balanced by the pressure gradient. In the vertical direction, hydrostatic balance is assumed. The momentum equation is written for geostrophic balance

$$ifq = -\frac{1}{\rho}\nabla p, \qquad \frac{\partial p}{\partial z} = -\rho g \qquad (5.12)$$

The solution is $q = -i(\rho f)^{-1}\nabla p$; in words, the flow direction is perpendicular to the pressure gradient (i.e., parallel with isobars) with higher pressure on the right when facing the direction of motion in the northern hemisphere. In the barotropic case, geostrophic flow is depth-independent, whereas in the baroclinic case the flow varies with depth. The geostrophic solution is horizontally non-divergent and therefore $w = 0$ everywhere. Consequently, the flow conserves its forcing, the pressure field. Taking the derivative with respect to the depth gives

$$\frac{\partial q}{\partial z} = -i\frac{g}{\rho f}\nabla \rho \qquad (5.13)$$

known as the thermal wind law because it was first found in meteorological research (Holton, 1979; Cushman-Roisin, 1994). If density changes in the horizontal direction, then the geostrophic current will vary in the vertical. This is the case in baroclinic situations; in barotropic situations the geostrophic current is independent of depth.

Geostrophic flow in the surface layer results from the inclination of the sea surface. If the tilt of the sea surface towards the east is β, the northward-directed flow speed $v = g\beta/f$. In practice $\beta \sim 10^{-6}$ ($= 1\,\text{cm}/10\,\text{km}$), thus resulting in $v = 10\,\text{cm/s}$. In a two-layer system, the flow in the surface layer v_1 is obtained from the tilt of the sea surface and the difference between the flow in the surface and lower layer is obtained from the density difference between these two layers according to the thermal wind law:

$$v_1 - v_2 = \frac{g}{f}\frac{\rho_1 - \rho_2}{\rho_2}\frac{\partial H_1}{\partial x} \tag{5.14}$$

where v_1 and v_2 are the flows in the surface and bottom layers; ρ_1 and ρ_2 are the corresponding densities; and H_1 is the thickness of the surface layer.

Determination of geostrophic flow includes one of the fundamental problems of physical oceanography: namely, the problem of finding a reference level where the pressure gradient is known. In the deep ocean it is usually assumed that below the permanent thermocline layer, at a depth of 1 km–2 km, the isobars are horizontal and thus geostrophic flow equals zero ("level of no motion"). In the Baltic Sea no such depth can be assumed, but the approach has been to consider the deep geostrophic flow as small ($\approx$ zero). This assumption is inaccurate because the near-bottom frictional layer is met at a depth where geostrophic flow is still significant. But even though absolute flows may be inaccurate, the relative currents inside the water body as obtained from the thermal wind law—Equation (5.13)—are good.

Exact determination of surface tilt would solve the problem concerning the reference level. However, the tilts of isobars are usually very small (10^{-6}), introducing a measurement problem. One of the first proofs of the relation between geostrophic flow and sea-level tilt is from the Great Belt. Comparisons of the results with measured flows in the strait are shown in Figure 5.6.

The combination of wind-driven currents and geostrophic currents

Consider the full horizontal equation of motion—Equation (5.2a)—for the combination of Ekman theory and geostrophic theory. Write current velocity as $q = q_E + q_G$, where q_E is the Ekman solution and q_G is the geostrophic current. It is assumed that the situation is stationary and that advection and horizontal friction are omitted, and Equation (5.2a) is then reduced to

$$if(q_E + q_G) = -\frac{1}{\rho}\nabla p + A_v \frac{\partial^2 (q_E + q_G)}{\partial z^2} \tag{5.15}$$

If it is furthermore assumed that in the last term the influence of geostrophic flow is negligible, the Ekman equation and the geostrophic equation can be simply summed together (i.e., the solution is the sum of the Ekman and geostrophic solutions). Especially in the barotropic case $q_G =$ constant and thus $\partial^2 u_G/\partial z^2 = 0$.

142 Circulation [Ch. 5

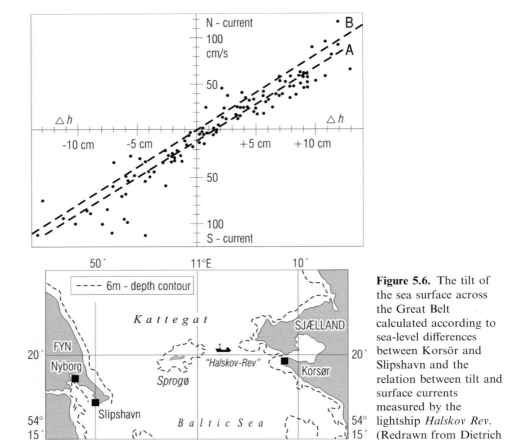

Figure 5.6. The tilt of the sea surface across the Great Belt calculated according to sea-level differences between Korsör and Slipshavn and the relation between tilt and surface currents measured by the lightship *Halskov Rev*. (Redrawn from Dietrich et al., 1963.)

5.2 GENERAL CIRCULATION BASED ON OBSERVATIONS

5.2.1 Surface currents

The first studies of Baltic Sea circulation were based on observations, which had been collected onboard lightships ever since the beginning of the 1900s (Witting, 1912; Palmén, 1930). The measurements were based on quite simple instrumentation[4] but they provided basic knowledge, and hence even 100 years ago an overall picture of surface circulation was available. These early measurements showed mean circulation in the Baltic Sea main basins to be cyclonic (i.e., counter-clockwise, Figure 5.7) and to be reflected in the transport processes in the Baltic Sea. Even though long-term mean

---

[4] Drifters or drifting buoys, sometimes called "flow crosses", were deployed and followed from lightships.

Sec. 5.2] General circulation based on observations 143

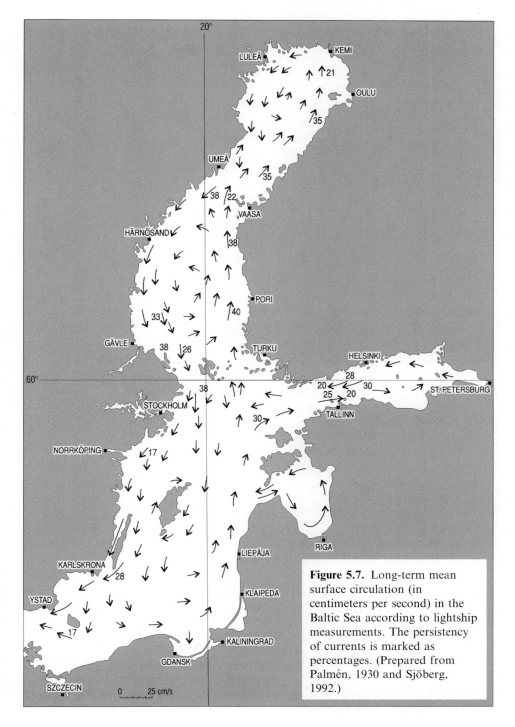

Figure 5.7. Long-term mean surface circulation (in centimeters per second) in the Baltic Sea according to lightship measurements. The persistency of currents is marked as percentages. (Prepared from Palmén, 1930 and Sjöberg, 1992.)

currents are weak with average speeds of some 5 cm/s, the persistency of the circulation system is in some areas relatively strong.

Based on early measurements, Witting (1912) and Palmén (1930) composed maps of mean surface currents in the Northern Baltic Sea. Observed mean circulation is a weak cyclonic system in the main basins (Gotland Sea, Gulf of Finland, Gulf of Riga, Sea of Bothnia, and Bay of Bothnia). However, in the Baltic Sea there are no permanent, stable current structures such as the Gulf Stream or the Kuroshio Current in the Atlantic and Pacific Oceans. Mean circulation transports salt and heat, and therefore the water is warmer and saltier in the eastern side of the basins than in the western parts. In the narrow, west–east oriented Gulf of Finland such differences, induced by mean circulation, are observed between the northern and southern side (Figure 5.7). During storms, wind drift currents can reach 50 cm/s, in straits up to 100 cm/s; on average, the speed of surface currents is 2%–3% of the wind speed, and direction is 20°–30° to the right of the wind direction. This is a realization of the Ekman solution at the surface.

Palmén (1930) defined the persistency of the direction of mean circulation (R) as the ratio

$$R = \frac{|\langle \mathbf{U} \rangle|}{\langle U \rangle} \qquad (5.16)$$

where vector $\langle \mathbf{U} \rangle$ is mean current velocity; and $\langle U \rangle$ is mean current speed. If the direction is constant, persistency is 100%. When persistency reaches zero, the mean flow and net transport of water are also equal to zero. In a symmetric, bimodal coastal current, persistency equals the difference between modes (e.g., if the flow is 75% to the west and 25% to the east with the same speeds, persistency is 50%). In practice, the persistency of surface circulation is 20%–40%, and in specific locations such as north of the longitudinal axis of the Gulf of Finland it can be even higher.

The general current pattern also becomes visible when investigating the distribution of currents. Figure 5.8 shows current velocity distributions in the Sea of Bothnia presented as so-called "current roses". Near the coast the currents are parallel to the shoreline and thus generally have a bimodal structure, whereas in the open sea the current system is more isotropic. In the open sea the steering effect of the coast is

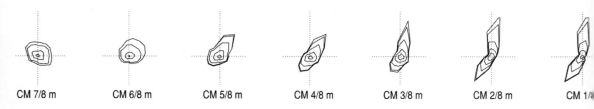

Figure 5.8. Current roses based on measurements outside Rauma, Sea of Bothnia during the International Gulf Bothnia Year 1991. The graphs describe the cumulative probability of current speeds at 5 cm/s spacing in different directions. Station CM1 is closest to the coast (10 km) whereas Station CM7 is 55 km from the coast. The x-axis directed to the east and the y-axis to the north. The measurement depth is 8 meters. (Murthy et al., 1993.)

missing, and thus the persistency of currents is lower; the current field in the open sea contains many mesoscale vortices.

5.2.2 Three-dimensional water circulation

The abovementioned quasi-permanent cyclonic circulation takes place in the upper layer of the Baltic Sea. In the lower layer, the water dynamics is rather different from that in the upper layer and the flow is steered by channels and restricted by various sills (Figure 3.2). Deepwater moves southwards through the Belt Sea and Öresund to Arkona Basin, from there via the Bornholm Channel to the Bornholm Deep. Then deepwater moves through the Stolpe Channel to the Eastern Gotland Deep. Note that the sill between the Bornholm Basin and the Western Gotland Basin is shallow and the water cannot directly reach the Western Gotland Basin (Figure 3.3). From the Eastern Gotland Basin some water flows into the Gdansk Deep but most flows northwards to the Gotland Deep. After that, water movement continues to the Fårö Deep and farther on to the Northern Gotland Basin. Due to the shallow connection, Eastern Gotland Basin deepwater cannot enter the Gulf of Riga.

In the north the Gotland Basin is bounded by the Archipelago Sea and the Åland Sea, restricting saline deepwater from the Gotland Sea from flowing into the Gulf of Bothnia. Deepwater from the Northern Gotland Basin flows partly to the southwest to the Western Gotland Basin and becomes deepwater for the Landsort Deep and the Norrköping Deep. Due to the absence of a sill towards the Gulf of Finland, water from the Northern Gotland Basin can gain entry there and add to the extremely complex estuarine dynamics in this basin.

The interaction between upper and lower layers is quite restricted in the Baltic Sea due to its strong stratification. In the Kattegat, dense North Sea water forms a deepwater pool, whereas fresher Baltic water is located in the surface layer. Deepwater circulation is characterized by dense bottom currents in the inflowing saline water at the mouth area of the Baltic Sea. Convection, mechanical mixing, entrainment, and vertical advection of watermasses lead to interactions between upper and lower layers in other parts of the Baltic Sea (Figure 5.9, see color section).

Döös *et al.* (2004) coined a new term "haline conveyor belt" to describe, at a general level, the Baltic Sea circulation system. Water is effectively recirculating in the Baltic Sea despite the existing low-permeable halocline. Overturning circulation is sometimes called the "Baltic Sea haline conveyor belt" (Döös *et al.*, 2004) in analogy to the deepwater conveyor belt of the World Ocean. A vertically overturning circulation consists of many important factors: gravity-driven dense bottom currents of inflowing water from the North Sea, the entrainment of ambient surface water, mixing due to diffusion, interleaving of inflowing watermasses into the deep at the level of neutral buoyancy, vertical advection due to the conservation and upward entrainment of deepwater into moving surface water in the northern Baltic Sea proper.

5.2.3 Examples of current dynamics

Spatiotemporal variability and mesoscale vortices can be superposed on relatively highly stable surface currents. If drifting buoys are deployed at sea at the same time their routes start to diverge (Figure 5.10). This diffusion is a general feature of large-scale turbulence. So, even a small difference in the initial positions of buoys leads to a situation in which the buoys are under the influence of slightly different wind and

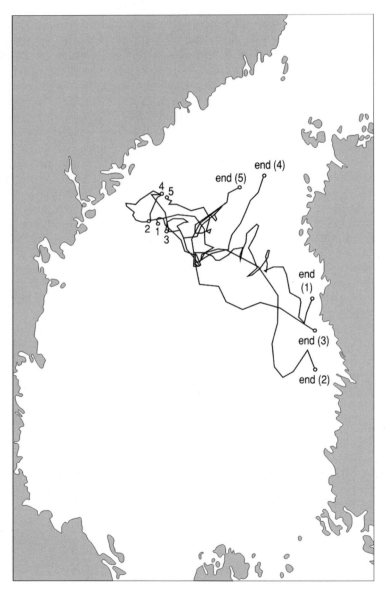

Figure 5.10. The movement of drifting buoys in the Gulf of Bothnia between July 7 and 18, 1991. The position of buoy deployments are marked by 1–5, the paths of the buoys are shown by continuous lines, and the end positions are marked as well. (Redrawn from Håkansson et al., 1994.)

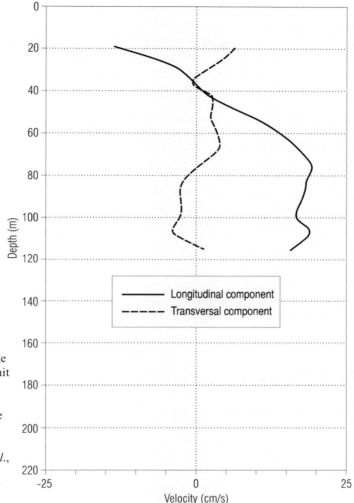

Figure 5.11. An example of an instantaneous strait flow in the Southern Quark at Station F33 (60°32'N, 18°57'E). The positive directions are northwards and eastwards. (Hietala *et al.*, 2007, with permission from *Journal of Marine Systems*.)

current conditions. The paths of the buoys diverge from each other in time and the differences in the local wind and currents felt by the buoys further increase.

The Baltic Sea consists of several basins joined by means of straits and channels. The flows in straits are usually guided by the geometry. Currents in the relatively narrow Southern Quark in the Sea of Åland on October 14, 2004 are shown in Figure 5.11. Currents parallel to the axis of the strait are on average remarkably stronger than currents directed across it. The steering effect of topography is thus obvious. The Southern Quark is relatively deep, more than 100 meters, and currents have a two-layer structure, as is the case with salinity and temperature. The upper layer is well mixed down to 40 meters, where at that time there was both a thermocline and a halocline. In the upper layer, currents are directed away from the Sea of Bothnia due to the cyclonic circulation system (Figure 5.7), whereas below 40 meters an inflow is observed. In the present case, observed current speeds are 10 cm/s–25 cm/s. Across the strait the flow is weak (0 cm/s–5 cm/s) and the vertical profile is variable.

Currents in the Baltic Sea normally have either a one-layer or two-layer vertical structure, which strongly interact(s) with hydrographical conditions. Figure 5.12 shows a time-series of currents in the Archipelago Sea in autumn 2002. The sea depth is 41 meters at the site, and hydrographic conditions were characterized at the outset by a two-layer structure. At that time, wind-induced surface currents had a speed between 5 cm/s and 15 cm/s, whereas currents close to the bottom were much weaker, 2 cm/s–5 cm/s. Near-bottom salinity and temperature remained nearly constant during the early autumn, whereas oxygen conditions deteriorated due to the lack of water exchange between bottom and upper layers. At the end of October, high current speeds were observed, and as a consequence turbulence became much stronger and turbulent mixing took place across the bottom boundary layer. Bottom temperature increased drastically and, correspondingly, near-bottom salinity decreased and oxygen conditions improved dramatically. Thereafter, near-bottom temperature decreased slowly due to overall cooling of the watermass, while near-bottom salinity remained nearly constant due to the absence of a layered structure. Oxygen conditions also remained unchanged due to the lack of stratification and oxygen consumption in wintertime being very small.

5.3 DEEPWATER CIRCULATION AND VENTILATION

Because the Baltic Sea is permanently two-layered, a key physical feature is deepwater circulation and its implications for overall dynamics (Figure 5.9, see color section). Today, there are still gaps in our understanding of the physics of deepwater dynamics. This subject has been studied and reviewed by several authors (e.g., Fonselius, 1969; Mälkki and Tamsalu, 1985; Rodhe, 1999; Stigebrandt, 2001; Meier et al., 2006; Matthäus, 2006; Elken and Matthäus, 2008). The main problems rest with different inflows and stagnation periods, water exchange between different basins, diapycnal mixing,[5] eddies, entrainment, and mixing.

5.3.1 Internal water cycle

Watermasses flowing into the Baltic Sea in the Belts and Öresund are driven barotropically due to along-strait sea-level tilt. Most frequently, salinity is between 12‰ and 16‰ and during major inflows values up to 22‰–25‰ are obtained. Watermasses are spread and transported further through channels and over sills where local conditions regulate the inflows depending on the densities of ambient water and new incoming water. Saline water further flows over the Drogden Sill in Öresund and over

---

[5] Mixing across a density gradient.

Sec. 5.3] Deepwater circulation and ventilation 149

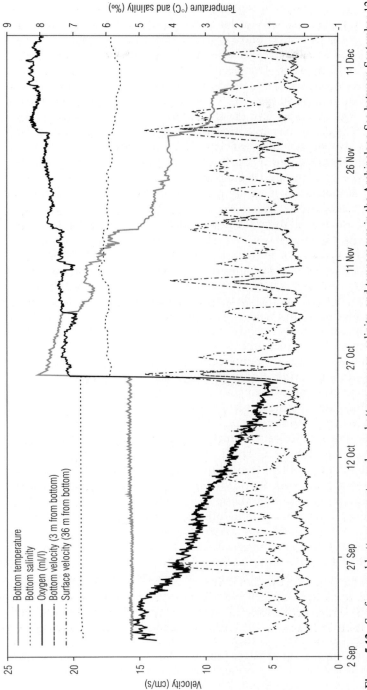

Figure 5.12. Surface and bottom currents, and near-bottom oxygen, salinity, and temperature in the Archipelago Sea between September 12 and December 14, 2002. The sea depth is 41 m. (Prepared by M.Sc. Riikka Hietala, printed with permission.)

the Darss Sill in the Belt Sea. Sinking watermasses entrain ambient surface water, which leads to reduced salinity of inflowing water with distance from the Kattegat (Figures 3.3, 3.10, 3.17, 3.18, and Figure 5.9, see color section).

Inflows reach the Arkona Basin first, which has a maximum depth of 53 meters. Inflowing water forms a 5 m–15 m deep bottom layer in which salinity can reach 24‰. Bottom salinity decreases due to wind-induced mixing. Gustafsson (2001) calculated from observations at Station BY2 in the Arkona Basin a mean inflow of 22,400 m^3 s^{-1} (mean inflow into the Baltic Sea in the Kattegat is 38,100 m^3 s^{-1}, see Section 4.1). The deepwater flow proceeds along the northern side of the basin as a baroclinic geostrophic boundary current (Liljebladh and Stigebrandt, 1996; Lass and Mohrholtz, 2003) to the Bornholm Basin through the Bornholm Channel where the sill locates between Hamrare (Bornholm) and Sandhammare (Skåne, Sweden). The sill depth is 45 meters. The salinity of the water flowing into the Bornholm Basin varies between 8‰ and 18.5‰.

From the Bornholm Basin the flow is directed east through the Stolpe Channel into the Eastern Gotland Basin. The Stolpe Channel, with a maximum depth of 90 meters, has a sill depth of 60 meters towards the Bornholm Basin and 80 meters towards the Gotland Basin. The flow system is very complicated. First, the Bornholm Basin works as a buffer where incoming salty water may be trapped. Elken and Matthäus (2008) took up the classical flow description by categorizing saltwater intrusion in three different modes: (a) regular inflow just below the primary halocline interleaving into the level of neutral buoyancy; (b) occasional inflow of more saline water, sinking to the bottom and exchanging with Bornholm Basin deepwater; and (c) major inflow of large amounts of saline water, filling the lower layer of the Bornholm Basin and further flowing through the Stolpe Sill to the Eastern Gotland Sea. The dynamics of the inflowing water contains many specific features like internal fronts with fine-scale intrusion and subsurface eddies (Piechura *et al.*, 1997; Zhurbas and Paka 1997, 1999; Zhurbas *et al.*, 2003; Meier *et al.*, 2006).

After passing the Stolpe Channel, the saline water flows northeast along the eastern end of the Hoburg Channel. The flow usually forms a complex cyclonic circulation cell in the Gotland Basin. Some of the water flows in to the Gdansk Deep (sill depth 100 m, maximum depth 114 m) making cyclonic loops there. Most water flows northwards to the Gotland Deep (maximum depth 249 m). Here, bottom water can be renewed only when major inflows take place from the Bornholm Basin. According to Elken (1996) and Elken and Matthäus (2008) the Gotland Deep regularly receives saline water interleaving at a depth of 80 m–130 m with a maximum northward flow across the basin reaching 2,500 m^3 s^{-1} around a depth of 100 meters. This ventilation process may well explain why hydrogen sulfide does not appear at depths above 140 m–150 m, even during long stagnation periods.

Farther north the deepwater is divided into two branches, northern and western. The northern branch moves over a sill (140 m) to the Fårö Deep (maximum depth 205 m) and farther on over another sill (115 m) to the Northern Gotland Basin. The western branch turns to the Western Gotland Basin. The Gulf of Riga, located east of the Eastern Gotland Basin is a rather isolated and shallow bay, and the main water exchange channel is the Irbe Strait. Watermasses from the lower layer of the Gotland

Basin cannot enter this strait, but surface water can where it forms the lower layer water of the Gulf of Riga.

In the Northern Gotland Basin, which is a dynamically active area, watermasses branch to the southwest, north, and east (i.e., to the Western Gotland Basin, the Gulf of Bothnia, and the Gulf of Finland). In the north the Gotland Basin is bounded by the Åland Sea where the sill depth is 70 meters. This is a very important sill in the Baltic Sea because it stops the saline deepwater of the Gotland Sea from flowing into the Gulf of Bothnia. Instead, the deepwater flows into the Western Gotland Basin and the Gulf of Finland.

Elken *et al.* (2003, 2006) carried out investigations into the large halocline variation and related mesoscale and basin-scale processes in the Northern Gotland Basin–Gulf of Finland system. The authors suggest that long-lasting pulses of southwesterly winds cause an increase in the water volume of the Gulf of Finland. The resulting increase in hydrostatic pressure in the gulf leads to an outflow of deepwater. Such counter-estuarine transport weakens the stratification of watermasses at the entrance to the Gulf of Finland. As a consequence, the same energy input leads to an intensified diapycnal mixing as compared with the classical situation at the entrance (strong upward vertical advection). Owing to the variable topography both in the Northern Gotland Basin and in the Gulf of Finland, basin-scale barotropic flows are converted into baroclinic mesoscale motions with a large isopycnal displacement (more than 20 m within a distance of 10 km–20 km), which causes intrahalocline current speeds greater than 20 cm/s. As a result, Elken *et al.* (2006) concluded that the near-bottom layers of the Gulf of Finland actively react to wind forcing, a reasoning that considerably modifies the traditional concept of the partially decoupled lower layer dynamics of the Baltic Sea. The multitude of processes at the entrance to the Gulf of Finland certainly makes the modeling of deepwater inflow extremely difficult. Internal wave activity is high, the production of strong eddies and topographically controlled local currents is frequent, and thus diapycnal mixing is intense.

Deepwater in the Gulf of Bothnia is mostly formed in such a way that the surface water of the Gotland Sea cools in winter and sinks over the Southern Åland Sill to the Åland Sea and farther to the Sea of Bothnia across the Middle Åland Sill at a depth of 70 m between Söderarm and Lågskär. The near-bottom layers of the Åland Sea are initially filled with salty water down to the depth of the Northern Åland Sill between the Åland Sea and the Sea of Bothnia (100 m).

During summer, deepwater spreading in the Gulf of Bothnia resembles that in the Baltic Sea proper: cyclonic flow along slopes either deeper along the sloping bottom or interleaving into the level of neutral buoyancy. However, during late autumn the water body is usually rather well mixed, as is the case in the eastern Gulf of Finland. In the Sea of Bothnia the water flows through the deepest route to the northeast guided by the shoal of Finngrundet and the east coast, and north up to the Ulvö Deep. Then, at the entrance to the Northern Quark, deepwater is forced to return south. The maximum depth of the Northern Quark is 65 meters, and there are two sill areas at the southern and northern sides of Holmöarna, both having a depth of 25 meters. The deepwater of the Bay of Bothnia is formed during winter cooling

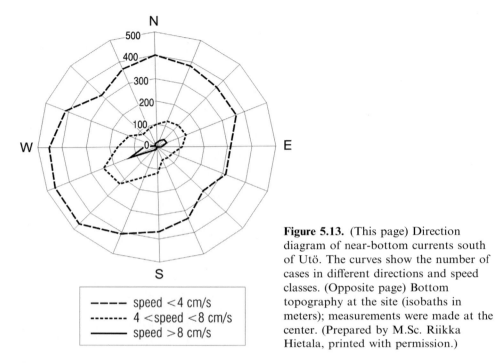

Figure 5.13. (This page) Direction diagram of near-bottom currents south of Utö. The curves show the number of cases in different directions and speed classes. (Opposite page) Bottom topography at the site (isobaths in meters); measurements were made at the center. (Prepared by M.Sc. Riikka Hietala, printed with permission.)

from the surface water of the Sea of Bothnia, which flows northeast at both sides of the Nordvalen shallows. The currents are often strong in this area.

An example of the strong topographic steering of near-bottom currents is given in Figure 5.13, which shows the situation at a deep located in the Archipelago Sea. It can be seen that, especially at high current speeds, the flow is directed nearly always parallel to the depth contours and hardly any flow is directed across the isobaths.

5.3.2 Internal mesoscale dynamics and mixing

Eddies

Eddies in the Baltic Sea have a large scatter in their horizontal size, about 5 km–50 km, and their vertical extension is some tens of meters. This scale usually exceeds the scale of the internal Rossby radius of deformation, which is 2 km–10 km (Fennel et al., 1991; Alenius et al., 2003). The highest velocities of eddies are roughly 30 cm/s. Eddies are cyclonically or anti-cyclonically rotating vortices, which can be observed at different depths in the sea. Their main role is to transport and mix watermasses. The overview of eddies given here is mostly based on the reviews by Stigebrandt (2001), Pavelson (2005), and Meier et al. (2006).

There are several mechanisms behind the birth of eddies, most of which are now clearly understood. The basic mechanisms are related to strong meandering currents

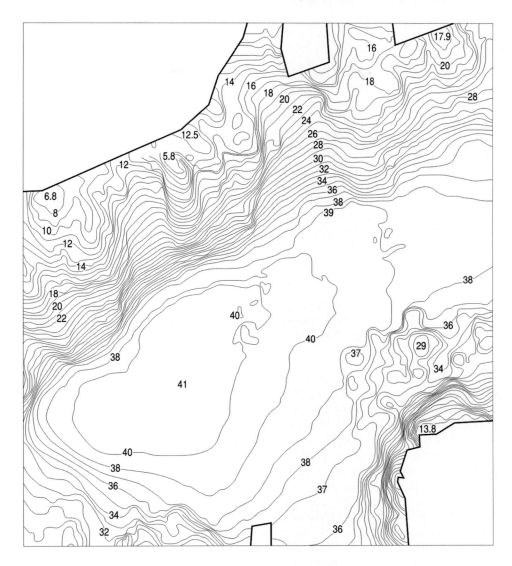

from which eddies become detached. This is a typical feature in the Gulf Stream, for example. The baroclinic instability of mean flow, influence of bottom topography on currents, and the effects of wind forcing furthermore affect the formation of eddies. They can have a long life history, as is the case with Mediterranean salt lenses, which propagate into the Atlantic (Kamenkovich et al., 1986). Such lens features have been observed in the Baltic as well (Kõuts, 1999, see Figure 5.14).

Systematic investigations of mesoscale eddies in the Baltic Sea had already begun in the 1970s by carrying out regular CTD measurements at a high enough resolution.

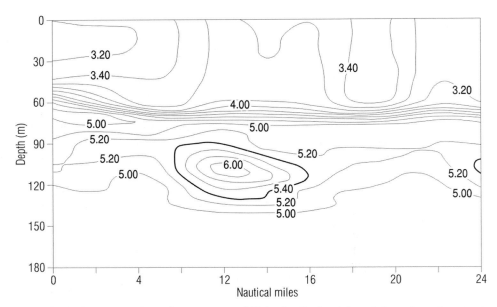

Figure 5.14. A cross-section of temperature (°C) in the Gotland Sea. A lens-shaped vortex is visible in the centre of the cross-section. (Redrawn from Kõuts, 1999.)

These early results were reported by Aitsam et al. (1984) among others. The PEX-86 experiment (Baltic Sea Patchiness Experiment) offered up new information concerning eddy dynamics. It was found that in the salinity field there is a lot of patchiness, in which eddies play a key role. Many eddies 10 kilometers in size were found, suggesting their relation to frontogenesis. Intra-halocline cyclonic and anticyclonic eddies have been reported in many projects, such as PEX-86 and DIAMIX (DIApycnal MIXing), in the Gotland Deep region. However, the role of eddies in deepwater mixing is still partly open (Elken et al., 1994; Stigebrandt et al., 2002). Eddies have also been observed in the southern Baltic: in the Stolpe Channel and in Gdansk Bay (Zhurbas et al., 2004) and in Arkona Basin (Lass and Mohrholtz, 2003). In the latter area, the geostrophic adjustment of eddies was suggested. Zhurbas et al. (2003) have concluded that cyclonic eddies are formed by vortex-stretching when the flow loses its potential vorticity from the Bornholm Basin via the Stolpe Channel into the southeastern Gotland Basin where stratification is weaker. Zhurbas et al. (2003) have also discussed the role of baroclinic jets in eddy formation. Eddies have also been reported in the Gulf of Finland (see Alenius et al., 1998; Pavelson, 2005).

There is additionally a number of experimental case studies of eddies in various parts of the Baltic Sea (see, Aitsam and Elken, 1982; Aitsam and Talpsepp, 1982; Aitsam et al., 1984; Elken et al., 1988, 1994; Kõuts, 1999; Elken, 1996; Reissman, 2002; Stigebrandt et al., 2002; Lass and Mohrholtz, 2003; Piechura and Beszczyńska-Möller, 2004). Eddies have also been confirmed in many numerical experiments (Lehmann, 1995; Elken, 1996; Zhurbas et al., 2003, 2004).

Thermohaline intrusions

The ventilation of deepwater in the Baltic Sea is very limited due to pronounced vertical stratification. In such a situation so-called "thermohaline intrusions" can play an important role. These intrusions are closely linked to fluxes of heat, salt, and momentum injected laterally into a vertical profile. They are ubiquitous in the ocean and occur in a variety of processes but often most strongly near ocean fronts. Though often expected to be built up by double-diffusive flux divergences, a number of processes can generate the lateral pressure gradient that is needed for interleaving. This includes differential mixing of temperature and salinity, as well as symmetric instability of gravity-driven current dynamics (Alford and Pinkel, 2000).

Zhurbas and Paka (1997, 1999) came up with two mechanisms to explain deepwater ventilation in the Baltic Sea proper: (1) continuously inflowing gravity-driven dense bottom flows filling up the deepest layers, and (2) intermittent eddies with cyclonic rotation within the halocline, eroded laterally by intrusions. Zhurbas and Paka (1997) found (based on measurements) that thermohaline intrusions play an important role in deepwater ventilation especially when coupled with similar effects caused by mesoscale eddies. The propagation of inflowing water farther into the Baltic Sea is accompanied by intensive intrusive layering in the permanent halocline.

For example, after the major inflow in April 1993 the permanent halocline in the Eastern Gotland Basin contained intensive thermohaline intrusions that lasted from several weeks to some months (see, Kõuts, 1999; Zhurbas and Paka, 1997, 1999). The well-defined front of the intrusive region was found to propagate northwards from the Stolpe Channel to the Gotland Deep at a velocity of a few centimeters per second. Such intrusions transport oxygen-rich and saline water northwards.

Stigebrandt (2001) stated that most inflowing dense water seems to be interleaved in the Baltic deepwater pool. Most interleaving occurs in the halocline region, which thus becomes well ventilated. This is also clearly illustrated by observed oxygen development in the Eastern Gotland Basin, showing that anoxic conditions in this basin only exist at depths greater than 120 meters below the sea surface. Kõuts (1999) summarized earlier findings of lens-shaped mesoscale eddies in and below the halocline, similar to the lenses of Mediterranean waters in the Atlantic Ocean. Some of these eddies seem to have their origin in pulsating inflows from the Bornholm Basin.

According to Zhurbas and Paka (1999), thermohaline intrusions gather in domains with a horizontal scale of 20 km–35 km (Figure 5.15). Each such domain, filled with intrusions, is separated from the ambient water by sharp thermohaline fronts, containing an eddy-like mesoscale baroclinic feature. In the permanent halocline Zhurbas and Paka (1999) found two types of intrusion-like structures: the so-called "odd" and "usual" intrusions. The odd ones are relatively rare phenomena characterized by a horizontal extension of about ten kilometres. Their most specific feature is the ability to cross isopycnals in such a manner that the along-layer density ratio is approximately equal to the vertical density ratio. Odd intrusions are not necessarily attributed to the process of double-diffusion driven interleaving. Usual intrusions have a smaller horizontal extension (about 1 km), and do not display

156 **Circulation** [Ch. 5

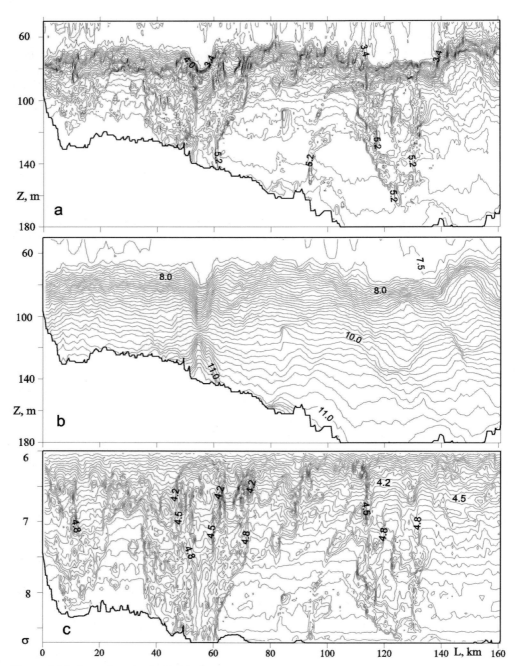

Figure 5.15. Intrusions in the Baltic Sea proper. Temperature (a) and salinity (b) vs. distance and depth, and (c) temperature vs. distance and σ (= density—1,000 kg m^{-3}) for a transect in the Eastern Gotland Basin. (Zhurbas and Paka, 1999, printed by permission from *Journal of Marine Systems*.)

a tendency to cross isopycnals. The existence of usual intrusions is partly explained by diffusive convection, and most likely inertial oscillations play a significant role, too. Recently, Kuzmina et al. (2005) found that most intrusions have a non-double-diffusion origin. These two different types of intrusions are likely driven by diffusive convection: relatively thin (3 m–5 m) and long (up to 8 km) intrusions inherent in high-baroclinicity regions, and relatively thick (about 10 m) and short (2 km–5 km) intrusions inherent in low-baroclinity regions.

Deepwater mixing

There is a long-term approximate advective–diffusive balance in deepwater (Stigebrandt, 2001). Advective supplies of new deepwater tend to increase and diffusive flows tend to decrease salinity. However, this is not in balance on shorter timescales due to the discontinuous character of the advective supply of deepwater. Since tides are usually small in the Baltic Sea, most of the energy sustaining turbulence in deepwater pools must be provided by the wind.

Stigebrandt (1987, 2001) concluded, according to results from the long-term modeling of large-scale vertical circulation in the Baltic Sea proper, that under contemporary conditions basin-wide vertical diapycnal diffusivity (or the diapycnal mixing coefficient) in deepwater pools can be reasonable well described by

$$\kappa = \min\left(\frac{\alpha}{N}, \kappa_{\max}\right) \quad (5.17)$$

where α and $\kappa_{\max}$ are constants; and N is the Brunt–Väisälä frequency. In his horizontally integrated model for the Baltic Sea proper Stigebrandt (1987) fine-tuned α to equal 2×10^{-7} m^2 s^{-2}. According to Meier et al. (2006) α depends on energy fluxes from local sources, such as wind-driven inertial currents, Kelvin waves, and other coastally trapped waves. This means that mixing near the coasts and near topographic slopes is more thorough than in the open sea. Axell (1998) found, based on measurements, that $\alpha = 1.5 \times 10^{-7}$ m^2 s^{-2} and that there is seasonal variability as well. For $N = 10^{-2}$ s^{-1} we have $\alpha/N \sim 1.5 \times 10^{-5}$ m^2 s^{-1}, while the normal level in the mixed layer is 10^{-3}–10^{-2} m^2 s^{-1}, which serves as a reference for $\kappa_{\max}$.

The processes involved in diapycnal mixing are not yet fully understood. A key question is to find the sources and paths for the energy that sustains the turbulence. It has been anticipated that internal waves and their dissipation play a key role in the transfer of energy down into deepwater. Several mechanisms may generate internal waves (see Chapter 6).

During the DIAMIX Project, Lass et al. (2003a) measured dissipation rates and stratification between 10 m and 120 m depths in this 9-day experiment in the Eastern Gotland Basin. The main finding was that there are two well-separated turbulent regimes. The turbulence in the surface layer, as expected, was closely connected with wind. However, in the strongly stratified deeper water turbulence was quite independent of meteorological forcing at the sea surface. The integrated production of turbulent kinetic energy exceeded the energy loss of inertial oscillations in the surface layer suggesting that the additional energy sinks might have been caused by inertial wave radiation during geostrophic adjustment of coastal jets and mesoscale eddies. The diapycnal mixing coefficient (κ) of Stigebrandt (1987) was estimated to be 7×10^{-7} m^2 s^{-2}.

5.4 BALTIC INFLOWS

Baltic Sea deepwater originates from the North Sea, and with distance from the Kattegat becomes more diluted. The inflow of this water renews lower layer watermasses, maintains strong salinity stratification, and is critical for the oxygen conditions in bottom waters. The inflow takes place in a continuous manner with moderate salinities and with aperiodic strong and highly saline pulses called *Major Baltic Inflows*. These major inflows are critically important for the ecological state of the Baltic Sea because only they can renew bottom waters. The time interval between major inflows is called a *stagnation period* since the oxygen level at this time decreases in the deepest basins.

Major Baltic Inflows have been investigated by a number of authors who have analyzed the measurements (see, Matthäus and Franck, 1990, 1992; Matthäus, 1993, 2006; Matthäus et al., 1993; Matthäus and Schinke, 1994, 1999; Håkansson et al., 1993; Dahlin et al., 1993; Lass and Matthäus 1996; Liljebladh and Stigebrandt, 1996; Jakobsen, 1995; Matthäus and Lass, 1995; Paka, 1996; Zhurbas and Paka, 1997; Feistel et al., 2003; Piechura and Beszczyńska-Möller, 2004). In this section these major inflows will be reviewed based on extensive analyses of measurements. The work of Matthäus (2006) and a number of other works form the basic material for this section. Numerical simulations of major inflows are discussed later in this chapter.

5.4.1 Characteristic features of Major Baltic Inflows

Normal water exchange with the North Sea is not effective enough to renew watermasses in the Baltic deeps. This is because the volume of water with a higher density crossing the sills is insufficient to displace bottom water and thus to markedly change hydrographic conditions. For bottom water renewal a Major Baltic Inflow is needed (Figure 5.16). Such an inflow takes place at present on average once in 10 years and brings salty, oxygen-rich, and dense water to Baltic main deeps. During such an event oxygen conditions clearly improve and the anoxic state ends.

Major inflows take place very irregularly since favorable conditions are rare and quite specific. Water exchange is forced by the difference in sea level and density between the Kattegat and the Arkona Basin. The density difference is almost constant, but the sea-level difference is strongly correlated with zonal winds. Hence, the exchange varies on the timescale of atmospheric circulation over Northern Europe because the narrow and shallow area linking the two seas largely prevents the continuous flow of saline water into the Baltic Sea. In addition, the circulation of deepwater in the Baltic Sea is restricted horizontally by bottom topography and vertically by permanent stratification.

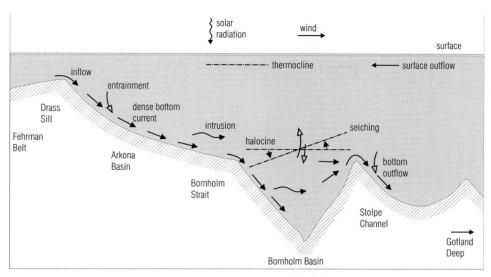

Figure 5.16. A sketch of the inflow into the Baltic Sea and determining processes. (Jakobsen, 1995, printed with permission from *Journal of Marine Systems*.)

The irregularity and long time intervals between subsequent Major Baltic Inflows are the reason stagnation conditions are present very often in the Baltic Sea. In such a situation, phosphate concentration increases and salinity and oxygen concentration decrease in deepwater. Sometimes considerable hydrogen sulfide concentrations are formed in the deep basins. Inflows of saline water through the Danish Straits are intermittent and usually characterized by their small volume and/or low salinity. Mixing with ambient water lowers the density of inflowing water along its way from the sills to the inner part of the Baltic Sea (Fonselius, 1981; Matthäus and Lass, 1995).

A Major Baltic Inflow takes place only if a given sequence of events takes place. Prior to the start of a Major Baltic Inflow, there is usually a period with easterly winds. This pushes the sea-level in the Baltic Sea down to a minimum level, which is important for the overall dynamics of inflow. Easterly winds enhance the outflow of surface water from the Baltic Sea, and in turn saline water flows in the lower layer close to the entrance to the Baltic Sea creating favorable conditions for the development of a major inflow (Dr. Andreas Lehmann, pers. commun.).

A Major Baltic Inflow consists of three main periods (see, Matthäus and Schinke, 1994): precursory period, main inflow period, and post-inflow period (Figure 5.17). The precursory period covers the time the Baltic sea-level is at a minimum until the start of the main inflow period. Meteorological conditions during the last two weeks of the precursory period play an important role in forming favorable conditions for the major inflow event (so-called pre-inflow period). After the precursory period, inflow only occurs if wind directions are such that wind forces North Sea water to pile up in the Danish Straits and simultaneously pushes Baltic Sea water eastwards, hence lowering the sea-level in the Darss Sill region. For this to happen strong westerly

160 Circulation [Ch. 5

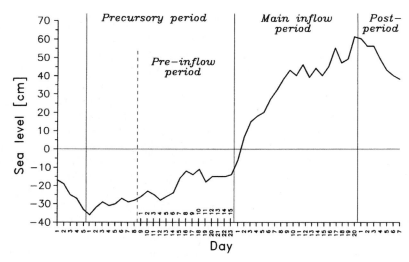

Figure 5.17. Main phases of a major inflow: precursory period, main inflow period, and post-inflow period, and the related changes in sea-level. (Matthäus and Schinke, 1994, printed by permission from Springer-Verlag.)

winds are a necessary condition. Just before the beginning of the main inflow period, water flowing over the Darss Sill has a relatively low salinity ($S < 17‰$), while saline watermasses ($S > 25‰$) start to flow in over the Drogden Sill.[6] Thus, during the main inflow period strong westerly winds prevail. In extreme cases the sea-level in the Baltic Sea rises as much as one meter, meaning that barotropic inflows dominate, driven as they are by sea-level difference between the Baltic Sea and the Kattegat. Watermasses with high salinity (and density) flow in over the Darss Sill and Drogden Sill and sink to the bottom layer. Wind needs to persist long enough for North Sea watermasses to reach at least the Bornholm Basin. It might also be possible for saline water to move later on from the Arkona Basin to the Bornholm Basin, if not flushed away. The post-inflow period starts when westerly winds weaken and North Sea water is no longer piled up at the Danish Straits, but as the Baltic sea-level has considerably increased during the inflow event, strong outflow takes place from the Baltic Sea for a while and the sea-level quickly drops. Some salty water flows out without having influenced the long-term salinity level and stratification in the Baltic Sea.

Relatively strong easterly winds dominate for several weeks over the entrance area of the Baltic Sea and are then replaced by durable, strong westerly winds; such conditions are not met very often. Major inflows have been observed to take place in September–April. Most frequently they occur between November and January, when according to Matthäus and Schinke (1994) about 60% of all cases take place. So, the tracks and intensity of cyclones are important factors for inflows, hence it is only in

---

[6] In major inflows the salinity of incoming water has to be more than 17‰ by definition (Franck et al., 1987).

winter that favorable conditions seem to occur. In other words, a strong zonal circulation the direction of which alternates favorably between the central North Atlantic Ocean and Eastern Europe, with only small fluctuations in the wind direction over the transition area, is a necessary prerequisite for a major inflow to happen.

In a usual major inflow case a gradual increase in wind speed takes place for several weeks during the precursory period (Figure 5.18). Saline waters ($S > 17‰$) start to flow across the Darss Sill about one day after the wind speed has reached its largest value. The geostrophic wind direction changes from west–southwest to west–northwest during the last two days before the main inflow period starts. This reflects the change in surface wind direction from southwest to west. The time variation of wind speed is determined by the east component, and the highest speed is about 15 m/s. The rule of thumb commonly used is that the higher the mean wind speed and the duration of high wind speed, the stronger the inflow (Matthäus and Schinke, 1994; Matthäus and Lass, 1995; Lass and Matthäus, 1996).

Inflows are restricted very much by the narrowness of the channels in the Danish Straits (Little Belt, Great Belt, and Öresund) and by the shallow sills between them and the Arkona Basin. The Darss Sill has a cross-section of only 0.8 km^2 and the sill depth is 18 meters. For the Drogden Sill the corresponding numbers are 0.1 km^2 and 7 m. There is no consensus as to whether the volume of highly saline water inflowing through the Drogden Sill is enough on its own to renew the bottom waters of the Baltic Sea; however, far more saline water penetrates through the Darss Sill. To constitute a major inflow, there needs to be a larger volume of water flowing across the Darss Sill, supported by inflow across the Drogden Sill.

The transformation of watermasses occurs mainly in the Belt Sea where entrainment of water from the deep layer into the surface layer takes place and *vice versa* (detrainment). During major inflows a complete mixing of the entire water body takes place. The transformation of watermasses then takes place in the Arkona Basin and Stolpe Channel as well. Wintertime inflows bring cold water to the Gotland Sea. The main routes of watermasses are given in Figure 5.19. It has been recently found that inflow across the sills into the Bornholm Basin and through the Stolpe Channel to the Gotland Sea has a complex internal dynamics. There are internal fronts with fine-scale intrusions and eddies (Hagen and Feistel, 2001) whereas recorded flows of higher salinity water over the Stolpe Sill are splash-like in nature (Piechura *et al.*, 1997; see also Matthäus and Lass 1995; and Matthäus, 2006). The propagation of inflowing water is additionally restricted by the bottom topography and by the mixing of the inflowing watermass with ambient water (Kõuts and Omstedt, 1993).

5.4.2 Discharge index and characteristic numbers

Attempts have been made to characterize the intensity of the major inflows by adopting an index. Franck *et al.* (1987) presented an index based on the duration of inflow (k) and the salinity of inflowing water S at the Darss Sill:

$$Q_{87} = 50\left(\frac{k-5}{25} + \frac{S-17}{7}\right), \quad k \geq 5; S \geq 17 \qquad (5.18)$$

162 Circulation [Ch. 5

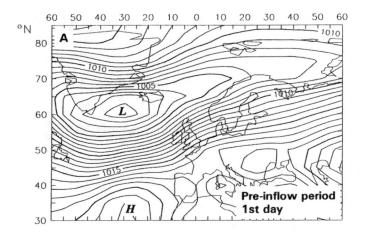

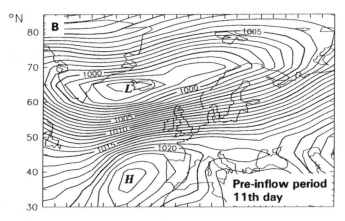

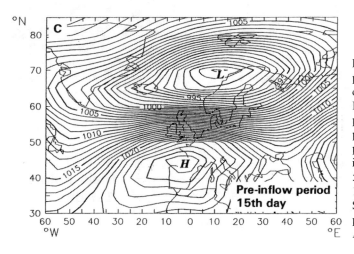

Figure 5.18. Mean sea-level pressure patterns (millibars) during a typical Major Baltic Inflow event. (A) Pre-inflow period 1st day; (B) pre-inflow period 11th day; (C) pre-inflow period 15th day; (D) main inflow 1st day; (E) main inflow 5th day; (F) post-inflow period 1st day. (Matthäus and Schinke, 1994, printed by permission from *Deutsche Hydrographische Zeitschrift*.)

Sec. 5.4] Baltic inflows 163

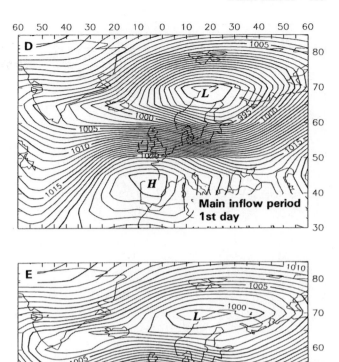

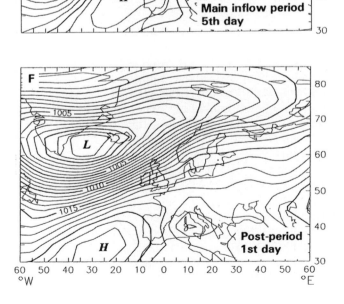

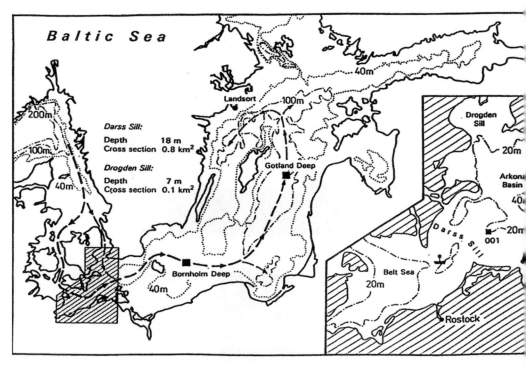

Figure 5.19. Main routes followed by Major Baltic Inflows, main sills, and channels. (Redrawn f Matthäus, 1993.)

where the unit of k is per day; and S is given in ‰. The index Q_{87} is taken between 0 ($k = 5$ days, $S = 17$‰) and 100 ($k = 30$ days, $S = 24$‰). This index has limitations in that the strength of the flow is not accounted for: 1 day increasess Q_{87} by 2 and 1‰ by 7. Inflows are classified according to their Q_{87}-index as weak ($Q_{87} < 15$), moderate ($15 < Q_{87} < 30$), strong ($30 < Q_{87} < 45$), and very strong ($Q_{87} > 45$) inflows. More than 50% of all inflows are weak. Their duration is less than 8 days, and average salinity does not exceed 18.6‰. Inflows with the highest observed indexes take place between November and January.

Inflow measurements had already started by the late 1800s, but there are data gaps due to World War I and World War II. Time-series analysis shows that the strongest inflows usually take place in clusters (17 cases) and more seldom as separate events (6 cases). In many cases the inflows frequently happen in 3–5 year time periods. Among the most important such periods were 1948–1952 (12 inflows) and 1968–1972 (10 inflows). The longest period without any inflow was 1983–1992. In addition, the periods 1927–1930 and 1956–1959 included no inflows.

Table 5.2 gives the Q_{87}-indexes for the most pronounced inflow events. The strongest inflow ever recorded was in 1951 with $Q_{87} = 79.1$. Thereafter, the highest

Table 5.2. Characteristic numbers of the 35 most intensive Major Baltic Inflows between 1880–2005 based on Q_{96} index. Q_{87} index is shown for comparison (according to Matthäus, 2006).

No.	Main inflow period at the Darss Sill	MBI	Q_{96}	Q_{87}
1	November 25–December 19, 1951	Very strong	51.7	79.1
2	December 16, 1921–January 6, 1922	Very strong	51.2	49.4
3	November 18–December 16, 1913	Very strong	38.0	76.6
4	December 1–20, 1886	Very strong	34.9	52.3
5	January 18–28, 1993	Very strong	34.0	21.2
6	November 20–December 4, 1897	Very strong	33.5	30.8
7	November 26–December 13, 1906	Very strong	30.3	38.0
8	October 30–November 8, 1965	Strong	29.5	16.9
9	October 29–November 25, 1969	Strong	29.2	54.8
10	January 17–31, 1921	Strong	28.7	46.6
11	December 7–22, 1898	Strong	27.9	29.1
12	January 7–22, 1902	Strong	27.7	31.1
13	January 24–February 6, 1938	Strong	27.5	27.3
14	November 13–29, 1973	Strong	27.0	41.4
15	December 22, 1975–January 14, 1976	Strong	25.6	60.0
16	December 18–28, 1900	Strong	25.5	21.3
17	January 20–February 7, 1898	Strong	25.0	34.4
18	November 30–December 10, 1960	Strong	24.4	11.8
19	March 7–16, 1926	Strong	24.3	24.5
20	January 3–13, 1925	Strong	23.5	35.7
21	November 10–20, 1930	Strong	22.9	37.3
22	October 12–22, 1938	Strong	22.6	25.9
23	February 8–20, 1939	Strong	21.5	24.9
24	December 2–15, 1914	Strong	20.3	30.0
25	January 16–22, 2003	Strong	20.3	14.1
26	December 26–31, 1902	Strong	20.3	6.2
27	November 15–22, 1964	Moderate	19.7	17.5
28	September 3–9, 1925	Moderate	19.7	10.6
29	September 18–26, 1948	Moderate	19.6	30.9
30	November 14–25, 1920	Moderate	19.5	21.9
31	December 26, 1931–January 2, 1932	Moderate	19.4	9.9
32	October 12–21, 1934	Moderate	18.9	25.9
33	January 24–31, 1901	Moderate	18.5	8.6
34	September 28–October 8, 1914	Moderate	18.0	17.0
35	December 26, 1910–January 3, 1911	Moderate	17.6	14.7

value is from the turn of the year 1975 to 1976 ($Q_{87} = 60$). During strong inflows the vertical average of salinity is around 20‰ in the Darss Sill. There is a large variability in imported seawater temperature since the inflows take place between late August and April. If the inflow comes during late August—December, the temperature of Baltic deepwater increases, whereas the inflows between January and April reduce it. The mean density of inflowing water is between 1,013 kg/m^3 and 1,017 kg/m^3 and the oxygen concentration is 7.5 mL/L–8.5 mL/L.

A revised version of the Q-index was introduced by Fischer and Matthäus (1996) because the 1993 inflow led to a re-assessment (see below). The Q_{87}-index used to be only a function of the duration of the event and the mean salinity of the water penetrating across the Darss Sill, with the role of Drogden Sill ignored. In reality, the significance of each sill varies considerably from event to event. The volumes crossing the Drogden Sill during major inflows are on average one-third of those crossing the Darss Sill. However, in some special cases the amount of salt transported across the sills is equal, and sometimes the amount crossing the Drogden Sill is even larger.

The amount of salt q_s (in 10^{12} kg) entering the Baltic Sea over both the Darss Sill (DS) and the Drogden Sill (DR) were used in the Q_{96}-index to establish the intensity of major inflows. In addition, to identify major inflows in the Darss Sill, Fischer and Matthäus (1996) used the following criteria to ascertain inflow days at the Drogden Sill that belonged to each identified event:

DR1   Surface salinity $S_0 \geq 17‰$ and the current must be toward the Baltic Sea.
DR2   All (but no more than 15) precursory days and all main inflow days that meet DR1 must be taken into account.

The intensity of a major inflow event corresponds to the amount of salt transported across the sills in watermasses with $S > 17‰$. In order to normalize the intensity to numerical values between 0 and 100, Fischer and Matthäus (1996) divided the calculated amount by $k_1 = 10^{11}$ kg. The Q_{96}-index accounts for the contributions of the Darss Sill ($q_{s,DS}$) and Drogden Sill ($q_{s,DR}$) as:

$$Q_{96} = \frac{q_{s,DS} + q_{s,DR}}{k_1} \tag{5.19}$$

Fischer and Matthäus (1996) adapted the major inflow classification introduced by Franck et al. (1987) and re-categorized the Q_{96}-index into weak ($Q_{96} \leq 10$), moderate ($10 < Q_{96} \leq 20$), strong ($20 < Q_{96} \leq 30$), and very strong ($Q_{96} > 30$). The major inflows identified between 1880 and 2005 are shown in Table 5.2 and Figure 5.20. The 1951 inflow is the largest according to both indexes ($Q_{87} = 79.1$, $Q_{96} = 51.7$) belonging to the category "very strong". The indexes do not always match each other. According to Q_{87} the 1993 inflow was only "moderate", but when using Q_{96} it is "very strong"! An opposite situation can be found with the inflow in winter 1975–1976. The old index indicates it was a "very strong" inflow, but the new index gives only "strong". The latest inflow in 2003 is "weak" according to Q_{87} but "strong" according to Q_{96}.

Sec. 5.4] **Baltic inflows** 167

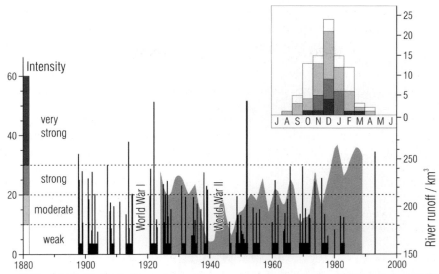

Figure 5.20. Major Baltic Inflows shown in accordance with the Q_{96}-index between 1880 and 2005 and their seasonal distribution (upper right) and lowpass-filtered annual river runoff to the Baltic Sea (inside the entrance sills) averaged from September to March (shaded). Black boxes on the time axis indicate Major Baltic Inflows arranged in clusters. (Redrawn from Fischer and Matthäus, 1996.)

Each major inflow is preceded by an inflow of less saline water formed in the Belt Sea by the mixing of relatively fresh water originating from the Baltic Sea with saltier waters from the North Sea (Matthäus and Franck, 1992). The duration of this so-called precursory period varies significantly—between 4 and 58 days—but usually it is 5–25 days. The average lengths of the precursory and main inflow periods are 22 and 10 days, respectively. Roughly, two-thirds of a major inflow forms the precursory period and one-third forms the main inflow period. Figure 5.21 shows the evolution of hydrographical characteristics in the Gotland Deep reflecting the influence of inflows and stagnation periods.

During an inflow event the average sea-level in the Baltic Sea varies between −60 cm and +70 cm (i.e., sea-level can change by over one meter). Matthäus and Franck (1990) estimated the water volumes penetrating the Baltic Sea during major inflows and calculated the frequency distribution of discharges during the precursory and inflow periods based on 90 major inflows between 1897 and 1976. Their estimate was based on the increase in total Baltic Sea water volume calculated from sea-level increase. During the precursory period the total volume was predominantly between 80 km^3 and 160 km^3 mainly originating from the Belt Sea. The amount of highly saline water (more than 17‰) was generally smaller. The total volume of inflow varied between 140 km^3 and 200 km^3 (Matthäus, 2006). A major inflow typically imports 2×10^{12} kg salt and 10^9 kg dissolved oxygen if it takes place in winter or spring (Feistel et al., 2006).

168 Circulation [Ch. 5

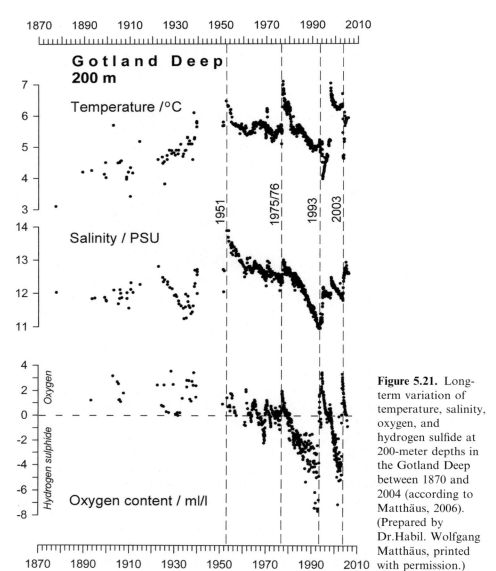

Figure 5.21. Long-term variation of temperature, salinity, oxygen, and hydrogen sulfide at 200-meter depths in the Gotland Deep between 1870 and 2004 (according to Matthäus, 2006). (Prepared by Dr.Habil. Wolfgang Matthäus, printed with permission.)

5.4.3 Major inflows in 1993 and 2003

The two most recent major inflows took place in 1993 (January) and 2003 (from January on). These inflows are the best documented as a result of the detailed measurements taken and have been widely studied by means of data analysis and numerical model simulations.

In January 1993 a major inflow of highly saline water took place after 17 years of stagnation during which hardly any North Sea water penetrated Gotland Basin deepwater. During the inflow a total of about 310 km^3 of water, 135 km^3 of it highly saline (more than 17‰) with a high oxygen content entered the Baltic Sea. The highly saline water crossing the Darss Sill was characterized by a mean salinity of about 19‰, a mean temperature of 3.5°C, and a mean oxygen concentration of 8.2 mL/L. According to the Q_{96}-index (34.0) the 1993 inflow is ranked the fifth largest inflow of all recorded cases (Table 5.2).

This very strong inflow had several specific features. Its duration was very short (total main inflow period 22 days). The average level of the Baltic Sea increased by 70 cm above the mean. Forcing over the Danish Straits was quite strong, because the inflow of highly saline water started when the sea-level had already risen to more than 30 cm above the mean level. This inflow was the first to be recorded in a comprehensive way. Temporal development of the flow over the sills and the flushing of near-bottom layers in the Arkona Basin with highly saline, oxygen-rich water was observed and analysed by Matthäus and Lass (1995) and by Liljebladh and Stigebrandt (1996).

The inflow of highly saline water through the Öresund across the Drogden Sill started on January 5, 1993 immediately after the currents had changed direction. Saline water was further advected into the Arkona Basin where a thin salty bottom layer formed. It is interesting to note that the inflow of saline water into the Arkona Basin via the Great Belt took place 1–2 weeks later. The salinity at Darss Sill exceeded 17‰ on January 18. The inflow was at its highest on January 26 (Figure 5.22). Owing to the large amounts of highly saline water that crossed the Darss Sill into the Arkona Basin, the halocline was lifted from 38 m to 10 m and the 20‰ isohaline was displaced from 42 m to 32 m. The halocline was inclined from the central Arkona Basin to Bornholm Channel. Even if the main inflow period was short, an unusually large amount of salt flowed into the Arkona Basin and lifted the halocline above the level of the Darss Sill depth. Consequently, a relatively large amount of salty water flowed back into the Belt Sea (Matthäus, 2006). It has been estimated (Matthäus and Lass, 1995) that half the salt that entered the Arkona Basin returned.

The remaining water flowed through the Bornholm Channel into the Bornholm Basin and replaced the old bottom water. Salinity in the bottom water increased from 15‰ to 20‰ and the oxygen concentration from about 1 mL/L to 7.5 mL/L between October 1992 and March 1993. The stagnant bottom water of the Bornholm Basin was lifted above the sill depth of the Stolpe Channel and flowed due to gravity into the Eastern Gotland Basin. The first sign of the renewal of bottom water in the Eastern Gotland Basin was observed in early April 1993. The deepwater of the Gotland Deep between 200 meters and the bottom was renewed by the middle of May. Weaker effects were also identified in the Northern Gotland Basin and most probably in the Gulf of Finland as well (Alenius *et al.*, 1998).

The 2003 inflow has been studied by several authors (see Feistel *et al.*, 2003; Piechura and Beszczyńska-Möller, 2004). About 200 km^3 of saline water flowed into the southwestern Baltic Sea. This strong inflow ($Q_{96} = 20.3$) was very interesting for many reasons. Due to a warm inflow in August 2002 and by another in November 2002, watermasses were already ventilated in the Southern Baltic (see Section 5.4.4) and were extremely warm in December 2002. The surface temperature was about 7°C

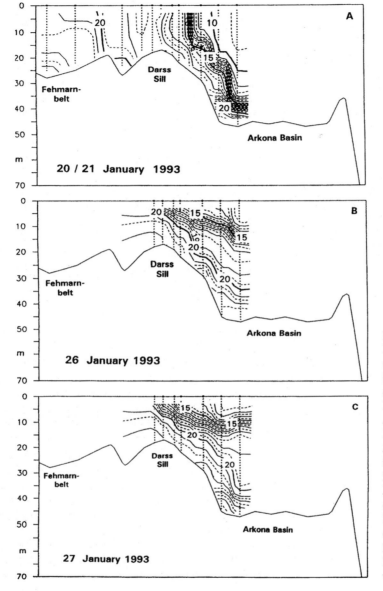

Figure 5.22. Longitudinal transects of salinity in the Darss Sill area during the Baltic Major Inflow in January 1993. (A) Inflow during January 20–21; (B) inflow on January 26; (C) inflow on January 27, 1993 (according to Matthäus, 2006). (Prepared by Dr.Habil. Wolfgang Matthäus, printed with permission.)

and the deep-water temperature was as high as $11°C–12°C$, whereas the salinity distribution was rather normal. During the inflow in January 2003 (actually a series of inflows), exceptionally cold water replaced the exceptionally warm deepwater (Piechura and Beszczyńska-Möller, 2004). The January inflow was followed by weaker inflows in March and May (Feistel et al., 2003).

The inflow started 10 years after the previous very strong inflow (Matthäus, 2006). Prior to the event in the beginning of January, an atmospheric high-pressure area over Scandinavia was associated with northeasterly winds. The sea-level fell to 80 cm below the mean. On January 11 the wind over the western Baltic Sea increased to 15 m/s, turned to the west, and triggered the inflow. The inflow continued until January 18 in the Drogden Sill and until January 22 in the Darss Sill, and the Baltic Sea level rose to 25 cm above the mean. In the Öresund very high salinities and low temperatures were reported on January 15 (26.6‰, 2.5°C) and January 18 (26.4‰, 2.2°C). In the Darss Sill, salinity was up to 21‰ at the bottom and 18‰ at the surface. Estimated salt transport across the Darss Sill (Feistel et al., 2003) was 1.18×10^{12} kg (58% of total transport) and through the Öresund 0.85×10^{12} kg (42%).

Inflowing water moved exceptionally fast into the Arkona Basin and through the Hamrare Strait into the Bornholm Basin. The estimated speed of the flow was 30 cm/s over a 12-day period. The inflowing water interacted with frequent baroclinic eddies, particularly in the Bornholm Deep. The Bornholm Basin was the main area where mixing of cold inflow water with local warm water took place. Farther east, no "pure" inflow water was detected. As a consequence of the inflow, colder and more saline, mixed water from the intermediate layers of the Bornholm Deep flowed into the Stolpe Channel over the Stolpe Sill and farther to the Gdansk Deep (Figure 5.23, see color section). It should be mentioned that both post-inflow periods in March and May 2003 enhanced this Major Baltic Inflow because the Bornholm Basin was already filled with saline water and favored the rapid eastward propagation of the following inflow water (Piechura and Beszczyńska-Möller, 2004; Matthäus, 2006).

5.4.4 Warm inflows and small-size and medium-size inflows

During recent years there has been a lot of discussion regarding warm inflows. Actually, inflows of warm water in late summer and autumn to the southern basins of the Baltic Sea occur regularly, as shown by the mean long-term annual temperature cycles in the deepwater of the Arkona, Bornholm, and Gdansk Basins. On average, temperatures higher than 12.5°C can be observed in the Arkona Basin in September–October, higher than 9°C in the western Bornholm Basin in October–November, and higher than 8°C in the central Bornholm Basin between November and January (Matthäus, 2006).

However, in addition to regular warm water inflows, some with exceptionally warm waters have recently occurred (2002 and 2003). According to Matthäus (2006) two types of such inflows have been observed and the dynamic mechanisms of their formation have been found quite specific. The first type is caused by heavy westerly gales, which pass over the Darss and Drogden Sills. The second type is a long-lasting baroclinic inflow, which only passes through the Darss Sill, caused by calm weather conditions over central Europe. In such a situation the inflows are driven by baroclinic pressure gradients, especially caused by horizontal salinity differences (Feistel et al., 2006).

Warm baroclinic inflows can transport large volumes of exceptionally warm water to the deeper layers of the Gotland Sea. However, these inflows in fact import oxygen-deficient water, but (Feistel et al., 2006) they seem to be important for the ventilation of intermediate layers in Eastern Gotland Basin deepwater through entrainment. On the other hand, warm-water inflows, whether baroclinic or barotropic, do transport less oxygen to the Baltic Sea than cold-water inflows, and higher temperatures increase the rate of oxygen consumption in deepwater and favors the formation of hydrogen sulfide (Matthäus, 2006). Although warm inflows do not fulfill the criteria for major inflows, they can still effectively influence the deepwater of the Bornholm, Gdansk, and Eastern Gotland Basins by their higher salinity and temperature.

In the summers of 2002 and 2003 an unusual sequence of inflows occurred. The first ever important baroclinic inflow took place in August 2002, followed by the 2003 major inflow, which was described earlier in this chapter. Surprisingly, a warm inflow again took place in August 2003 (Feistel et al., 2006). This warm summer inflow in conjunction with the preceding warm inflow of 2002 is only comparable with the processes observed in summer 1959 (Feistel et al., 2004).

Let us take a brief look at these two rare features (Figure 5.24, see color section). Between the end of June and mid-September 2002 moderate easterly winds prevailed over the western Baltic Sea. A near-bottom current with a mean layer thickness of about 5 meters and salinity more than 14‰ transported warm, saline water across the Darss Sill to the Arkona Basin. This inflow also caused the ventilation of deepwater in the Gdansk Deep and a reduction of hydrogen sulfide concentrations in the Gotland Deep (Matthäus, 2006). In summer 2003 a further warm inflow took place (Feistel et al., 2004, 2006; Piechura and Beszczyńska-Möller, 2004). This inflow resulted in an increase in temperature and salinity of Gotland Basin deepwater in 2004. The density of inflowing water was large enough for the ambient water to be lifted up. A temperature of 7.2°C and salinity of 13.2‰ were measured in the near-bottom layer of the Gotland Deep in February 2004. The 2002 and 2003 inflows had after-effects that lasted until 2005 (Feistel et al., 2006).

Small-size and medium-size inflows take place several times a year. They have a high enough capacity to ventilate certain parts of the intermediate layers of the Baltic Sea proper's halocline, which is subject to oxygen renewal. During such events, the density of inflowing watermasses is reduced to a large extent due to the turbulent mixing that is already going on in the Arkona Basin. This is why quantification of watermass transformation in this basin is essential to understanding the key processes in the internal dynamics of the Baltic Sea. In particular, the role of climate change and the human impact on watermass transformation should be studied further (see, Meier et al., 2006 for details). Burchard et al. (2005) have as an example discussed the role of offshore winds in the Arkona Basin in relation to mixing conditions and found the effect of wind farms to be significant.

The fine structure of inflowing water to the Arkona Basin has been studied by Lass et al. (2005). Detailed investigations of medium-scale saltwater inflows (during 1998) revealed a complex fine-scale structure of the saltwater inflow with division of the watermass into different branches.

5.4.5 Stagnation periods

During recent decades Major Baltic Inflows have occurred very infrequently. In fact, there are many potential reasons for their absence but no single factor can give a full explanation. In this section the main reasons are discussed. The reader may like to consult the works, based on observations, of Fonselius (1969), Lass and Matthäus (1996), Samuelsson (1996), Schinke and Matthäus (1998), Matthäus and Schinke (1999), Zorita and Laine (2000), Winsor et al. (2001), Rodhe and Winsor (2002), Stigebrandt (2003), Matthäus (2006), and Meier et al. (2006).

One of the main reasons behind long stagnation periods may be variability in river runoff to the Baltic Sea (Figure 5.25). There seems to be a good fit between minimum deepwater salinity in the Baltic Sea and maximum river runoff if a time shift of 6 years between salinity and runoff is performed (see Launiainen and Vihma, 1990). There are at least two main mechanisms driven by river runoff variability and counteracting inflows (Matthäus and Schinke, 1999).

First, mixing of low-salinity outflowing water takes place in the surface layer with saline lower layer water which penetrates the Baltic Sea in the near-bottom layer of the sill areas. Increased water supply from rivers reduces the salinity of outflowing water and strongly dilutes inflowing water. Second, an increase in the freshwater supply to the Baltic Sea causes a greater outflow, reduces or impedes the inflow of

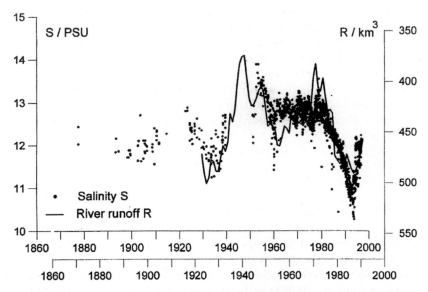

Figure 5.25. Long-term variations in river runoff into the Baltic Sea compared with salinity (‰) in central Baltic deepwater (Gotland Deep 200 m; Fårö Deep 150 m; Landsort Deep 400 m; Karlsö Deep 100 m). Salinity in the Gotland Deep shows real values. The values at other stations are changed by +0.8‰ (Fårö Deep), by 1.8‰ (Landsort Deep), and by 2.6‰ (Karlsö Deep). The time axis for salinity is shifted forwards by 6 years. (Matthäus and Schinke, 1999, printed with permission from Springer-Verlag.)

saline water, and gives rise to unfavorable conditions for major inflows. Freshwater supply depends on precipitation, evaporation, changes in water storage, vegetation, and land use in the drainage area. It also may be affected by human impacts, through changes in the annual river runoff cycle by river regulation. This has been a special factor during the last 20 years and caused unfavorable conditions for saline water inflows since regulation increases wintertime runoffs.

An explanation for the decreasing frequency of major inflows can also be related to changes in meteorological patterns (Matthäus and Schinke, 1994; Zorita and Laine, 2000; Meier et al., 2006). Currently, there are strong wintertime westerly winds (a high NAO index) which have many links to Baltic Sea physics. Westerly winds intensively transport moist airmasses from the North Atlantic to Europe which results in intensified precipitation in the Baltic Sea region, lower evaporation, and increased river runoff. In consequence, above-normal Baltic Sea levels occur frequently for long periods which hampers saltwater inflows. Stronger than usual zonal atmospheric circulation also increases oxygen intake. This could, according to Zorita and Laine (2000), be in relation to weakened stratification.

One potential reason for today's increasing stagnation periods may also be the high salinity of bottom waters. According to Meier et al. (2006) the period with high bottom salinities in the 1950s and 1960s may have been caused by the major inflow in 1951. This filled the Baltic deeps with relatively highly saline water, and made difficult their replacement by later inflows.

There are some indications, as a result of direct measurements, of changes in Baltic Sea wind climatology. Lass and Matthäus (1996) found an anomalous west wind component at the Kap Arkona station between August and October for seasons without a Major Baltic Inflow compared with years with such inflows between 1951 and 1990. In years without a major inflow the period when easterly winds predominate is shortened. Such changes in local wind patterns may thus cause variations in long-term salinity, which cannot be explained by accumulated freshwater inflow or by low-frequency variability of the zonal wind.

5.5 NUMERICAL MODELING

5.5.1 General

The presentation of numerical modeling in this chapter is mainly focused on the current results of three-dimensional models at studying circulation dynamics and related processes in the Baltic Sea. These models are full models in that they attempt to solve the full ocean circulation problem; they include seven quantities: horizontal velocity $U = (u, v)$; vertical velocity w; pressure p; density ρ; temperature T; and salinity S. The fundamental equations are the conservation laws of momentum and mass, the equation of state of seawater, and the conservation laws of heat and salt. The system is written in general form as—see Equations (4.7), (4.14), and (5.2)

$$\frac{\partial \boldsymbol{U}}{\partial t} + \boldsymbol{U} \cdot \nabla_H \boldsymbol{U} + f\boldsymbol{k} \times \boldsymbol{U} = -\frac{1}{\rho}\nabla_H p + A_H \nabla_H^2 \boldsymbol{U} + \frac{\partial}{\partial z}\left(A_V \frac{\partial \boldsymbol{U}}{\partial z}\right) \quad (5.20a)$$

$$\frac{\partial w}{\partial t} = -\frac{1}{\rho}\frac{\partial p}{\partial z} - g \quad (5.20b)$$

$$\nabla_H \cdot \boldsymbol{U} + \frac{\partial w}{\partial z} = 0 \quad (5.20c)$$

$$\rho = \rho(T, S, p) \quad (5.20d)$$

$$\frac{\partial T}{\partial t} + \boldsymbol{U} \cdot \nabla_H T + w\frac{\partial T}{\partial z} = K_{TH} \nabla_H^2 T + K_{TV} \frac{\partial^2 T}{\partial z^2} + q \quad (5.20e)$$

$$\frac{\partial S}{\partial t} + \boldsymbol{U} \cdot \nabla_H S + w\frac{\partial S}{\partial z} = K_{SH} \nabla_H^2 S + K_{SV} \frac{\partial^2 S}{\partial z^2} \quad (5.20f)$$

This is the system of equations for an incompressible, turbulent, geophysical flow with horizontal and vertical motions separated due to their anisotropy. The vertical equation of motion—Equation (5.20b)—is in most models taken in hydrostatic form (left-hand side equal to zero). Otherwise, the main differences in the physics of the models are in the formulations of eddy diffusivities and boundary conditions. The dynamic part of this system—Equations (5.20a–c)—was discussed in Section 5.1, the equation of state—Equation (5.20d)—was discussed in Section 3.2.1, and the conservation laws of heat and salt—Equations (5.20e–f)—were discussed in Sections 4.3.1 and 4.2.1, respectively.

Numerical solutions are mainly based on finite difference schemes. The horizontal grid size in current models is about 1 km–10 km, and in the vertical direction 10–100 layers are taken. Both geometric and σ-coordinate formulations have been employed in different studies.

Three-dimensional modeling of the Baltic Sea started as early as the 1970s. Kuzin and Tamsalu (1974) studied the baroclinic circulation of the Baltic Sea with a flat bottom. Sarkisyan et al. (1975) developed a diagnostic model to study the interaction of circulation and bottom relief. Wind-driven circulation as a result of topography was investigated by Simons (1978) and by Kielmann (1981). In the study of Lehmann (1995) the full problem was studied in which both wind-driven and baroclinic circulations as a result of topography were taken into account. Recently, new applications have been carried out. For example, Meier (2007) studied, with the help of a three-dimensional model, residence times with respect to freshwater and inflowing saline water for the entire Baltic Sea for the first time.

Three-dimensional circulation modeling is an important tool to help improve our knowledge of circulation dynamics. Three-dimensional models have been applied, for example, to study the mean thermohaline and wind-driven circulation in the Baltic Sea. Three-dimensional research-oriented modeling of the Baltic Sea is based on many different modeling approaches (see Omstedt et al., 2004 for details). Several modeling groups use a GFDL-type (Geophysical Fluid Dynamics Laboratory)

z-coordinate ocean model (Bryan, 1969; Killworth *et al.*, 1991) with free surface and its modular updated versions MOM 1, MOM 2, and MOM 3 (Pacanowski and Griffies, 1998), or further versions of the OCCAM model (Ocean Circulation Climate Advanced Modeling project; Webb *et al.*, 1997). Another family of three-dimensional models is based on the BSH model (Bundesamt für Seeschifffahrt und Hydrographie; Kleine, 1994), which forms the basis of HIROMB (High Resolution Operational Model for the Baltic Sea) and DMI (Danish Meteorological Institute) operational models (see Chapter 9).

Many modeling attempts are based on σ-coordinate models (e.g., Blumberg and Mellor, 1987) such as the POM (Princeton Ocean Model) and non-hydrostatic approach (Marshall *et al.*, 1997a, b). The COHERENS model (Luyten *et al.*, 1999) is also based on the application of σ-coordinates.

Several three-dimensional models have been developed inside the Baltic Sea community. Various versions of MIKE models (DHI Water and Environment, 2000) have been used in various applications. We can also mention two models developed by Russian scientists, OAAS (Andrejev and Sokolov, 1989) and SPBIO (Neelov, 1982), the Finnish–Estonian model FinEst (Tamsalu and Myrberg, 1995), and the Estonian–Russian model FRESCO (Finnish–Russian–Estonian Cooperation; Zalesnyi *et al.*, 2004).

Thermohaline and wind-driven circulation have been studied by Simons (1978), Kielmann (1981), Lehmann and Hinrichsen (2000a, b), Lehmann *et al.* (2002), Meier and Kauker (2003a), Stipa (2004), and Döös *et al.* (2004). Specific regional studies have been carried out, for example, by Andrejev *et al.* (2004a, b) focusing on investigation into the Gulf of Finland, and by Myrberg and Andrejev (2006) studying the mean circulation of the Gulf of Bothnia.

Pure ocean models have been considered in this section. Coupled ice–ocean models are discussed in Chapter 7. Two-dimensional ocean models, which have been used for forecasting sea-level change, are discussed in Chapter 8. The climate change section of Chapter 10 will introduce modeling results of long-term (decadal) change in the overall stratification conditions of the Baltic Sea. Today operational modeling is a rapidly developing field of ocean modeling. The present state of the art in the Baltic Sea is also discussed in Chapter 9.

5.5.2 Modeling of general circulation dynamics

Today's three-dimensional numerical models of the entire Baltic Sea have a horizontal resolution of 1 km–5 km and a vertical structure described by about 20–100 layers. In local applications, resolutions can be even finer. For modeling purposes there are now commonly used forcing datasets for the Baltic Sea. For long-term hindcast simulations SMHI gridded data at a $1° \times 1°$ resolution and, for example, European Centre for Medium Range Weather Forecasting (ECMWF) analyses are available. Current meteorological models (like the High Resolution Limited Area Model, HIRLAM) have a horizontal resolution of 10 km–20 km. They produce operational forecasts of meteorological fields, which are frequently used as input for marine models. River runoff data on a monthly basis are available, for example,

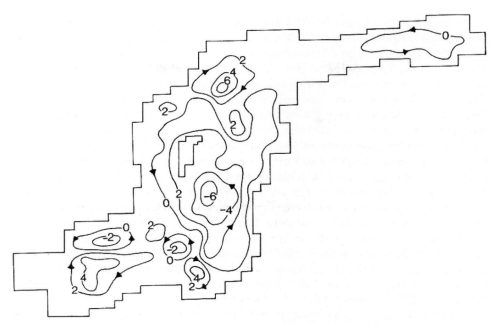

Figure 5.26. Baroclinic circulation in the Baltic Sea based on a diagnostic model. (Redrawn from Sarkisyan *et al.*, 1975.)

from the BALTEX Hydrological Data Centre (Bergström and Carlsson, 1994), and in recent years hydrological model results have also been available. The Baltic Environmental Database (BED) has improved the possibilities for scientists to construct initial conditions for their models and to find verification data when the Data Assimilation System (DAS) came into common use (Sokolov *et al.*, 1997). The ICES and SHARK databases are also available through the BALTEX Ocean Data Centre at SMHI. Observations from tide gauges can be used for open boundary conditions or used for verification and assimilation. In addition the data from buoys and individual measurement campaigns also are a help to modelers.

The first paper in which the cyclonic circulation of the Baltic Sea was produced by numerical modeling was by Sarkisyan *et al.* (1975). They used a diagnostic model (Figure 5.26). Later on, circulation characteristics were studied widely using present-day high-resolution modeling tools; a short summary of the main findings is now given. Lehmann and Hinrichsen (2000b) studied the variability in mean circulation using an eddy-resolving model. They concluded that in early spring when the wind forcing is weak, the circulation is mostly determined by the baroclinic field, whereas in autumn and winter strong westerly winds dominate and the circulation is controlled by Ekman dynamics, and basin-wide bottom topography superimposing the baroclinic current. However, annual mean barotropic circulation shows minor fluctuations when comparing different years. Generally, a classical cyclonic circulation was found in each sub-basin. The main differences in circulation patterns between different years concerned the strength of circulation cells. The surface velocity field strongly differed from the barotropic current field, reflecting a surface circulation determined by Ekman dynamics.

The annual average of near-bottom velocity reflects stronger currents in shallow areas and less intense bottom currents in deeper parts of the sub-basins, a feature that coincides with areas where sediment deposition occurs (Krauss and Brügge, 1991). Lehmann and Hinrichsen (2000b) found areas where high near-bottom current velocities exist, such as the Danish Straits, Bornholm Strait, and Stolpe Channel. reflecting their important role in watermass exchange and thermohaline circulation.

Atmospheric conditions play the leading role in the variability of water exchange between different basins (Lehmann and Hinrichsen, 2002). It was found that in the Arkona Basin the strength of upper layer flow, which is on average directed to the west, opposite to the mean wind direction, is compensated by a flow in deeper layers to the east. Increased upper layer flow results in increased lower layer flow in the opposite direction and *vice versa*. Annual mean flow is only weakly correlated with annual mean discharge to the Baltic Sea. In accordance with mean circulation the flow through the Bornholm Channel is on average directed to the east. South of the Bornholm, the flow is directed to the west indicating an import of heat and salt to the Bornholm Basin through the Bornholm Channel and an export south of Bornholm. The flow of the lower, more saline layer is increased downstream due to entrainment (Kõuts and Omstedt, 1993).

Flux characteristics change farther downstream in the Stolpe Channel. The volume flow in the upper layer shows a strong seasonal signal. During autumn to spring the flow is mainly directed to the east whereas in summer the flow direction is reversed. Flow in the westerly direction is related to increased lower layer flow in easterly directions. This exchange mechanism is controlled by prevailing wind directions and has been described by Krauss and Brügge (1991) and by Lehmann *et al.* (2002). On average the net flow through the Stolpe Channel is directed to the east in accordance with mean circulation. The flow through the Stolpe Channel is part of a huge circulation cell comprising the Eastern and Western Gotland Basins (Lehmann and Hinrichsen, 2000a). The difference in flow characteristics between the Arkona Basin and the Stolpe Channel is due to the fact that in the Stolpe Channel river runoff has no direct influence on upper layer flow. Thus, the flow can change its direction subject to prevailing wind forcing. Calculated fluxes show high intra-annual and interannual variability with no obvious trend.

The mean circulation of the entire Baltic Sea was modeled recently by Meier (2007). The results confirm the main characteristics of the early findings of Palmén (1930) but also gives new fine-scale characteristics. The mean transport above and below the halocline confirm the existence of strong cyclonic gyres both in the Baltic Sea proper and in the Sea of Bothnia; these circulations have a high stability (Figure 5.27a). In the Eastern Gotland Basin high transport is found around the Gotland Deep. It turns out that the strength and persistency of currents is lower in the Gulf of Riga, Gulf of Finland, and Bay of Bothnia than in the Baltic Sea proper. This might be due to the impact of ice during the winter. Close to the Swedish coast an intense southward-directed flow becomes visible as part of the cyclonic gyre of the Baltic Sea

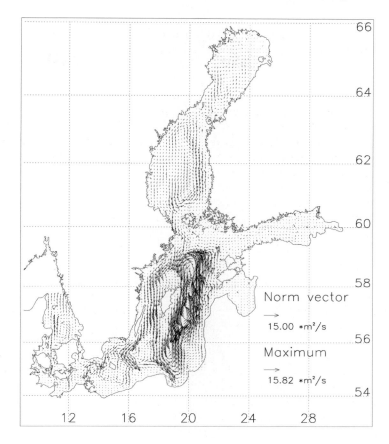

Figure 5.27a. Annual mean transport per unit length ($m^2 s^{-1}$) for 1981–2004: upper layer. Only every third gridpoint of the model is shown. (Meier, 2007, printed by permission from *Estuarine, Coastal and Shelf Science*.)

proper. The flow is directed into the Bornholm Basins and to the Arkona Basin. The main flow crosses the central Arkona Basin and bifurcates north of Rügen Island. One branch leaves the Baltic Sea at the Darss Sill and the flow continues through the Belt Sea and the Great Belt. The other branch recirculates and forms a cyclonic gyre in the Arkona Basin. Also a flow follows the Swedish coast into the Öresund and Kattegat. In the lower layer (Figure 5.27b) the flow follows the topography closely, from the Darss Sill into the Arkona Basin and farther towards the Bornholm Channel passing Rügen Island. The deepwater flows farther to the Bornholm Deep and into the Stolpe Channel with high persistency. East of the Stolpe Channel the main flow is directed along the southwestern slope of the Gdańsk Deep. In the Gotland Deep the flow is characterized by cyclonic gyres. The watermasses furthermore have a cyclonic gyre which finally leads some of the water to flow into the western Gotland Basin.

Lehmann *et al.* (2002) studied the relation between the NAO index and Baltic Sea circulation. To relate the local wind field over the Baltic Sea to large-scale atmospheric circulation, the authors defined a Baltic Sea Index (BSI), which is the difference in normalized sea-level pressures between Oslo in Norway and Szczecin in

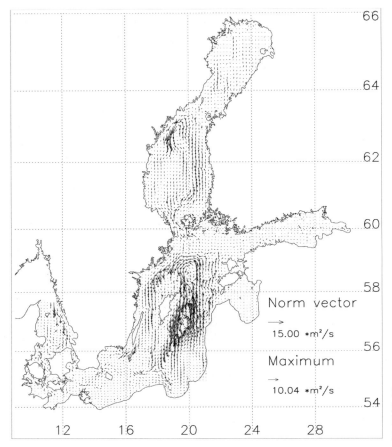

Figure 5.27b. As for Figure 5.27a, but for lower layer.

Poland. The BSI is significantly related to the NAO index and furthermore highly correlated with the mean sea-level of the Baltic Sea and water exchange through the Danish Straits. It is shown that the different phases of the BSI during winter result in major changes in horizontal water transport in the deep basins of the Baltic Sea and in upwelling along the coasts as well as in the interior of the basins. During NAO$^+$ phases (positive NAO index), strong Ekman currents are produced with increased upwelling and downwelling along the coasts and associated coastal jets. During NAO$^-$ phases the winds are weak and therefore Ekman drift and upwelling are strongly reduced, and the flow field can almost entirely be described by the barotropic stream function.

Circulation characteristics in the Northern Baltic Sea

For local detailed investigations, mean circulation in the Gulf of Finland and in the Gulf of Bothnia has been modeled, respectively, by Myrberg (1997, 1998), Andrejev *et al.* (2004a, b), and Myrberg and Andrejev (2006). Cyclonic mean circulation in the Gulf of Finland (Figure 5.28, see color section) generally is discernible but the resulting patterns and the persistency of the currents according to Andrejev *et al.* (2004a) deviate to some extent from the classical analyses of Witting (1912) and Palmén (1930). Both the mean and instantaneous circulation patterns in the Gulf of Finland contain numerous mesoscale eddies with a typical size clearly exceeding the internal Rossby radius. Modeled circulation patterns (resolution 1×1 nautical miles) reveal certain nontrivial and temporally and spatially varying vertical structures. The Neva Bight in the easternmost part of the gulf is characterized by strong and persistent north- to northwest-directed currents caused by the voluminous runoff from the River Neva. The surface layer (0 m–2.5 m) flow pattern is characterized mainly by an Ekman-type drift in the rest of the gulf. There is an inflow near the southern coast that is visible at all depths with a high degree of persistency (up to 50 %). A compensating outflow exists in the rest of the Gulf. It becomes stronger and is highly persistent (up to 80%) in the sub-surface layers (2.5 m–7.5 m and downwards) slightly north of the axis of the Gulf (about 30 km offshore from the Finnish coast) and weakens considerably near the bottom.

Up until very recently, there have been few water exchange estimates between different basins, mostly based on measurements, but now model-based calculations have also been carried out. The earliest well-known estimates of water exchange between the Gulf of Finland and the Gotland Basin across the Hanko–Osmussaar line were given by Witting (1910). He estimated the magnitude of the inflow and outflow as $480 \, \text{km}^3 \, \text{yr}^{-1}$ and $600 \, \text{km}^3 \, \text{yr}^{-1}$, respectively. This corresponds to a mean flow of only $1 \, \text{cm} \, \text{s}^{-1}$, which is clearly less than observed values (see Mikhailov and Chernyshova, 1997). Lehmann and Hinrichsen (2000b) also found, using numerical modeling tools, that the difference between inflows and outflows is around $130 \, \text{km}^3 \, \text{yr}^{-1}$. However, their absolute numbers were much higher than those of Witting (1910).

It was found that inflow and outflow strongly depend on the time period used for averaging (Andrejev *et al.*, 2004a). When water exchange was summed at every time step (about 30 minutes) during a 5-year model run (1987–1992), the dynamics of the short-term mesoscale circulation system at the entrance to the Gulf of Finland was properly taken into account and average inflows and outflows of $3{,}154 \, \text{km}^3 \, \text{yr}^{-1}$ and $3{,}273 \, \text{km}^3 \, \text{yr}^{-1}$ were obtained. When water exchange was estimated from average velocities for the 5-year period, short-term random variability had no influence but the quasi-stationary mesoscale circulation pattern at the western end did, and average inflows and outflows became $1{,}417 \, \text{km}^3 \, \text{yr}^{-1}$ and $1{,}532 \, \text{km}^3 \, \text{yr}^{-1}$. Net water exchange became $119 \, \text{km}^3 \, \text{yr}^{-1}$ or $115 \, \text{km}^3 \, \text{yr}^{-1}$, respectively, using full data or just average inflows and outflows (the difference was due to numerical errors in budget calculations). This estimate is close to total river runoff into the Gulf of Finland. Andrejev *et al.* (2004a) also studied the mean current field at the mouth of the Gulf of Finland across a north–south transect where a multi-layer velocity structure frequently occurs. Mean velocities show an inflow adjacent to the Estonian coast across the entire water column with typical velocities of $1 \, \text{cm} \, \text{s}^{-1}$–$4 \, \text{cm} \, \text{s}^{-1}$. The most intense flow ($7 \, \text{cm} \, \text{s}^{-1}$–$10 \, \text{cm} \, \text{s}^{-1}$) occurs near the surface.

Somewhat counter-intuitively, there is comparatively intense transport at 8 cm/s as a result of Ekman drift entering the Gulf of Finland on the Finnish side in the uppermost model layer (0 m–2.5 m). However, this side is primarily dominated by an offshore outflow down to a depth of 40 m–50 m (mean velocity is 8 cm s^{-1}). Similar results have also been obtained, for example, by Meier (1999). This outflow can be interpreted as a persistent buoyancy current stabilized by certain specific features concerning baroclinic instability (Stipa, 2004). Generally, baroclinic instability leads to disintegration of a flow but the relatively gently sloping bottom at the Finnish coast causes an internal baroclinic adjustment of the flow making it quasi-stable. In other words, the outflow can be interpreted as a current flowing along bottom isobaths. Such currents (analogous to nearly zonal flows in the β-plane) are generally more pronounced in baroclinic than in barotropic conditions, although they are somewhat more inclined to meander across the slope (Soomere, 1995, 1996).

Numerical simulation of the circulation of the whole Baltic Sea was performed for the period 1991–2000 with a special focus on the Gulf of Bothnia by Myrberg and Andrejev (2006). Their results supported the traditional view of cyclonic mean circulation in this basin. Its persistency ranges from 20% to 60% and it is at its largest close to coasts, as demonstrated many years ago from observations by Witting (1912) and Palmén (1930). Calculations using the barotropic model of Myrberg and Andrejev (2006), at a fairly high horizontal resolution (3.4 × 3.4 km), supported the idea that the main features of circulation can be reproduced using a barotropic, wind-driven model. However, mean current velocities were clearly larger than those according to the Witting–Palmén results. The difference apparently comes from the lower resolution of early measurements that were unable to resolve mesoscale features, such as the quite pronounced differences in speed and direction between coastal and open sea currents. The persistency of mean circulation, especially in the Sea of Bothnia, was close to the results by Lehmann and Hinrichsen (2000a), who also used a barotropic model but for a different period.

Water exchange estimates at the mouth of the Gulf of Bothnia are mostly based on water budgets (Fonselius, 1971; Dahlin, 1976; Ehlin and Ambjörn, 1978). This approach gives, as a first approximation, useful information and can be easily compared with the results of three-dimensional models. There exist large discrepancies in estimates of the inflow and outflow magnitudes for the Gulf of Bothnia. According to Dahlin (1976), inflow to the Gulf of Bothnia is 1,380 km^3 yr^{-1} whereas outflow equals 1,570 km^3 yr^{-1}. Ehlin and Ambjörn (1978) gave much higher numbers, 2,200 km^3 yr^{-1} inflow and 2,400 km^3 yr^{-1} outflow, while Fonselius (1971) ended up with 900 km^3 yr^{-1} (inflow) and 1,100 km^3 yr^{-1} (outflow). The large scatter in these estimates is embedded in the various datasets used and in the different averaging periods, but net exchange comes close to river runoff to the Gulf of Bothnia: approximately 180 km^3 yr^{-1}–190 km^3 yr^{-1} (Fonselius, 1995). According to three-dimensional modeling results, Lehmann and Hinrichsen (2000b) found the corresponding differences to be from 51 km^3 yr^{-1} (in 1994) to 221 km^3 yr^{-1} (in 1993).

Thus, net water exchange between the Gulf of Bothnia and the Gotland Sea is known quite well but the exact amounts of inflow and outflow need further investigations. Myrberg and Andrejev (2006) calculated, using their three-dimensional model, water exchange across latitude 60°02′ (section: Åland Sea–Åland–Archipelago Sea), which is approximately the southern boundary of the Gulf of Bothnia. Two different approaches were used for integration as in the case of the Gulf of Finland (p. 181). According to the model, the difference between inflows and outflows was about 170 km^3 yr^{-1} for the first and second cases, both values being reasonably close to total river discharge debouching into the Gulf of Bothnia.

When analysing short-term (daily) variability of currents, it becomes clear that in the Åland Sea, at the entrance to the Gulf of Bothnia, there are quasi-permanent anticyclonic circulation cells, with a continuous inflow–outflow system across the section where water exchange was calculated by the model. The discrepancy between budget estimates and model estimates is very large, and it was concluded that standard budget calculations (i.e., those roughly based on inflowing and outflowing watermasses) can hardly be used in practice because of the associated uncertainties.

5.5.3 Inflow modeling

A large number of Baltic Sea inflow modeling studies have been made since the early 1990s. The reader can consult papers concerning major inflows and related dynamics by Krauss and Brügge (1991), Lehmann (1992, 1993, 1994a, b, 1995), Kleine (1993), Meier (1996), Sayin and Krauss (1996), Gustafsson and Andersson (2001), Andrejev et al. (2002), and Meier and Kauker (2003b). Recent warm inflows have been modeled by Meier et al. (2003, 2004a) and Lehmann et al. (2004). Reviews of inflow modeling can be found in Meier et al. (2006) and Matthäus (2006).

One of the first model integrations of inflow-related dynamics was carried out in the early 1990s by Krauss and Brügge (1991). The authors stated that strong westerly winds over the western Baltic Sea yield an inclination in sea-level from the Skagerrak to the Arkona Basin and a strong inflow of saline and oxygen-rich water takes place into the western Baltic Sea penetrating as far as the Bornholm Basin. However, model runs indicated that northerly and easterly winds are necessary to transport these waters from the Bornholm Basin to the Gotland Sea.

The 1993 inflow has been the subject of numerous modeling studies. Lehmann (1995) employed a model with a horizontal resolution of 5 km and 21 vertical levels using realistic wind forcing (Figure 5.29). The model reproduced a realistic salinity distribution but the depth of the mixed layer was underestimated and thus the vertical gradient of salinity across the halocline was too weak. Good agreement between modeled and observed volume and salt transport for the 1993 inflow was obtained, and the importance of the Drogden Sill in major inflows was shown by Meier (1996). The important role of slope convection in the inflow process was confirmed by, for example, Andrejev et al. (2002). The phenomenon of splitting bottom intrusions of saline waters into two branches when entering the Eastern Gotland Basin from the Stolpe Channel was indicated by Zhurbas et al. (2003). One branch went northeast towards the Gotland Deep, and the other moved southeast towards the Gdansk Deep. Meier et al. (2003) showed, while studying the 1993 inflow, that river runoff and sea-level in the Kattegat play a major role in the inflow process.

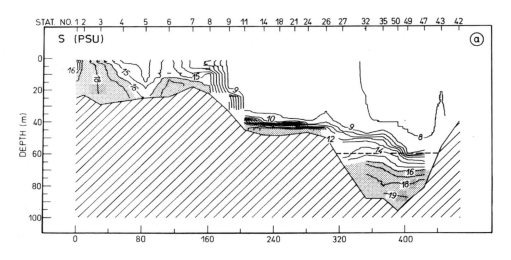
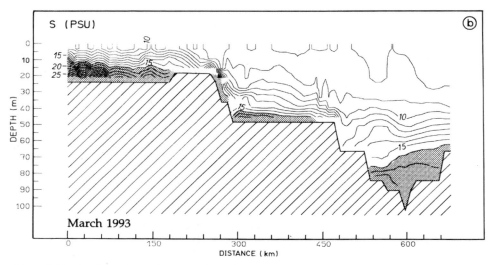

Figure 5.29. Section of salinity (‰) through the Belt Sea, Arkona Sea, and Bornholm Sea, March 1993, (a) observation, (b) model. (Lehmann, 1995, printed with permission from *Tellus.*)

A long-term hindcast simulation was carried out by Meier and Kauker (2003b) for the period 1902–1998 showing that increased runoff and precipitation reduce the intensity of major inflows but cannot explain the stagnation periods in Gotland Basin deepwater. During stagnation periods anomalous strong westerly winds cause increased sea-levels and reduced salt transport into the Baltic Sea. Increased runoff and precipitation changed the baroclinic pressure gradient on a slower timescale and so salt transport was further reduced. The series of inflows in 2002–2003 was also recently confirmed successfully by numerical models (Meier et al., 2004a; Lehmann et al., 2004). However, the mechanism behind warm inflows is still unknown and thus modeling work is still in progress.

5.5.4 Renewal times of watermasses

In Chapter 4 simple water renewal estimates for the Baltic Sea were given. However, in general the water entering the Baltic Sea may flow out again within a short time and hence has no effect on the main water body of the Baltic Sea. So, the summing of individual inflow and outflow events, calculated from sea-level differences along the straits, gives little information about actual water renewal. Flow in the straits is often separated by salinity and the term "inflow" usually covers water that forms the deep layers below the halocline (i.e., water entering the Arkona Basin with salinity greater than 8‰–9‰). This treatment also includes Baltic Sea water that is entrained to the inflow. There are a large number of salinity-weighted inflow estimates (see Elken and Matthäus, 2008). Calculated inflows vary from 19,000 $m^3 s^{-1}$ to 43,000 $m^3 s^{-1}$ mainly depending on how methods or models are used and how salt fluxes are adjusted. These estimates give water residence times from 11 to 22 years.

The main question that arises is how long do water parcels inflowing to the Baltic Sea with different origins remain in the sea. This question has many environmental dimensions: the longer the renewal time the more vulnerable the Baltic Sea is to human activity being just one of them. This topical subject has been investigated by several authors in recent decades (Bolin and Rodhe, 1973; Engqvist, 1996; Andrejev et al., 2004b; Meier, 2005; and Myrberg and Andrejev, 2006). Recently Meier (2007) carried out a comprehensive review of the subject.

There are two main approaches to solving the problem: one Lagrangian the other Eulerian. Döös et al. (2004) calculated, from three-dimensional circulation model results, the residence times of Baltic Sea watermasses using Lagrangian particles released either at sills in the Baltic Sea entrance area or at the mouth of the River Neva. They found for both sites that e-folding times were between 26 and 29 years. These numbers are relatively close to the traditional assumption of water renewal time (about 30 years). One potential problem with the Lagrangian method is that subgrid-scale mixing is not included, whereas Eulerian tracers do not have this problem. The concept of age distribution and transit time has been applied in which a passive tracer was added to the model to characterize the average age of seawater in the reservoir with prescribed values at open boundaries or at the sea surface.

Another approach where the age of seawater is defined as the time elapsed since a water particle left the sea surface was used by Meier (2005). He calculated the mean age of bottom water as 1 year in the Bornholm Deep, 5 years in the Gotland Deep, and up to 7 years in the Landsort Deep. A slightly different approach is to make the assumption that the age of inflowing water at the source regions (i.e., lateral open boundaries towards the Baltic Sea proper and river mouths) is set to zero. In such a way Andrejev et al. (2004b) and Myrberg and Andrejev (2006) found maximum water ages of around 2 years in the Gulf of Finland and 7.4 years in the Gulf of Bothnia.

A common factor in both studies was that the age of seawater was estimated from the equation of pure water (Deleersnijder et al., 2001).

There also exist techniques that are able to separate the ages of different watermasses (Meier, 2007). For example, one of them marks various watermasses with passive tracers and calculates the associated ages of specific water volumes. At time t and location r, the concentration of a tracer $C(t,r)$ would obey the advection–diffusion equation:

$$\frac{\partial C}{\partial t} + \nabla \cdot (\boldsymbol{U}C - \boldsymbol{K} \cdot \nabla C) = 0 \qquad (5.21)$$

where $\boldsymbol{K}$ is the diffusion tensor. The age concentration $\alpha(t,r)$ of the watermass (see Deleersnijder et al., 2001) under study is the solution of

$$\frac{\partial \alpha}{\partial t} + \nabla \cdot (\boldsymbol{U}\alpha - \boldsymbol{K} \cdot \nabla \alpha) = C \qquad (5.22)$$

Finally, the age is then given as a ratio $a(t,r) = \alpha(t,r)/C(t,r)$. The initial tracer and age concentrations are set to zero and in case of inflow the tracers and associated ages are relaxed to zero on the timescale of a day.

Model simulations (Meier, 2007) were based on the idea of passive tracers marking inflowing water (Figure 5.30a, see color section) and in another case tracers marking the discharge from all Baltic rivers (Figure 5.30b, see color section). Figure 5.30 shows the results of a 96-year run (using the atmospheric forcing of 1980–2004) with tracers marking inflowing waters at the Darss Sill and the Drogden Sill.

In the case of inflowing waters from the Kattegat there are pronounced vertical and horizontal age gradients between the uppermost layers and near-bottom layers. Spatial distribution also shows large differences in the water ages between the mouth area of the Baltic and the Bay of Bothnia. At the sea surface the age of Belt Sea water is younger than 14 years whereas in the northernmost Baltic Sea the water age is up to 40 years. In bottom layers, water is in general younger than at the surface (e.g., in the Arkona Basin it is less than 10 years). The halocline separates the watermasses of the upper and lower layers that have in the Gotland Basins associated ages younger or older than 26 years. The west–east cross-section in the Gotland Sea confirms that surface water is older than bottom water and that the Western Gotland Basin is characterized by older ages than the Eastern Gotland Basin due to the cyclonic circulation system.

For tracers marking freshwater from all rivers, vertical age gradients in the Gotland Deep are much smaller than in the case of inflowing water from the Kattegat. This indicates an efficient re-circulation of freshwater in the Baltic Sea. This finding is most probably because in the model downward tracer flux across the halocline caused by the entrainment of surface water into deepwater balanced by an upward tracer flux in the interplay between vertical advection and diffusion. Consequently, the upward tracer flux caused by the entrainment of deepwater into the surface layer flow is less important. According to Meier (2007) the oldest surface water mean ages, more than 30 years, are found in the central Gotland Basin and Belt Sea. Young water is found only at narrow coastal areas and in river mouths. At the bottom the mean ages are oldest in the western Gotland Deep (about 36 years). At the halocline depth, age distribution is rather homogeneous in the Baltic Sea proper—an opposite situation in comparison with the age calculation based on inflowing waters where high spatial gradients are found.

In this chapter the main characteristics of the circulation in the Baltic Sea have been given. Currents in the Baltic Sea are wind- and density-driven, and both barotropic and baroclinic processes are important. Long-term surface water circulation is cyclonic, whereas in the near-bottom layer the dynamics are strongly coupled with inflow processes and topographic steering is important. Due to the strong stratification of the Baltic Sea, mixing between the upper and lower layers is a complicated matter. Irregular major inflows are important for the renewal of the near-bottom watermasses and the transport of oxygen-rich water. At present, stagnation periods, when no major inflows take place, are more frequent than ever before; this may be due to the influence of climate change on the physics of the Baltic Sea. The dynamics of inflows are still not fully understood. Today, numerical modeling is a useful tool in inflow studies, as are simulations of general circulation.

6

Waves

Professor Günter Dietrich (1911–1972) studied mathematics, physics, oceanography, and meteorology at the University of Berlin from 1931 to 1935, taking his doctor's degree in 1935. He was appointed assistant professor at the Institute of Oceanography in Berlin in the following year. During the Second World War, he served as an oceanographer with the German Navy. In 1959 he was appointed Director of the Institute for Marine Sciences at the University of Kiel on the southwest corner of the Baltic Sea. He played a major role in the development of the institute into one of the world's centers of excellence in the field of marine science. In 1957 he published his famous book *Allgemeine Meereskunde*, translated in English in 1963. (© Archives of Leibniz-Institut für Meereswissenschaften an der Universität Kiel.)

6.1 PHYSICS OF WAVE PHENOMENA

This chapter is a treatise on wave dynamics with special emphasis on the Baltic Sea. First, a general theoretical framework is given based mainly on linear theory. Waves are periodic solutions of Navier–Stokes equations examined with special techniques, as will be illustrated. The main division of waves in oceanography is into shallow-water waves and deepwater waves; the former lead to basin eigenoscillations,

Figure 6.1. As is well known, waves are a common feature on the sea surface. They are part of ocean dynamics, they influence the air–sea interaction, and they drive bottom erosion. Here wind-generated waves are breaking offshore Warnemünde, Germany in the Southwestern Baltic Sea. (Photograph by Anna-Riikka Leppäranta, printed with permission.)

astronomical tides, and internal waves, while the latter lead to wind-generated surface waves. All these types of waves are described and their role in the Baltic Sea is discussed (Figure 6.1).

Wave phenomena in the classical linear framework are periodical processes in time and space. The primary mathematical model is the plane sine wave, presented in real and complex forms, respectively, as

$$\psi(\boldsymbol{x}, t) = A \sin(\boldsymbol{K} \cdot \boldsymbol{x} - \omega t + \phi) \tag{6.1a}$$

$$\psi(\boldsymbol{x}, t) = A \exp[i(\boldsymbol{K} \cdot \boldsymbol{x} - \omega t + \phi)] \tag{6.1b}$$

where ψ is the physical quantity under investigation; A is amplitude; $\boldsymbol{K} = k\mathbf{i} + l\mathbf{j} + m\mathbf{k}$ is the wave vector in three-dimensional space; $K = |\boldsymbol{K}|$ is the wave number; ω is angular frequency; and ϕ is the phase. The wave vector gives the direction of wave motion and the wave number—its length—characterizes the wavelength as $\lambda = 2\pi/K$. The wave period is $T = 2\pi/\omega$, and the phase velocity is ω/K. Wave amplitude is the maximum deviation from the physical quantity under consideration (e.g., water surface) from its resting position (e.g., calm sea-level). Wave height equals twice the amplitude; that is, for surface waves it is the vertical distance between the sea-level at the wave bottom (trough) and crest. The complex representation is very useful in mathematical analysis, because differentiation and integration are then made simple.

The presence of waves reflects certain important properties of the physics of the system (in particular, regularity and consequent predictability). The basic linear wave equation is written as $\partial^2 \psi / \partial t^2 = c^2 \nabla^2 \psi$, where c is the wave speed. Sine waves turn out to be basic solutions, and different solutions can be superposed for processes with many periods or wave numbers; and, inversely, any periodic process can be decomposed into a sum of sine waves. For any equation, the existence of a wave solution can be examined by inserting the sine wave—Equation (6.1)—as a trial solution (frequently called *ansatz*) into this equation to see whether linear wave solutions are possible. In positive cases one ends up with condition $\omega = \omega(\boldsymbol{K})$, the *dispersion relation*. The group velocity of waves is $d\omega/d\boldsymbol{K}$, which shows the velocity of energy transfer. If phase velocity is independent of wave number, as is the case in the above basic linear wave equation, the waves are called non-dispersive; otherwise, they are dispersive. Seemingly similar waves may exhibit completely different properties depending of their scale. For example, long surface waves in shallow water are non-dispersive whereas waves in deep water experience "normal" dispersion (longer waves propagate faster than shorter waves). "Abnormal" dispersion (shorter waves propagating faster than longer waves) is present in capillary wave fields.

In dynamical oceanography, waves are observed in a wide range of size and period: from short-period wind-generated capillary waves to long planetary waves (e.g., Pond and Pickard, 1983). Specific features in the Baltic Sea are the absence of long planetary waves,[1] the extreme rarity of tsunami,[2] and the very small level of tidal energy. There is geological evidence of a tsunami in pre-historical time in the Baltic Sea (Mörner, 1995). Wind-generated short waves, long gravity waves including gyroscopic effects, and internal waves are the most important waves, which show up in the elevation of the sea-surface or (internal wave) density interfaces. The associated velocities of water particles follow wave behavior. The paths of these particles are circular or elliptic and the particles experience no net displacement unless a wave becomes highly non-linear and breaks. Inertial oscillations are horizontal waves of water particles (they were treated in connection with current dynamics in Chapter 5).

Wave motion starts from an initial disturbance and continues under the influence of a restoring force and inertia. Here the principal restoring force is gravity. In long waves the Earth's rotation together with gravity acts as the restoring force. In very small disturbances, wavelengths less than 1.7 cm, surface tension is a stronger restoring force than gravity. The resulting waves are called capillary waves, which are not

---

[1] Planetary waves have wavelengths of the order of Earth's radius (e.g., Rossby waves due to variations in the Coriolis parameter with latitude).
[2] *Tsunami* (Japanese) is literally "harbor wave", but in physical oceanography it is understood as waves created by seismic events in the sea floor.

192 **Waves**

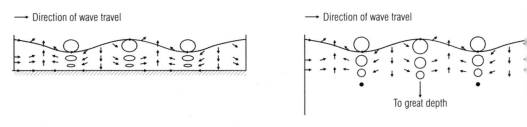

Figure 6.2. Schematic illustration of shallow-water and deepwater waves.

discussed further in this chapter. Wave motion is also influenced by friction, which causes the damping and disappearance of waves in the course of time.

It can be shown that in linear theory a small amplitude wave has the phase speed

$$c = \sqrt{\frac{g\lambda}{2\pi} \tanh\left(\frac{2\pi H}{\lambda}\right)} = \sqrt{\frac{g}{k} \tanh(kH)} \qquad (6.2)$$

where H is sea depth. If $\lambda < 2H$, then $\tanh(kH) \approx 1$ and the phase speed is almost independent of depth; if $\lambda > 20H$, then $\tanh(kH) \approx kH$ and the phase speed is almost independent of wavelength. Consequently, waves in natural water bodies are divided into three categories based on water depth H and wavelength λ (Figure 6.2):

Shallow-water waves	$\lambda > 20H$	Waves touch the bottom
Deep-water waves	$\lambda < 2H$	Waves do not reach the bottom
Intermediate regime	$2H < \lambda < 20H$	Mixture of the two above

Shallow-water waves and deepwater waves are also called long waves and short waves, respectively. *Seiches*[3] are examples of shallow-water waves, while wind-generated surface waves in the open ocean represent deepwater waves (depth must be more than half a wavelength). Waves may be free, natural oscillations depending on the geometry of the basin, or they may be forced by a cyclic forcing into a given frequency. Seiches are free oscillations, while tidal waves are externally forced.

Real marine waves are an extremely complex system of motions. To a first approximation, they are interpreted as consisting of a large number of sine waves. Superposition of linear waves leads to formation of independently propagating groups of wave crests. Such groups of waves travel at a *group speed*, which is obtained

---

[3] Eigenoscillation of a water body, analogous to pendulum motion. The word *seiche* is Swiss French dialect and means "to sway back and forth"; they were first studied and explained (1869) in Lake Geneva, Switzerland.

from the dispersion relation as $c_g = d\omega/dk$. Equation (6.2) then gives

$$c_g = \frac{c}{2}\left(1 + \frac{2kH}{\sinh 2kH}\right) \tag{6.3}$$

For long waves ($kH \ll 1$) $c_g = c$, and for short waves ($kH \gg 1$) $c_g = \tfrac{1}{2}c$.

In stratified conditions *internal waves* also occur. Their characteristics are determined on the basis of density stratification, and in two-layer situations they travel along the boundary layer surface. The restoring force is the difference in gravity between the layers, and therefore these waves are easily generated and may have a large amplitude (the gravity difference is small and thus little force is needed for vertical displacements of water parcels).

6.2 SHALLOW-WATER WAVES

6.2.1 General model

Shallow-water waves move in the horizontal plane. The wave vector is therefore two-dimensional, $\boldsymbol{K} = k\mathbf{i} + l\mathbf{j}$. Water particles move in three-dimensional space, but each trajectory is confined to a vertical plane. The theory is based on the equations for a frictionless fluid (or Euler's equation) with Coriolis acceleration and the continuity equation of an incompressible fluid; that is, Equations (5.2) with $A_H = A_v = 0$ (e.g., Pedlosky, 1979; Cushman-Roisin, 1994). The density of seawater is taken as constant and the hydrostatic law is used for the vertical direction. In linear theory the momentum advection term is neglected, which is a good approximation as long as advection is much smaller than pressure gradient:

$$\frac{U^2}{L} \ll \frac{gH}{L} \quad \text{or} \quad Fr = \frac{U}{\sqrt{gH}} \ll 1 \tag{6.4}$$

where Fr stands for the Froude number. The system of equations becomes

$$\frac{\partial \boldsymbol{U}}{\partial t} - f\mathbf{k} \times \boldsymbol{U} = -\frac{1}{\rho}\nabla p \tag{6.5a}$$

$$\frac{\partial p}{\partial z} = -v\rho \tag{6.5b}$$

$$\frac{\partial w}{\partial z} = -\nabla \cdot \boldsymbol{U} \tag{6.5c}$$

Vertical velocity and consequent variations in surface level follow from the continuity equation, when the hydrostatic theory is used. The boundary conditions are:

$$\text{Surface:} \quad w = \frac{\partial \xi}{\partial z} \tag{6.5d}$$

$$\text{Bottom:} \quad \boldsymbol{u} \cdot \mathbf{n} = 0 \tag{6.5e}$$

where **n** is a unit vector perpendicular to the sea bottom. These conditions state that vertical velocity equals the rate of change in sea-level elevation at the surface, while at the bottom velocity must be parallel to the bottom.

Since density is constant, the horizontal pressure gradient is independent of depth and proportional to sea-level slope, $\nabla p = g\rho\nabla\xi$, where ξ is sea surface elevation. Thus, the pressure field is known as soon as the surface elevation is known. Since forcing is vertically constant, horizontal velocity will also be vertically constant. Furthermore, the *small-amplitude assumption*, $\xi \ll H$ (i.e., $H + \xi \approx H$) is employed. Now we have the classical linear shallow-water model:

$$\frac{\partial \mathbf{U}}{\partial t} - f\mathbf{k} \times \mathbf{U} = -g\nabla\xi \qquad (6.6a)$$

$$\frac{\partial \xi}{\partial t} = -H\nabla \cdot \mathbf{U} \qquad (6.6b)$$

Equation (6.6a) is first operated by $\partial/\partial t\, \nabla\cdot$ and then by $f\nabla \times |_z$, the results are added together again using Equation (6.6b), and we come up with the sea level equation (Pedlosky, 1979)

$$\frac{\partial}{\partial t}\left[\left(\frac{\partial^2}{\partial t^2} + f^2\right)\xi - \nabla \cdot (c_0^2\nabla\xi)\right] = gf[\nabla H \times \nabla\xi]_z \qquad (6.7)$$

where $c_0 = \sqrt{gH}$ is the speed of shallow-water waves in a non-rotating system. Here the wave equation is on a rotating plane (in brackets on the left-hand side) and is modified by interactions with bottom topography (on the right-hand side). Now when the sea-level has been ascertained, velocity is obtained from the original Equation (6.6a) (see Pedlosky, 1979).

Equation (6.7) has an interesting stationary wave solution. It only exists when $\nabla H \times \nabla\xi = 0$; that is, the sea-level contours are parallel to the isobaths and consequently the geostrophic flow goes along the isobaths. The direction of the flow is, however, indeterminate. In Baltic Sea basins the direction of circulation is known to be counter-clockwise (see Section 5.2).

6.2.2 Free waves in ideal basins and in the Baltic Sea

If the depth is constant, the cross-product in Equation (6.7) is zero, and in an infinite plane there are free-wave solutions. The trial solution produces the dispersion relation as $\omega(\omega^2 - f^2 - gHK^2) = 0$, and the roots are

$$\omega_1 = 0, \qquad \omega_{2,3} = \pm\sqrt{f^2 + gHK^2} \qquad (6.8)$$

where K is the absolute value of the wave vector. The first root gives the steady-state geostrophic current, and the other roots give free waves traveling in opposite directions. The wave speed is $c = [c_0 + (f/K)^2]^{1/2}$; if $Kc_0 \gg f$, the non-rotational solution is a good approximation.

Table 6.1. Theoretical uninodal seiche periods in Baltic Sea basins. These periods are based on the Merian formula, which assumes basins to be one-dimensional channels with constant depth.

	Length (km)	Depth (m)	Period (h)
Baltic Sea proper–Gulf of Bothnia	1,500	56	35.6
Baltic Sea–Gulf of Finland	1,000	56	23.7
Gotland Sea, longitudinal	600	64	13.3
Gulf of Riga	100	28	3.4
Gulf of Finland, longitudinal	300	38	8.6
Gulf of Finland, transverse	80	38	2.3

In classical theory the sum of two similar waves traveling in opposite directions produces standing waves or seiches. They can be found in lakes, bays, harbors, and in different enclosed or semi-enclosed basins, and their period depends on the size and depth of the basin. In a one-dimensional channel with constant depth the wave speed equals c_0, and the period is given by the Merian formula[4]

$$T_n = \frac{2L}{n\sqrt{gH}} \tag{6.9}$$

where L is the length of the basin; and n is the number of nodes. Uninodal oscillation ($n = 1$, the node in the center of the basin) represents the longest possible free oscillation in an enclosed basin. If the basin is semi-enclosed, one node is at the open boundary since outer water masses also take part in the oscillation, and the period becomes twice that in a closed basin. Table 6.1 shows the theoretical uninodal seiche periods based on the Merian formula for several basins of the Baltic Sea. Oscillations of the whole Baltic Sea have periods of 1–1½ day, while in smaller bays the period is some hours (e.g., 3.4 hours in the Gulf of Riga).

For $H \sim 50$ m, we have $f/K \sim c_0$ at $K^{-1} \sim 175$ km corresponding to a basin size (half-wavelength) of 550 km, and therefore in basin-scale oscillations of the whole Baltic Sea, Gotland Sea, and Gulf of Bothnia, Coriolis acceleration is significant, but in smaller basins the non-rotational model is a good approximation.

Seiches play an important role in the dynamics of the Baltic Sea. They are observed in sea-level data, and they may be dominant after wind forcing ceases. Seiches may even cause very high sea-levels when favorably coupled with wind-driven sea-level variations. For example, in the eastern Gulf of Finland the sea-level may rise almost 2 m in westerly storm winds, and at the same time the sea-level becomes low in the southern Baltic Sea. When the wind situation changes or wind ceases, sea surface tilt starts to come back and oscillations form, damping out due to frictional effects.

---

[4] Presented by J. R. Merian in 1828 (see Dietrich et al., 1963).

196 Waves [Ch. 6

Table 6.2. Oscillation modes (in hours) of the whole Baltic Sea based on absence of a Coriolis effect ($f = 0$) and with a Coriolis effect ($f \neq 0$) in a two-dimensional numerical model (Wübber and Krauss, 1979).

Mode	1	2	3	4	5	6
Period ($f = 0$)	40.5	27.7	23.7	21.4	17.9	13.0
Period ($f \neq 0$)	31.0	26.4	22.4	19.8	17.1	13.0
Mode	7	8	9	10		
Period ($f = 0$)	10.5	8.7	7.6	7.2		
Period ($f \neq 0$)	10.5	8.7	7.8	7.3		

The e-folding decay timescale of seiche amplitude is about two days in the Baltic Sea (Mälkki and Tamsalu, 1985).

The actual seiche period has received much attention. The best-known seiche is the Baltic Sea proper–Gulf of Finland oscillation in the Baltic Sea. In the uninodal case the node is at Gotland and antinodes are at St. Petersburg and the Danish coast. Using a channel model with variable depth, Witting (1911) obtained the period as 28–31 hours and Neumann (1941) as 27.5 hours, while the Merian formula gives 23.7 hours (Table 6.1). One-dimensional models have produced 27.7 hours.

Two-dimensional models were later applied to examine the oscillation modes (Wübber and Krauss, 1979). The second mode turned out to have the same period as the one-dimensional seiche model when Coriolis acceleration was not included and 26.4 hours when included (Table 6.2). The latter value is also close to what has been observed (Figure 6.3). Wave progression is not unidirectional in a two-dimensional basin such as the Gotland Sea, since Coriolis acceleration tends to curve the direction of propagation towards the right.

Observations also indicate the existence of 17-hour to 19-hour periods. The higher modes have local significance. The Baltic Sea proper–Gulf of Bothnia system has periods as long as 39 hours. Long waves show different results irrespective of whether the Coriolis term is included or not, but when the period becomes half a day or less the difference is very small as indicated by the analytical analysis above.

6.2.3 Kelvin waves and Poincaré waves

The properties of very long waves in large rotating basins are considerably modified by the rotation of the Earth. These rotational waves are represented by Kelvin and Poincaré waves. The phenomena first become evident for waves propagating along the coast. Consider an infinite channel of width L and constant depth H (see Pedlosky, 1979). The governing equation is (6.7), but now there is the additional constraint that velocity must be parallel with the channel walls at channel boundaries.

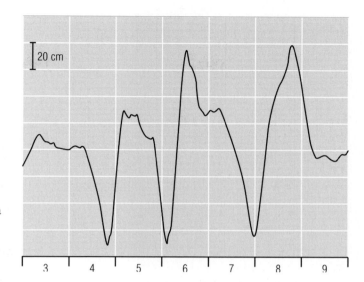

Figure 6.3. Sea-level variations at Ystad, southern Sweden during one week in December 1975. The graph shows a seiche with a period slightly above one day. (Redrawn from Sjöberg, 1992.)

From Equations (6.7) and (6.6a) we then have, taking $H = $ constant,

$$\frac{\partial}{\partial t}\left[\left(\frac{\partial^2}{\partial t^2}+f^2\right)\xi - c_0^2\nabla^2\xi\right] = 0 \tag{6.10a}$$

$$\frac{\partial^2\xi}{\partial y\,\partial t} - f\frac{\partial\xi}{\partial x} = 0, \qquad y = 0, L \tag{6.10b}$$

Periodic wave solutions along the x-axis take the form $\xi = \xi_x(y)\exp[i(kx - \omega t)]$, where $\xi_x(y)$ is amplitude along the y-axis. Inserting this into Equations (6.10) gives the eigenvalue problem

$$\frac{d^2\xi_x}{dy^2} + \alpha^2\xi_x = 0, \qquad \alpha^2 = \left(\frac{\omega^2 - f^2}{c_0^2} - k^2\right) \tag{6.11a}$$

$$\frac{d\xi_x}{dy} + f\frac{k}{\omega}\xi_x = 0, \qquad y = 0, L \tag{6.11b}$$

The general solution of Equation (6.11a) is $\xi_x = A\sin(\alpha y) + B\cos(\alpha y)$, and boundary conditions restrict the solutions to two categories (Figure 6.4):

(i) $\sin(\alpha L) = 0$ or $\alpha L = n\pi/L$ $(n = 1, 2, 3, \ldots)$. These waves are called *Poincaré waves*, and the dispersion relation is

$$\omega = \pm\sqrt{f^2 + c_0^2\left(k^2 + \frac{n^2\pi^2}{L^2}\right)}, \qquad n = 1, 2, 3, \ldots \tag{6.12}$$

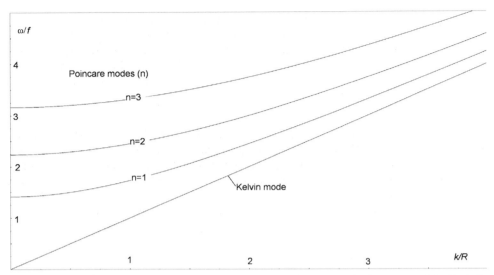

Figure 6.4. Dispersion relation of Kelvin and Poincaré waves; ω is frequency, f is Coriolis parameter, k is wave number, and R is barotropic Rossby radius.

These waves have a frequency greater than inertial frequency and they can travel in both directions along the channel.

(ii) The second root set is the pair $\omega = \pm c_0 k$ (i.e., free shallow-water waves traveling along the channel in both directions). These waves are called *Kelvin waves* (Figure 6.5). For waves traveling in the positive direction, $\omega = c_0 k$, the sea-level solution becomes

$$\xi = \xi_0 \exp\left(-\frac{fy}{c_0}\right) \cos([k(x - c_0 t) + \phi]) \tag{6.13}$$

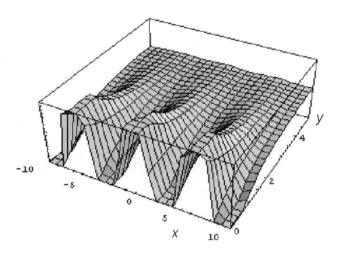

Figure 6.5. Theoretically calculated coastal profile of Kelvin waves. Unit is wave number in *x*-axis, barotropic Rossby radius in *y*-axis, and arbitrary in vertical axis.

Flow is in geostrophic balance in the x-direction, and cross-channel wave amplitude has an exponential profile with high amplitude on the right-hand side when looking in the direction of wave motion. The e-folding lengthscale of wave amplitude along the wave crest is $R = c_0 f^{-1}$, the Rossby radius of deformation. Kelvin waves need at least one boundary (on the right when looking in the direction of motion) for their existence.

Kelvin waves are observed on a daily basis in the English Channel as the incoming tidal wave from the Atlantic Ocean propagates farther up the channel, causing an extremely large tidal range along the French coast. There are just a few reports concerning the existence of Kelvin waves in the Baltic Sea (see Fennel and Seifert, 1995 and Section 8.3 "Upwelling"). These waves propagate in the northern hemisphere so that the shore is to the right of the direction of propagation.

6.2.4 Topographic waves and eddies

Topographic waves may occur in geophysical flows along a sloping bottom. They are also called "topographic Rossby waves" since the bottom slope has a similar role there to the Coriolis parameter in planetary Rossby waves.

Consider a channel of width L aligned in the x-direction and depth sloping as

$$H = H_0\left(1 + \frac{sy}{L}\right), \quad s \ll 1 \tag{6.14}$$

As shown before—see Equation (6.7) and the text below it—in such a topography the steady-state geostrophic current goes along the x-axis. In this fundamental wave equation (6.7), we have

$$c = \sqrt{gH_0\left(1 + \frac{sy}{L}\right)} \quad \text{and} \quad [\nabla H \times \nabla \xi]_z = -H_0 \frac{s}{L} \frac{\partial \xi}{\partial x} \tag{6.15}$$

Let us look at wave solutions $\xi = \xi_x(y)\exp[i(kx - \omega t)]$ along the x-direction. The wave equation with boundary conditions becomes (see, Pedlosky, 1979)

$$\left(1 + s\frac{y}{L}\right)\frac{d^2\xi_x}{dy^2} + \frac{s}{L}\frac{d\xi_x}{dy} + \xi_x\left[\frac{\omega^2 - f^2}{gH_0} - k^2\left(1 + s\frac{y}{L}\right) + \frac{fs}{L\omega}k\right] = 0 \tag{6.16a}$$

$$\frac{d\xi_x}{dy^2} + \frac{fk}{\omega}\xi_x = 0, \quad y = 0, L \tag{6.16b}$$

This system is similar to the case of uniform depth except for the terms that include bottom slope s. With $1 + sy/L \approx 1$, we have the terms proportional to s/L added into the basic equation. Solutions are of the form $\xi_x = \exp(xy/2L)[A\sin(\alpha'y) + B\cos(\alpha'y)]$, where α' is obtained by trial and error from Equation (6.16a). Employing boundary conditions, we get the eigenvalue problem. Solutions to this problem include Kelvin waves, as in the uniform depth case, and slightly modified Poincaré modes. In addition to them there is a low-frequency solution

$$\omega = s\frac{f}{L} \cdot \frac{k}{k^2 + (n\pi/L)^2 + (f/c)^2} \qquad (6.17a)$$

The frequency is greatest when $k^2 = k_n^2 = (n\pi/L)^2 + (f/c)^2$, and then

$$\omega = \omega_{max} = \frac{s}{2}\frac{f}{\sqrt{(n\pi)^2 + (fL/c)^2}} \qquad (6.17b)$$

These waves are called topographic Rossby waves. It can be seen that as $s \ll 1$, we have $|\omega| < |f|$ (i.e., these are low-frequency waves). In the x-direction the phase speed is $C_x = \omega/k < 0$. Thus, these waves travel in a positive x-direction or, in other words, such that the shallower area is to the right of the direction of motion. For higher wave numbers the frequency goes down, which is opposite to the behavior of Poincaré waves. Rossby waves are transverse waves: the motion of water particles is perpendicular to the wave propagation direction. The wave becomes evident in the form of a striped current field, the velocity of which varies sinusoidally in the wave propagation direction. Superposition of two such waves propagating in different directions results in a pattern of currents resembling the one formed by a system of eddies of alternating signs. For this reason oceanic eddies (including topographic eddies) are sometimes interpreted in terms of crossing waves although much more frequently they are created by other mechanisms and radiate Rossby waves instead.

Topographic eddies have been found in the Baltic Sea from observations and from model simulations (see Mälkki and Tamsalu, 1985). They propagate along topographic contours such that shallow depths are to the right of the direction of propagation, and they are often seen in observations along the coast of the Bornholm Basin and the Gotland Sea. Clockwise eddies are equivalent to atmospheric cyclones and lift watermasses in their center; for that reason they usually have a cold center in the thermocline and a salty center in the halocline and *vice versa* in the case of anticlockwise eddies (Aitsam and Elken, 1982). The lengthscale of such topographically "bottom-trapped" waves is typically 30 km–40 km, and the theoretical migration speed is estimated as 1.9 cm s^{-1}, while observations suggest 1.5 cm s^{-1} or 15 km per day. Talpsepp (1983) found from observed data that the strongest periodic variability was at 6–8 days in regions where the bottom slope was steep.

For the Muuga Bay, Gulf of Finland there are reports of trapped waves (Soomere *et al.*, 2008a). There is a general anticyclonic flow in the upper layer and a cyclonic flow in the lower layer substantially affected by the local topography giving rise to horizontal and vertical circulation cells (Raudsepp, 1998). The observed dynamics of the coastal jet was interpreted in terms of traveling coastally trapped waves (as had been found in many other parts of the Baltic Sea earlier, Talpsepp, 2006) although only a quarter of the eastward-traveling wave period was covered by measurements. The estimated wave period was around 40 days. The wave supposedly reversed the cross-shore flow direction while alongshore velocity was preserved (Raudsepp, 1998). Further analysis of current measurements in this framework displayed the existence of a periodic variability in the velocity of the current between 5 cm/s and 25 cm/s and with a period of 3–4 days (Talpsepp, 2006).

The numerical model investigations of Kielmann (1978) showed similar eddies in the Bornholm Basin with a migration speed of about 20 km per day. One region of eddy generation is the mouth of Stolpe Channel in the Bornholm Basin. In the open sea, winds may create eddies in regions with variable bottom topography. When near-bottom water layers are isolated from the direct influence of wind, the formation mechanism of topographic waves may be baroclinic instability.

In the Finnish side of the Sea of Bothnia there is a slightly curved 200 km long coastline, and the bottom slope is about 2×10^{-3} across the 60 km wide coastal zone. Alenius and Mälkki (1978) found a wide peak in current velocity spectra at periods of 57–65 hours (Figure 6.6), which they interpreted as barotropic topographic waves with wavelength 390 km. With several current meter stations, the phase, coherence, and decay of amplitude could be analysed and the result agreed with the characteristics of topographic waves.

6.3 TIDES

Tides are a global-scale oceanic phenomenon forced by the gravity of the Moon and Sun; they are actually the difference between the gravitational attraction of the Moon and Sun across the oceans. In smaller basins this differential force is not enough to generate strong tidal motions of their own, but penetration of tides into them is possible if the oceanic opening is wide and the amplitude of the oceanic tide at the opening is large. Thus, the region of tidal influence covers many marginal seas, but intracontinental seas are more sheltered. In the Baltic Sea neither condition for tide penetration is satisfied, and consequently tides are weak there. Tides in the Skagerrak are not strong, and the Kattegat and Danish Straits are too narrow and shallow for tides to get through them to the Baltic Sea (Defant, 1961; Feistel et al., 2008). In the Kattegat the amplitude of tidal sea-level variations is 10 cm–30 cm on the Danish coast and 4 cm on the Swedish coast (Defant, 1961).

Witting (1911) examined tides in the Baltic Sea. In the Belt Sea the tidal amplitude in sea-level is 10 cm, and over most of the Baltic Sea it is 2 cm–5 cm. In the eastern Gulf of Finland amplitudes greater than 10 cm have been reported. In the Kattegat and Skagerrak the tides are dominantly semidiurnal, but in the Baltic Sea they are made up of diurnal and semidiurnal components, both showing up in sea-level spectra (Figure 6.7). The main components are principal lunar (M_2), principal solar (S_2), luni-solar diurnal (K_1), and principal lunar diurnal (O_1).[5] Because of the weakness of tides, no notable tidal ecosystem has developed in the Baltic Sea.

Figure 6.8 shows the co-tidal lines of M_2 and K_1 tides in the Baltic Sea. Semidiurnal tides show three amphidromies,[6] all counter-clockwise, while diurnal tides

---
[5] Tidal variations contain several slightly different semidiurnal and diurnal periods with standardized notations (e.g., Neumann and Pierson, 1966).
[6] A system of tidal action in which tides progress around a point with little or no tide.

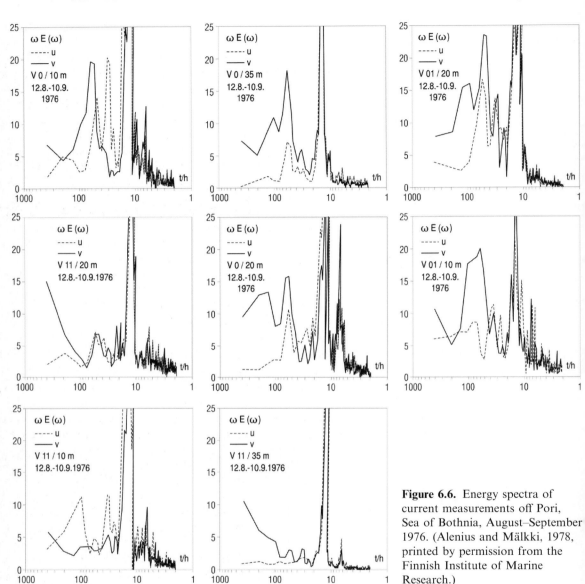

Figure 6.6. Energy spectra of current measurements off Pori, Sea of Bothnia, August–September 1976. (Alenius and Mälkki, 1978, printed by permission from the Finnish Institute of Marine Research.)

have one amphidromy in the central Gotland Sea. According to Defant (1961), diurnal tides are generated within the Baltic Sea and semidiurnal tides are dominantly co-oscillating with the tide in the Kattegat.

Tides in the Baltic Sea have not gained much attention since the investigations of Witting (1911), despite their significance especially in the Southwestern Baltic Sea (see Feistel *et al*., 2008). Very recently though, Müller-Navarra and Lange (2004) modeled the effects of tides in the Baltic Sea.

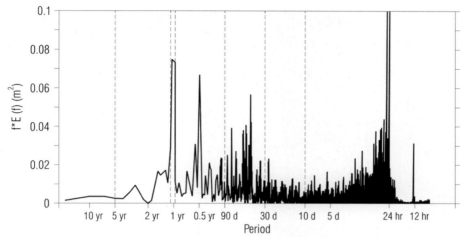

Figure 6.7. Energy spectrum of sea-level elevation in Helsinki mareograph from the period 1980 to 1999. (Redrawn from Johansson et al., 2001.)

6.4 WIND-GENERATED WAVES

6.4.1 Wave generation

Atmospheric shear stress together with air pressure fluctuations bring about wave motion at the sea-surface (e.g., Neumann and Pierson, 1966). As the wind begins to blow above a calm sea, water starts to flow and shear friction induces vorticity in the water. Vorticity in turn creates pressure differences, which deform the smooth surface. Bulges and depressions tend to return and thus oscillation is set in motion. The restoring force for the shortest waves is the combination of surface tension and gravity, and the resulting waves are called capillary waves, with periods less than 0.1 s. The role of these two forces is equal at wavelengths about 1.7 cm. When waves increase in size, gravity takes over as the restoring force. Waves can grow to the saturation limit where the energy input from the wind equals dissipation. The properties of the saturation state depend on wind stress, duration of the wind, fetch,[7] and depth of the sea.

Wind-generated waves are subject to deepwater theory or shallow-water theory depending on the wavelength (see Section 6.1). Deepwater waves are independent of the depth, and the deepwater approach is valid when the depth is at least half the wavelength, $H > \frac{1}{2}\lambda$. The longest wind-generated waves are 100 m requiring a sea depth of 50 m for the deepwater theory to be valid. The dispersion relation of gravity-restoring deepwater waves is

$$c = \sqrt{\frac{g\lambda}{2\pi}} \qquad (6.18)$$

---
[7] The distance along open water across which winds blow.

204 Waves [Ch. 6

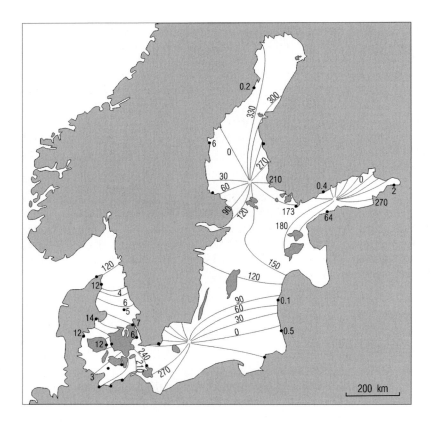

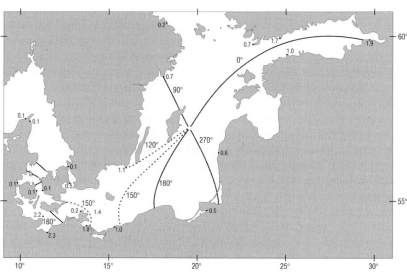

Figure 6.8. (Top) Co-tidal lines of the M_2 tide in the Kattegat and Baltic Sea; and (bottom) co-tidal lines of the K_1 tide the Kattegat and Baltic Sea. Phase in degrees and amplitude in centimeters. (Redrawn from Defant, 1961 and Witting, 1911, respectively.)

These waves are dispersive (i.e., wave speed depends on wavelength). The period of the waves is $T = \lambda/c = \sqrt{2\pi\lambda/g}$. Long waves move faster than short waves (for capillary waves the opposite is true).

In the Baltic Sea, wind-driven surface waves have special features that differ from those of oceans. There are large shallow areas where deepwater wave theory is not applicable. Central Baltic Sea basins are always deep enough but in the coastal regions mixed shallow-water/deepwater waves occur. Baltic Sea basins are small and wind events often short, consequently the growth of waves is limited by the duration of wind events and fetch. In addition, the shape of a basin has a notable influence on the evolution of a wave field.

Wave models can be used to consider the evolution of the one-dimensional power spectrum of wave height $S(\omega)$, where ω is angular frequency (Figure 6.9), or of the two-dimensional spectrum $S(\omega, \theta)$, where θ is the direction of waves. The energy spectrum describes how energy from the entire wave field is distributed between different sine wave components of different frequencies, and for the two-directional spectrum how energy is propagated in different directions. Earlier works considered the form of wave spectra and the statistical properties of wave fields. Similarity

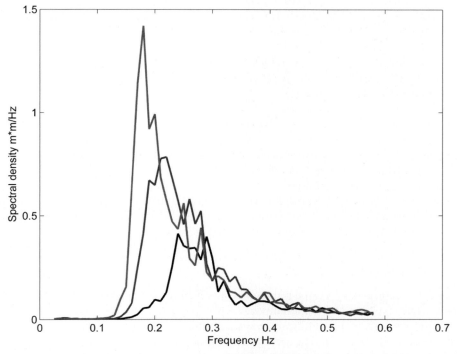

Figure 6.9. One-dimensional wave spectra from the Baltic Sea showing local waves and swell. The different spectra show a wave field developing as a result of a southwesterly wind, increasing from 7 m/s to 11 m/s. The frequency of 0.2 Hz is typical under Baltic Sea conditions. (Prepared by Dr. Heidi Pettersson, printed with permission.)

models, employed since the 1960s, attempt to explain the general physical relationships between wind and basin geometry. They are based on the assumption that the characteristics of a saturated wave field depend only on wind speed and fetch L. This assumption allows us to introduce the dimensionless fetch and energy of the wave field as

$$\tilde{L} = \frac{Lg}{U_a^2}, \qquad \tilde{E} = \frac{Eg^2}{U_a^4} \tag{6.19}$$

Energy is given by the wave spectrum as $E = \int_0^\infty S(\omega)\,d\omega$.

Contemporary wave models (so-called "third-generation" wave models, which are widely used today) are based on the energy balance across the entire wave spectrum. The wave energy spectrum at a particular location changes as a result of advection by the group velocity of waves (c_g), energy from the wind G_{in}, non-linear interactions G_{nl}, and dissipation G_{ds} (e.g., Komen et al., 1994). The deepwater equation can be written

$$\frac{\partial S(\omega, \theta)}{\partial t} + c_g \cdot \nabla S(\omega, \theta) = G_{in} + G_{nl} + G_{ds} \tag{6.20}$$

There are several debatable issues in the practical forecast of wave fields. The formula for wind input has been relatively well established, but non-linear interactions and dissipation are less understood. For example, the gustiness of wind has not been accounted for. Although the nature of non-linear interactions is theoretically well understood, owing to computational difficulties only a rough approximation is usually included in numerical models. The role of dissipation is less understood and is usually assumed to be the result of waves breaking. In shallow water, Equation (6.20) needs further modification.

Close to the shore and shallow sea areas, wind-generated waves change from deepwater waves to shallow-water waves. This is the time that wave speed gradually slows and reaches the value shown by shallow-water theory ($c^2 = gH$). Depending on bathymetrical features, reflection, refraction, and diffraction may take place. Due to the conservation of energy flux, as depth decreases the wave height first decreases slightly but then grows and finally waves break when the wave height/sea depth ratio is greater than about 0.78. Surface waves then have an effect on the bottom in shallow areas: sediments and sand may remain only in highly protected zones, and consequently the distribution of vegetation is affected by wave processes.

6.4.2 Wave statistics

A wave field consists of a large number of waves of different sizes. The characteristic wave height is defined in contemporary wave science on the basis of the wave spectrum. This gives the significant wave height h_s as four times the standard deviation of sea-level elevation. The traditional characterization, based on the analysis of single wave properties, was made using the average, $h_{1/3}$, of the upper one-third fractile of wave height distribution (Neumann and Pierson, 1966). The outcomes of these two definitions are in fact close to each other, $h_s \approx h_{1/3}$. The significant wave height corresponds well with observational estimates. The estimator of the maximum wave height in a storm is generally taken as twice the significant wave height; waves higher than 2.2 times h_s are frequently called rogue or freak waves. The overall wave climate of the Baltic Sea is quite mild. Figure 6.10 shows data from the Northern Gotland Basin: in 90% of cases the significant wave height is less than 2 m.

Wave periods are relatively short, usually not exceeding 7–8 s except in very strong storms a few times a year; thus long-period swell is almost absent here. The joint distribution of wave heights and periods is concentrated along sea states roughly corresponding to saturated waves but shifted somewhat toward shorter waves. Rough seas therefore contain a large portion of relatively short and steep waves (Figure 6.11). This feature makes Baltic Sea wave fields similar to those on large lakes, and to some extent increases the danger to shipping in this water body.

High waves have been recorded in the Baltic Sea. The Northern and Eastern Gotland Basins have the highest waves in the Baltic Sea. On December 22, 2004, during Storm Rafael the significant wave height was 7.7 m and the maximum wave height was 14 m according to measurements (*http://www.fimr.fi*). Only once has such a height been recorded before: in January 1984 at Almagrundet off the Swedish coast southeast of Stockholm. Significant wave height, estimated from the tenth highest wave, assuming wave heights follow the Rayleigh distribution, reached 7.82 m on the night of January 13–14, 1984 (Broman *et al.*, 2006). The classical estimate of significant wave height was, however, slightly smaller than the observed value. There is evidence that significant wave height was about 9.5 m offshore Saaremaa during Storm Gudrun on January 9, 2005 (Soomere *et al.*, 2008b).

In the Gulf of Finland, at Helsinki, a significant wave height of 5.2 m and maximum wave height of 9 m were measured on November 15, 2004. Westerly and easterly winds may be responsible for these wave heights. Record significant and maximum wave heights are, respectively, 5.5 m and 10 m for the Sea of Bothnia and 3.1 m and 5.6 m for the Bay of Bothnia.

In large oceanic basins, significant wave height can be 10 m–15 m, and maximum wave height can be as large as 30 m (Holthuijsen, 2007). The highest ever significant wave in the North Atlantic (17 m, highest single wave 27.7 m) was recorded near the Norwegian coast on January 11, 2006 (Norwegian Meteorological Service/Magnus Reistad, pers. commun. via Academician Tarmo Soomere). The roughest wave conditions ever measured occurred during Hurricane Ivan in the Gulf of Mexico: $h_s = 17.9$ m and highest single wave equal to 27.9 m (Wang *et al.*, 2005).

6.4.3 Similarity law of Kitaigorodskii

In the study of fetch-limited waves, the similarity law of Kitaigorodskii (1962) is often used. This is a deepwater model that considers the situation when offshore winds start to blow over a calm sea, wind speed is constant, and waves become limited by fetch. Spectral density is continuous and contains one peak.

Kahma (1981a, b) examined fetch-limited wave conditions from measurements in the Sea of Bothnia. The dimensionless total energy of the wave field ($\hat{E}$) and the peak

208 Waves [Ch. 6

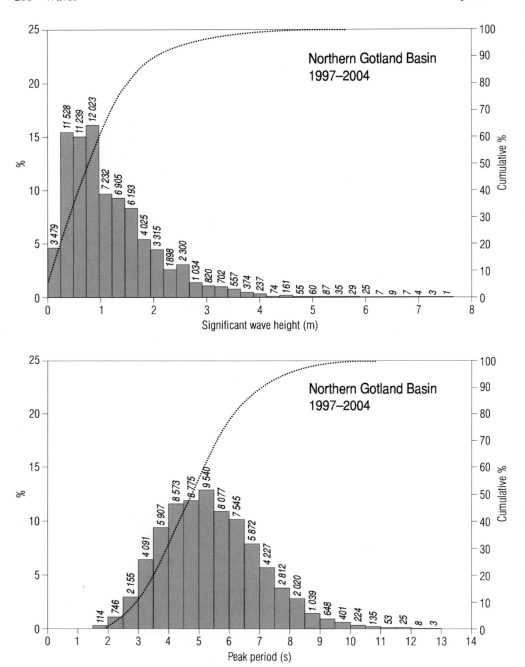

Figure 6.10. Histograms of significant wave height (top) and period of the peak wave (bottom) based on buoy measurements in the Northern Gotland Basin 1997–2004. (Source: Finnish Institute of Marine Research.)

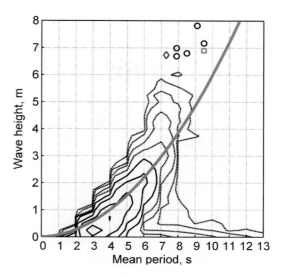

Figure 6.11. Scatter diagram of wave heights and periods at Almagrundet in the Western Gotland Basin southeast of Stockholm 1978–1995. The bin size is 0.25 m for the wave height and 1 s for the period. Isolines of probability of 0.0033%, 0.01%, 0.033%, 0.1% (dashed lines), 0.33%, 1%, 3.3%, and 10% (solid lines) are plotted. Wave conditions for storms with $h_{1/3} > 6.5$ m are shown as: circles—January 1984; diamonds—January 1988; and squares—August 1989. The bold line indicates saturated wave conditions with a Pierson–Moskovitz spectrum. (Broman et al., 2006, printed by permission from *Oceanology*.)

frequency of the wave spectrum (ω_p) were found to be

$$\tilde{E} = 3.6 \times 10^{-7}\tilde{L}, \qquad \tilde{\omega}_p = 20\tilde{L}^{-1/3} \tag{6.21}$$

where the dimensionless peak frequency is $\omega_p U_a g^{-1}$. These estimates somewhat differ from classical ones (cf. Komen et al., 1994) and may reflect certain specific features of wind fields over the Baltic Sea. One-dimensional wave spectra[8] were found to follow the law

$$S(\omega) = \alpha U_a g \omega^{-4} \tag{6.22}$$

where the constant α equals 4.5×10^{-3}. Based on this theory Kahma (1986) constructed a nomogram for significant wave height as a function of time, fetch, and wind speed (Figure 6.12). This nomogram also shows when wave growth is fetch-limited and when it is duration-limited. Pettersson (2004) examined wave growth in narrow bays from the Baltic Sea and, for comparison, from the Gulf of Lyon in the Mediterranean Sea (Figure 6.13, see color section).

The evolution and saturation state of a wave field in principle depend on air and water densities, but these can be taken as constants in this problem; generally, density ratio would become an additional non-dimensional quantity. The salinity and temperature of water slightly influence density and viscosity, but in the evolution of a wave field they have no significance. When wind calms down, there is no more energy input into short waves. Owing to the partial breaking of waves (whitecapping) and energy transfer to longer waves via non-linear interactions, the proportion of longer waves gradually increases. Wave systems that are not directly supported by wind are called "swells", which is also a general name for waves generated in remote sea areas.

---

[8] Wave spectrum at a point integrated over all directions of wave propagation.

210 Waves [Ch. 6

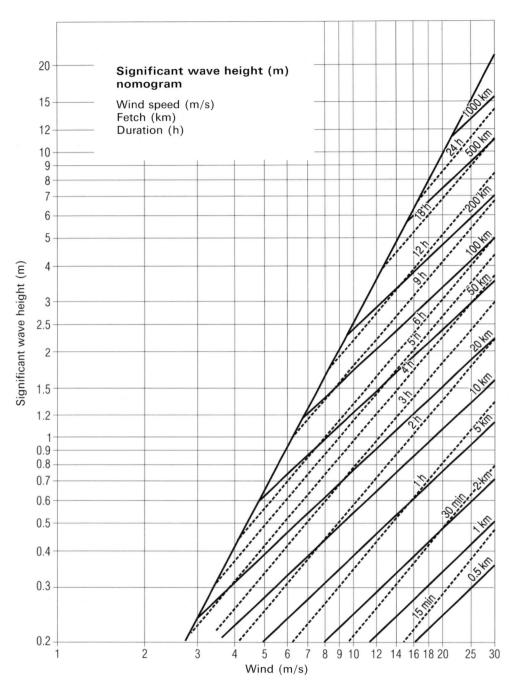

Figure 6.12. Significant wave height as a function of wind speed, duration (- - -), and fetch (———). (Kahma, 1986, printed by permission from the Finnish Institute of Marine Research.)

6.4.4 Wave forecasting

Wave-forecasting methods are used in the Baltic Sea for operational navigation and the design of shipping routes. In the Baltic Sea the size and form of basins introduce particular requirements for the construction of forecasting models. In wintertime, ice extent with reflection and refraction of waves at the ice edge also need to be considered. Models are used to produce hindcasts for investigations of accident cases and statistics for planning of routes and harbors. As an example, in the Finnish Institute of Marine Research operative forecasts are produced four times a day based on the WAM model forced by the HIRLAM model of the Finnish Meteorological Institute. Other leading wave forecast providers in this area are the Danish Meteorological Institute, SMHI (Sveriges Meteorologiska och Hydrologiska Institut, the Swedish equivalent), and Deutsche Wetterdienst (see Section 9.5 "Operational oceanography" and the short description of their models in Soomere et al., 2008b).

6.4.5 From linear to non-linear waves

The form of linear waves can be described by two dimensionless parameters: wave steepness h/λ and relative water depth H/λ (see Massel, 1989). The latter indicates whether waves are dispersive and whether celerity, length, and height are influenced by water depth. Wave steepness implicitly indicates whether the linear assumption is valid. The ratio h/H, relative wave height, also indicates whether the small-amplitude assumption, a basic feature of linear waves, is acceptable. Non-linear waves have recently been examined by Rannat (2007).

An appropriate non-dimensional parameter to characterize the nature of surface waves in shallow areas is the Ursell number—a combination of relative wave height and relative water depth (e.g., Massel, 1989)

$$Ur = \left(\frac{\lambda}{H}\right)^2 \frac{h}{H} = \frac{\lambda^2 h}{H^3} \qquad (6.23)$$

The Ursell number is often used to select an adequate wave theory according to typical λ (or T) and h in a given water depth H. High values of Ur correspond to large, finite-amplitude, long waves in shallow water and suggest the use of a non-linear wave theory could be appropriate.

For many coastal engineering tasks non-linear models must be used, as the waves cannot always be approximated as linear. For example, when waves of appreciable height travel through shallow water, higher order wave theories are often required to describe their behavior. First-order correction to the linear wave model is provided by Stokes theory (Stokes, 1847; see Massel, 1989). Stokes wave theory can be used up to $Ur \approx 75$ in some cases. For even larger Ursell numbers, Stokes wave theory is generally incorrect (Massel, 1989).

The necessity to switch from the linear and/or Stokes description to another framework becomes evident in the analysis of properties, behavior, and influence of the long components of surface waves. For storm waves in the Baltic Sea and in the Gulf of Finland with typical periods of 5–6 s (Kahma et al., 2003; Pettersson, 2001;

Broman *et al.*, 2006), the threshold $Ur \approx 75$ occurs for 1 m high waves at a depth of 1 m–2 m ($Ur \approx 25$ at depth 3 m–4 m), and for a 3 m high wave at a depth of 3 m–4 m ($Ur \approx 25$ at depth 6 m–7 m). Therefore, for storm waves, Stokes theory can be applied to a depth at which wave height is around 80% of water depth (i.e., practically up to the point at which waves break).

An appropriate model for long-plane, finite-amplitude surface waves in shallow water is the Korteweg–de Vries equation (Massel, 1989). Its periodic solution can be brought about by using so-called "cnoidal waves", because they can be explicitly expressed in terms of so-called cnoidal (Jacobi elliptic) functions.

A very important consequence of using Stokes theory is the incomplete closure of particle paths (i.e., there is net movement of water parcels in the direction of wave propagation). This resulting velocity is called *Stokes drift*. When considering wind-driven surface drift, Stokes drift should be added to Ekman drift forced by the shear stress of wind (Essen, 1993). However, it is very difficult to separate the Ekman and Stokes drift components from drifter observations, because the semi-empirical parameters underlying these components are inadequately known.

6.5 INTERNAL WAVES

6.5.1 Background

Internal waves may be generated in a stratified water body. In the case of distinct layers waves may form at the pycnocline at the layer boundary, and in the case of continuous stratification waves may form at density surfaces. The force disturbing a stratified fluid may be surface pressure gradient, diverging or converging surface layer motion, sea-level variations, flow over rough topography, etc. The restoring force is the difference between buoyancy and gravity, or net gravity. The presence of stratification is a critical background factor. Since the stratification of the Baltic Sea is stable, internal waves are a common feature, even though there has not been such a large number of studies into such waves. Internal wave generation does not need much energy, because the density differences between the water layers are small compared with the density differences between the air and sea. Therefore, the amplitude of internal waves can be one order of magnitude greater than the amplitude of surface waves. Internal waves do not show up in sea-level elevation but due to the movement of water particles in wave motion there are convergence and divergence bands in the sea-surface. Floating material consequently is accumulated at the convergence bands marking the existence of internal waves. However, not all such convergence zones are due to internal waves (Figure 6.14).[9] Further evidence of an internal wave field may be apparent changes in the reflectivity of the water surface (slicks) due to changes in capillary wave properties owing to velocity fields induced by internal waves.

---

[9] Langmuir circulation cells in the surface layer and atmospheric roll vortices collect floating particles into bands parallel to the wind direction.

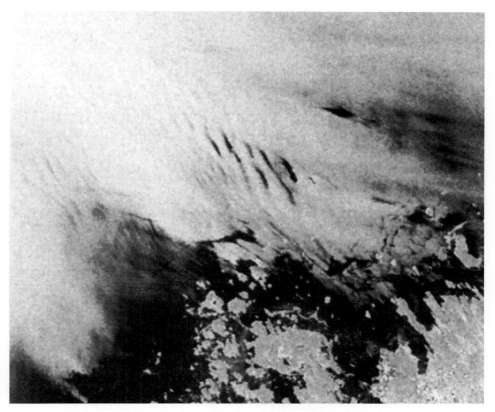

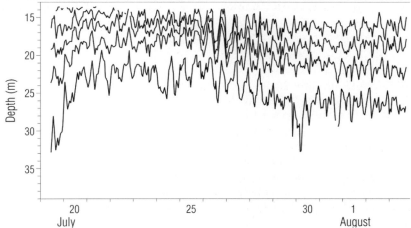

Figure 6.14. (Top) Convergence bands due to internal waves on the sea surface in the Bay of Bothnia offshore Kokkola seen in an ERS-1 SAR image (image width 40 km); (bottom) isotherms for the 16-day period in July–August 1978 in the Sea of Bothnia. (Herlevi and Leppäranta, 1994 and redrawn from Alenius, 1985, respectively.)

Internal waves may be initiated by disturbance at the surface, internal current fields, or flows along an uneven bottom. Animals and ships may also be the initiating factor (Alenius, 1985). So-called "dead water", related to the drastic increase in wave resistance owing to ship-induced internal waves, has been known for a long time to seamen. No reports of dead water cases are known to us from the Baltic Sea, but favorable conditions do arise for this phenomenon to occur. Internal waves are damped by turbulent viscosity, the fine structure of density stratification, wave breaking, and the absorption of waves at the bottom.

6.5.2 Theory

In a stratified fluid with water particles experiencing oscillations, density can be written as $\rho = \rho_0 + \underline{\rho}(z) + \rho'(x,y,z,t)$, where $\rho_0 =$ constant, $\underline{\rho}$ is the ambient equilibrium stratification, and ρ' is fluctuation due to the wave motion, with $\rho_0 \gg \underline{\rho} \gg \rho'$ (Cushman-Roisin, 1994). The Brunt–Väisälä frequency is given for ambient stratification as

$$N = \sqrt{-\frac{g}{\rho_0}\frac{\partial \bar{\rho}}{\partial z}} \qquad (6.24)$$

In the case of internal waves, vertical motions are forced by density fluctuations and the hydrostatic law no longer applies. The vertical component of the momentum equation is written as

$$\frac{\partial w}{\partial t} = -\frac{1}{\rho_0}\frac{\partial p'}{\partial z} - \frac{1}{\rho_0}g\rho' \qquad (6.25)$$

Looking for a wave solution of the form $\exp[i(kx + ly + mz)]$ results in the dispersion relation

$$\omega^2 = N^2 \frac{k^2 + l^2}{k^2 + l^2 + m^2} \qquad (6.26)$$

It can immediately be seen that $\omega \leq N$. It can also be seen that frequency does not depend on wave number magnitude but on direction $\omega = \pm N \cos\theta$, where θ is the angle of the wave number vector with respect to the vertical. The lower boundary of the frequency of internal waves is the Coriolis frequency, similarly to the case of shallow-water waves. Thus, $f \leq \omega \leq N$.

While the dynamics of internal waves is complex (Miropolsky, 1981), internal waves in a two-layer medium are very similar to surface waves. In a two-layer system the dispersion relation can be derived from the equation of motion, basically as in the case of shallow-water waves above. Consider a two-layer water body with layer thicknesses and densities of h_n and ρ_n ($n = 1, 2$), respectively. Assuming $\Delta\rho = \rho_2 - \rho_1 \ll \rho_2$, the dispersion relation becomes

$$(\omega^2 - gk \tanh[k(h_1 + h_2)])\left[\omega^2 - \frac{gk\Delta\rho}{\rho_2 \coth kh_2 + \rho_1 \coth kh_1}\right] = 0 \qquad (6.27)$$

The first factor gives gravity waves for the whole water body at total depth $h_1 + h_2$; they are not influenced by stratification. The latter factor represents internal waves. For deepwater internal waves, the hyperbolic cotangents are equal to 1, and we have the dispersion relation and phase speed as

$$\omega^2 = \left(\frac{\Delta\rho}{\rho_2 + \rho_1}\right)gk, \quad C^2 = \frac{g\Delta\rho}{k(\rho_2 + \rho_1)} \tag{6.28}$$

These waves are dispersive and have a group speed equal to half the phase speed, and because the density difference between the layers is small, the phase speed is much smaller than in the case of surface gravity waves. For $\Delta\rho/\rho \sim 2 \times 10^{-3}$, the ratio of phase speeds is about 1:30.

As long as the amplitudes of internal waves are small, the above linear theory works well (e.g., Cushman-Roisin, 1994). The maximum horizontal displacement is U/ω, whereas the horizontal wavelength is $2\pi/(k^2 + l^2)^{1/2}$. Taking the scale for vertical wavelength as sea depth H, we have the small-amplitude requirement as

$$Fr = \frac{U}{NH} \ll 1 \tag{6.29}$$

If this inequality is not satisfied, non-linear effects arise. Non-linear interactions spread energy in a continuous spectrum, and the result may lead to wave breaking in similar fashion to that of surface waves.

In the analysis of non-linear, dispersive internal waves the Korteweg–de Vries equation based on perturbation theory is normally employed. It is valid for small-amplitude long waves: the assumptions are $H/\lambda \ll 1$ and $a/H \ll 1$, where H is sea depth, λ is typical wavelength, and a is wave amplitude. The equation reads

$$\frac{\partial \eta}{\partial t} + c\frac{\partial \eta}{\partial x} + \alpha\eta\frac{\partial \eta}{\partial x} + \beta\frac{\partial^3 \eta}{\partial x^3} = 0 \tag{6.30}$$

where $\eta(x,t)$ is pycnocline depth; and α, β, and c are parameters describing non-linearity, dispersion, and phase speed. These parameters can be determined on the basis of the vertical distribution of density. The phase speed of long internal waves (c) and changes in the vertical position of the pycnocline $\phi(z)$ are obtained from the equation

$$\frac{d^2}{dz^2}\phi + \frac{N^2(z)}{c^2}\phi = 0, \quad \phi(-H) = \phi(0) = 0 \tag{6.31}$$

based on the solution to the eigenvalue problem. According to oceanic investigations the linear phase speed parameters (β, c) are influenced by sea depth while the non-linearity parameter α is influenced by stratification.

6.5.3 Observations

Measurements show fluctuations in current velocities and motions of isotherms on the following timescales. Periods of 1–30 minutes have been observed in Kiel Bight, while periods of 5–6 hours have been reported from the Gulf of Finland, Arkona

Basin, and Darss Sill (Krauss and Magaard, 1961; Hollan, 1966). Resulting temperature variations can be quite large (Figure 6.14, bottom). According to Alenius (1985), amplitudes are around one meter for waves with periods more than 2 hours.

The largest changes are found at the pycnocline, where the Brunt–Väisälä frequency reaches its maximum. When there is both a thermocline and a halocline, two distinct internal wave structures may be observed in the resulting three-layer structure.

Internal waves have a role in the vertical transfer of kinetic energy, because in many cases their group velocity substantially propagates in the vertical direction. Below the pycnocline, however, turbulence and other agents of vertical energy and momentum transfer seem to be occasional (intermittent), and it is likely that internal waves are able to transfer the energy to feed lower layer turbulence. In consequence, internal waves cause modifications to the distributions of temperature and salinity.

Figure 6.15 shows the vertical distribution of Brunt–Väisälä frequency based on salinity and temperature observations. The largest values ($0.05\,s^{-1}$) are found close to the pycnocline, while in the upper layer (top 60 m) the frequency is $0.004\,s^{-1}$. In shallower regions the minima are around $0.0012\,s^{-1}$.

In the Baltic Sea several kinds of internal waves exist because forcing factors and bottom topography vary to a large degree. Non-linear internal waves are thought to be physically the most important since they are the most effective at modifying watermasses.

Wave phenomena in the Baltic Sea have now been presented based on linear wave theory and observation material. The theory is well established and elegant, but real data from the Baltic Sea are limited. The seiches of Baltic Sea basins are common

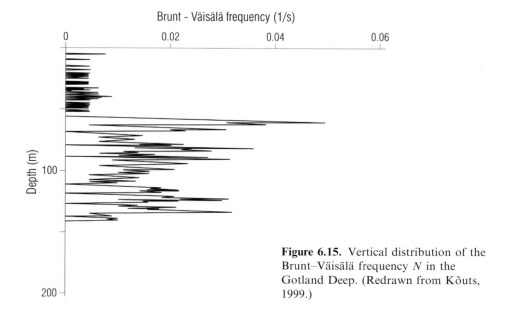

Figure 6.15. Vertical distribution of the Brunt–Väisälä frequency N in the Gotland Deep. (Redrawn from Kõuts, 1999.)

and well documented. Different periods exist due to its complex morphology. Topographic waves have been studied much but the results are still quite scattered. Tides are weak. Wind-generated surface waves are very important for practical applications, and forecast systems are already operational. Chapter 6 together with Chapter 5 constitute the dynamics of the Baltic Sea, and thus classical oceanography has been completely covered for the Baltic Sea. There are in addition particular fields of marine research that need to be studied to get a full picture of the physics of the basin. Chapter 7 presents the sea ice of the Baltic Sea, and its evolution from its interaction with the water body beneath and that forced by the atmosphere.

7

The ice of the Baltic Sea

Erkki Palosuo (1912–2007) commenced his sea ice science career as an ice reconnaissance pilot in the Finnish Air Force during World War II. He worked in the Ice Service of the Finnish Institute of Marine Research during 1947–1972 and presented his Ph.D thesis about severe ice seasons in the Baltic Sea, utilizing material collected during those flights (Palosuo, 1953). Thereafter, he served five years as a professor in geophysics at the University of Helsinki. He was a pioneering scientist in many branches of Baltic Sea ice research, and his best-known study fields were the crystal structure and salinity of ice, the physics of landfast ice, pressure ridges, and ice climatology. (Courtesy of Palosuo family archives.)

7.1 ICE SEASON AND ITS CLIMATOLOGY

Sea ice forms in the Baltic Sea annually, can be found there for seven months of the year, and has a very important role in the annual course of physical and ecological conditions in the basin. Therefore, a treatise on the Baltic Sea needs a chapter on ice, a topic that is usually less familiar to physical oceanographers. Section 7.1 introduces the ice season, and the following sections then progress from small to large scales. Section 7.2, for example, contains fine-scale questions and shows that brackish ice is structurally like normal sea ice; it also presents ice crystal structure, phase diagram,

Figure 7.1. Drift ice fields in the Baltic Sea make up its winter landscape: a field of ice floes with leads and pressure ridges as the most striking features.

and sediments. Section 7.3 focuses on the small scale considering the growth and melting of sea ice, using analytical and numerical models. Section 7.4 looks at meso-scale and large-scale physics and introduces drift ice, its material structure, kinematics, and dynamics. Sea ice is a peculiar, granular, compressible, non-linear medium. Section 7.5 presents full sea ice models (i.e., basin-scale systems for the growth, drift, and decay of ice with vertical thermodynamics and horizontal dynamics).

The Baltic Sea is a part of the seasonal sea ice zone of the World Ocean, where climate variations show up drastically in the ice conditions. The southern edge of this zone borders sub-Arctic semi-enclosed basins from the Baltic Sea to the Sea of Okhotsk and the Bohai Sea, to the Bering Sea, and to Hudson Bay and the Gulf of St. Lawrence.

Sea ice—or brackish ice—forms every year in the Baltic Sea. It is one of the key physical characteristics of this basin and has a significant role in the North European climate system (Figure 7.1). Ice acts as a buffer to surface water temperature right up to freezing point. The impact of ice melting is significant for surface water salinity as it kick-starts stable spring stratification which has further consequences for the spring algal bloom, but the influence of the salt flux as a result of water freezing is too weak to force convection across the halocline. The presence of ice has a pronounced influence on the transfer of momentum and light to the water and on the air–sea exchange of water and heat. Ice and snow have high albedos and cold surfaces, which reduces the amount of solar radiation getting into the ice and water and the turbulent exchange of sensible and latent heat. Atmospheric fallout accumulates in the ice sheet and is released during the short melting phase. Ice also has a remarkable role in human living conditions in the sea area, in particular regarding sea traffic. At the time of sailboats, winter shipping was cut off in the ice season, while presently a fleet of 20–25 icebreakers ensures there is a workable marine transportation system giving access to all the main harbors of the Baltic Sea.

In the fall the Baltic Sea cools due to radiational and turbulent heat losses, and the surface layer temperature goes down at the rate of $3°C–4°C$ per month (see Figure 3.14). The ice season begins on average in the middle of November on the northern coast of the Bay of Bothnia, and the freezing front then progresses southward (SMHI and FIMR, 1982). During the 20th century, for Kemi the earliest, average, and latest freezing dates were October 6, November 10, and December 23, the range covering as much as 2.5 months (Jevrejeva et al., 2002). In the central basins the freezing date is much later than coastal sites due to the large heat content in the water body (SMHI and FIMR, 1982). The Bay of Bothnia freezes over on average in mid-January, and in normal winters the Sea of Bothnia, the Gulf of Finland, and the Gulf of Riga freeze over one month later. In mild winters only the Bay of Bothnia and the eastern part of the Gulf of Finland freeze over. The three mildest winters in the last 100 years were 2008, 1989, and 1961.

The freezing point (T_f) of Baltic Sea brackish water is just a little below 0°C, while the temperature of maximum density (T_D) is 2°C–3°C—see Equations (3.3) and (3.4). At low salinities—see Equation (3.4)—adding 1‰ to salinity lowers the freezing point by 0.055°C and the temperature of maximum density by 0.215°C; for $S = 6‰$, we have $T_f = -0.33°C$ and $T_D = 2.7°C$. A boreal brackish water basin homogeneous in salinity would behave as a dimictic[1] freshwater lake with fall and spring turnover of the watermass and an inverse winter thermocline.

Then, compared with freshwater basins, one would expect that the freezing date would be delayed due to the lower temperatures of freezing point and maximum density. However, an additional significant feature is embedded in the equation of state of seawater. The density maximum is weaker in brackish water than in freshwater, and therefore shallow winter thermoclines break down easily in brackish water basins resulting in a more or less homogeneous upper layer. Thus, the existence of a density maximum above freezing point does not influence the freezing date to any great degree, with the possible exception of small bays along the coast.

Example. Let us examine the freezing date problem for a homogeneous brackish basin using the analytic models discussed in Section 4.4.1—see Equation (4.34) and the examples below it. Take the fall atmospheric cooling rate as 5°C/month. Then, for water salinity of 6‰, the timing of the maximum density would be eight days later than freshwater lakes. On the other hand, with full mixing until the freezing day, freezing point depression would give two days for the delay. Since the truth is somewhere between these two extremes, we may estimate that the delay would be around five days. ∎

Salinity in the Baltic Sea is, however, stratified as shown in Chapter 3. In fall and winter there is a homogeneous upper layer, a halocline at a depth of between 40 m and 80 m, and stratification in the lower layer is continuous. Fall mixing of the cooling watermass just reaches the halocline. Consequently, stratified deep basins cool faster than corresponding freshwater basins. The analytic slab model shown in Section 4.4.1 gives a simple answer. If lower layer thickness were around 30 m–40 m, the stratified basin would freeze one month earlier than a corresponding non-stratified basin.

---

[1] Turnover of the water body twice a year (spring and fall).

222 The ice of the Baltic Sea [Ch. 7

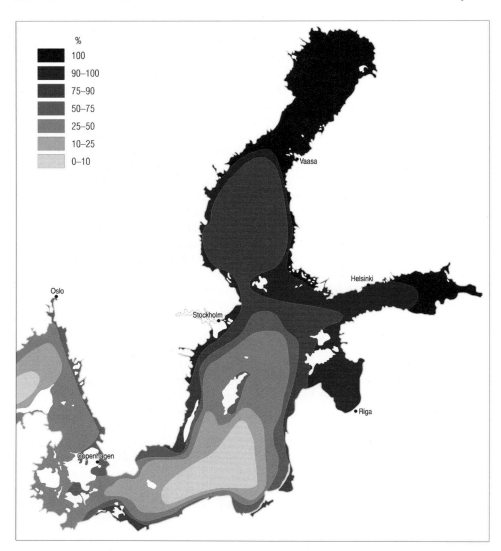

Figure 7.2. The probability of annual ice occurrence in the Baltic Sea. (Redrawn from SMHI and FIMR, 1982.)

Thus, the main reasons the Baltic Sea freezes are that the halocline limits the convection and renewal of the heat content of the upper layer from the lower layer or the North Sea is too slow to compensate for atmospheric cooling and radiation losses.

Annual ice extent is largest between mid-February and mid-March. On average, the ice-covered area is then 45% of the total area of the Baltic Sea, and ranges from 12.5% to 100% in different years (Figure 7.2). In normal winters the ice edge crosses the Gotland Sea in the north at the latitude of Stockholm and then turns south along the mouths of the Gulf of Finland and the Gulf of Riga. Farther south, ice occurs only in shallow coastal areas in normal winters, such as Curonian Lagoon, Vistula Lagoon, Puck Bay, Szczecin Lagoon, and parts of the Danish Straits.

The Gotland Sea takes a long time to freeze because the halocline is at great depth, 60 m–80 m. Its southern part is the last spot to freeze and only does so in very severe winters. The most recent freeze-over of the Baltic Sea dates to 1947 (complete freeze-overs took place in 1940 and 1942 as well, and in 1987 the coverage was 96% of the area of the Baltic Sea). Even in these very severe winters the ice in the Gotland Sea is thin and breaks easily. An excellent presentation of the ice conditions in very severe winters is given by Palosuo (1953).

On coastal and archipelago areas the ice appears as *(land)fast ice* and elsewhere as *drift ice* (Figure 7.3). Fast ice is a solid and even sheet of ice and remains immobile apart from very early and late in the ice season. On average, the outer boundary of the fast ice zone lies at the 10 m isobath (Leppäranta, 1981b). Freezing, growth of fast ice, and melting of fast ice closely follow changes in air temperature, as does ice in the lakes in Finland. The maximum annual thickness of fast ice in Kemi is 50 cm–110 cm, and the record value is from Tornio, 122 cm in winter 1985.

Beyond the fast ice boundary there is *drift ice*. A "drift ice landscape" consists of leads and ice floes with ridges, hummocks, and other variable morphological characteristics. *Ice types* are defined according to practical shipping activities in ice-covered waters (WMO, 1970). They are based on how ice appears to an observer on a ship or in an aircraft. The formation mechanism, aging, and deformation have an effect on this appearance, which furthermore contains information of ice thickness, seldom known in direct measurements.

The melting of ice begins when the radiation balance turns positive. In the Baltic Sea springtime, turbulent air–ice fluxes are small as is the heat flux from the water body to the ice bottom. The radiation balance is $Q_R = (1 - \alpha)Q_s + Q_{La} - Q_{Lo}$ (see Section 4.1). Mean net longwave radiation ($Q_{La} - Q_{Lo}$) is at a level of $-50\,\mathrm{W\,m^{-2}}$, while daily shortwave radiation Q_s increases from almost zero in January to $200\,\mathrm{W\,m^{-2}}$ in May. By March solar radiation peaks at $500\,\mathrm{W\,m^{-2}}$ on clear days, and this is large enough to start melting. As a result of liquid water formation and recrystallization of snow grains the albedo starts to decrease, and as incoming radiation increases day by day the melting rate accelerates.

Melting begins in the south in early March at the same time as new ice is forming in the north. Melting progresses in the central basins due to the absorption of solar radiation in leads and due to the decrease in ice compactness, and somewhat later melting starts from the shoreline due to the shallow sea depth and the proximity of warm land. In the 20th century, the mean date of ice break-up was May 21 in Kemi, with the extremes being April 16 and June 27 (Jevrejeva et al., 2002). Thus, the length of the melting season is 45–60 days in Kemi and the average melting rate is 1 cm–2 cm per day. Melting is an accelerating process in which the melt rate increases as the melting season progresses, and as the porosity of ice has achieved 30%–50% the ice breaks in the surface layer and the resulting remnants quickly melt.

224 The ice of the Baltic Sea [Ch. 7]

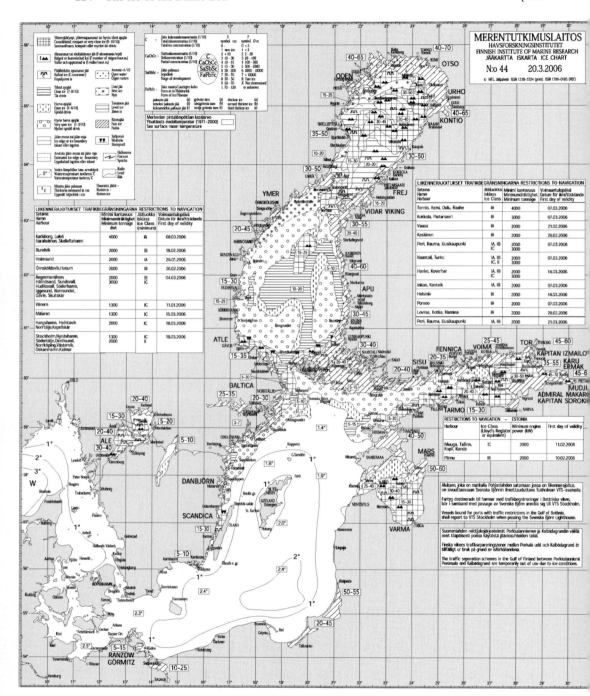

Figure 7.3. Baltic Sea ice chart on March 20, 2006. (Source: Finnish Institute of Marine Research.)

The outer fast ice boundary with large and grounded ridges is the last outpost of ice at the end of the ice season. Remnants of old ridges are seen in June in some years in the Bay of Bothnia. In 1867, the last very cold, famine year in Finland, ice was reported on July 17.

Ice cover is a stiff, thin lid between the atmosphere and the sea. It has a direct influence on the exchange of momentum, heat, moisture, and matter between the atmosphere and the sea. Because of surface roughness, in free-drifting ice fields the sea may even receive more momentum from the wind than in open water conditions. But due to the small size of Baltic Sea basins, much work needs to be done to overcome internal friction and momentum transfer to the water body beneath the ice is then small. Ice and snow on ice are good insulators, and therefore the heat exchange between the atmosphere and ocean weakens as ice grows. Ice growth slows down and the annual ice thickness does not grow thicker than about 1 m. Precipitation that accumulates on the ice is released to the water body during the short melting season. Because the salinity of ice is low, the surface layer receives a major freshwater flux as the ice melts. This shows up in the salinity climatology.

7.2 SMALL-SCALE ICE STRUCTURE AND PROPERTIES

7.2.1 Crystal structure

Ice crystals formed of seawater are thin crystal plates or needles consisting of water molecules. They are optically uniaxial, with the optical axis or the c-axis[2] perpendicular to the plate plane. Overlain together the crystals form *macrocrystals*, which are optically like single crystals but are structurally multiple crystals. In the geophysics of sea ice, the term *ice crystal* normally refers to these macrocrystals. The scale of their size is 10^{-4} m–10^{-1} m.

Sea ice (and brackish ice) crystals are in size and shape much like freshwater ice crystals in lakes and rivers, and external conditions are determining factors for their size and shape. But there are two important differences caused by dissolved substances in seawater (Weeks, 1998). In sea ice the crystal boundaries are irregular, jagged, and inside macrocrystals between the single crystal plates there are brine inclusions (Figure 7.4, see color section). Transition between the freshwater ice structure and the sea ice structure is in natural conditions at 1‰–2‰ of the salinity of the parent water. This has been shown in laboratory experiments, and under natural conditions in the Baltic Sea (Palosuo, 1961; Weeks *et al.*, 1990; Kawamura *et al.*, 2001). Consequently, apart from river mouths, the ice formed of brackish Baltic Sea water is similar to sea ice in general. It is low-salinity first-year sea ice, much like the ice in the Sea of Azov and in Ob Bay.

---

[2] Light passes unchanged in the direction of the optical axis, as through glass, while birefraction resulting in primary and secondary rays takes place in other directions. These rays produce interference colors, which are utilized in the analysis of crystal structure (Figure 7.4, see color section).

Ice formation in the Baltic Sea is based on three different mechanisms, which produce congelation ice, frazil ice, and superimposed ice. The most common form is *congelation ice*, which grows down from the ice–water interface (Weeks, 1998). The resulting crystals are columnar, of diameter 0.5 cm–5 cm, and of height 5 cm–50 cm (Figure 7.4, see color section). At the top the crystals are small; their *c*-axes are vertical if the ice formed under calm conditions, or randomly oriented if the ice formed under disturbed conditions (water flow, waves, wind, snowfall). Across a transition layer of 10 cm–20 cm, the crystals grow larger and their axes turn horizontally. The size of ice crystals increases with depth. Congelation ice is the governing ice type in lakes of the Baltic Sea drainage basin and in the Arctic Ocean as well.

The bottom layer of ice, which is about 1 cm thick, consists of separate crystal plates penetrating the saline water. This layer, called the *skeleton layer*, thus consists of seawater and crystal platelets and has no significant strength. In normal sea ice the skeleton layer is thicker (up to 5 cm) while freshwater ice has no such layer.

Frazil ice forms in open water areas. The crystals are small (1 mm or less) and granular; they may drift free in turbulent flow and attach later to the bottom of existing ice or join together into a solid sheet at the surface. In shallow and well-mixed waters frazil ice may also attach to the sea bottom to form *anchor ice*. There is not much information on the amount of frazil ice in the Baltic Sea, but it is known to exist there. In rivers frazil ice is generated in fast-flowing streams, which have an open surface for most of the winter. In Antarctic seas, frazil ice is the dominant ice type.

Superimposed ice is a more complicated form of ice.[3] As the term implies, superimposed ice forms on top of the ice in meltwater or rainwater pools or in slush layers generated by snow and liquid water from flooding, liquid precipitation, or snow meltwater. In the Baltic Sea the most common variant of superimposed ice is frozen slush, also called *snow ice*. Slush formation in flooded seawater and snow is possible when the weight of the snow forces the ice sheet below sea-level; that is, when

$$\frac{h_s}{h_i} > \frac{\rho_w - \rho_i}{\rho_s} \tag{7.1}$$

where h_s and h_i are the snow and ice thicknesses; and ρ_w, ρ_s, and ρ_i are the water, snow, and ice densities. Since $(\rho_w - \rho_i)/\rho_s \approx 1/3$, the thickness of snow needs to be at least one-third of the thickness of ice for flooding to occur. Sea ice is normally porous so that when forced beneath the sea-level seawater can filter through.

By means of melt–freeze cycles all snow can in principle change into superimposed ice, and with liquid precipitation the only limiting factor is the amount that accumulates. When a superimposed layer is produced purely from atmospheric precipitation, it is called *meteoric ice*. Snow ice formed by flooding is therefore partly (some 50%) meteoric ice. Snow ice is always granular and fine-grained like frazil ice, with a crystal size of 1 mm–5 mm (Figure 7.4, see color section).

The thickness of superimposed ice is monitored by means of stakes driven into the ice at the beginning of the ice season (Palosuo, 1965). This procedure is part of the

---

[3] In glaciology superimposed ice is the ice formed in the surface layer from liquid water or slush.

Table 7.1. Statistics on annual maximum ice thickness and the proportion of superimposed ice during 1961–1990 for Baltic Sea landfast ice (in centimeters) (Seinä and Peltola, 1991).

Station	Total ice thickness					Snow ice thickness					
	Min	Mean	Max	SD	n	Min	Mean	Max	SD	n	
Kemi	58	76	111	11	30	4	25	43	12	23	GB, north
Valassaaret	18	51	80	16	25	0	21	40	14	12	GB, central
Kumlinge	0	37	75	19	17	0	7	20	5	17	GB, south
Utö	0	21	69	22	19	0	8	50	17	10	GS, north
Kotka	16	50	90	18	26	4	18	35	9	19	GF, east

GB = Gulf of Bothnia, GS = Gotland Sea, GF = Gulf of Finland, SD = standard deviation.

routine observation system in Finland. The origin of superimposed ice is not clear from thickness measurements, but the oxygen isotope ratio gives the proportion of meteoric ice (Kawamura *et al.*, 2001; Granskog *et al.*, 2004). The volume of superimposed ice is 10%–50% in the Baltic Sea ice sheet. Mean annual maximum ice thickness increases toward the north and east (Table 7.1). Variance is largest in the south where the sea is ice-free in mild winters. Landfast ice in the Baltic Sea usually has two layers—congelation ice and snow ice, like the lake ice in Finland—but drift ice has a more complicated stratigraphy due to mechanical processes and frazil ice layers muddying the picture.

7.2.2 Ice salinity

When saline water freezes, ice crystals form from water molecules (i.e., freezing tends to separate dissolved substances out of the solid phase of water). However, due to constitutional supercooling, in the molecular diffusion of salt and heat in a liquid, a cellular ice–water interface forms that can close liquid brine pockets between crystal platelets (Weeks, 1998). In freshwater ice, in contrast, the ice–water interface is planar. As discussed above, transition in the structure of the interface is within 1‰–2‰ of the salinity of parent water (Palosuo, 1961; Weeks *et al.*, 1990; Kawamura *et al.*, 2001). These brine pockets also serve as a habitat for ice algae. In cold ice, solid salt crystals form from the brine.

In the Baltic Sea the salinity of new ice is a fraction μ—the segregation coefficient—of the salinity of parent water: $S_i = \mu S_w$; the numerical value is $\mu \approx 0.25$–0.4 (i.e., $S_i \approx 1‰$–$3‰$). The segregation coefficient depends on the speed of growth, so that a higher growth rate captures more salts. Since the speed of growth in general decreases when ice gets thicker, the top layer of congelation ice has the maximum salinity. In mid-winter, salinity decreases a little due to brine pocket expulsion and

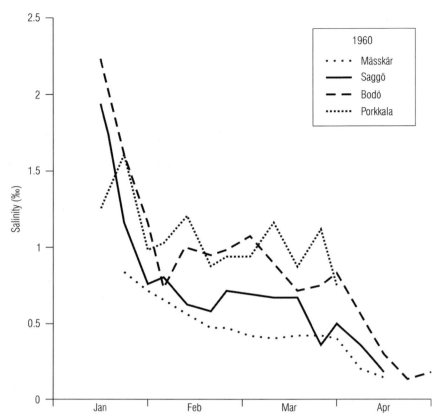

Figure 7.5. Evolution of mean vertical salinity of ice in the Finnish coast of the Baltic Sea. The raw data are from Palosuo (1963).

consequent drainage, but it stays at the fraction of 0.2–0.3 of water salinity (Figure 7.5).

The vertical salinity profile develops as a result of entrapment and advection of brine into a C-shape (maximums at the top and bottom, minimum in the middle) in midwinter, changing into an I-shape as the ice becomes warmer. In spring in the melting season the brine pockets expand to form a drainage network, and the salinity profile turns to an inverse C-shape (minimums on the top and bottom). Later on, salinity drops and the residual level is of the order of 0.1‰ in the Gulf of Finland.

The composition of sea ice at a given temperature can be read from a phase diagram (Figure 7.6). This diagram shows the fractions of pure ice, liquid brine, and solid salt crystals. As temperature decreases, salts start to crystallize from the brine, each at its eutectic temperature point. The most important points in temperatures above $-25°C$ are sodium chloride ($-22.9°C$) and magnesium sulfate ($-8.2°C$). The influence salinity has on the properties of sea ice is principally due to brine volume. Ice and brine pockets form a porous binary system, and the properties of sea ice are

Sec. 7.2] Small-scale Ice structure and properties 229

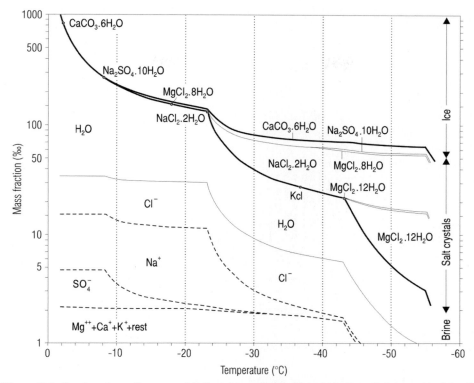

Figure 7.6. Sea ice phase diagram with the proportions of ice, main ions, brine, and solid salts as a function of temperature. The chemical salt crystal symbols show the eutectic points of the main salts. (Redrawn from Assur, 1958.)

determined from its components. In very cold ice, solid salt crystals scatter light and therefore influence the optical properties of sea ice.

The salinity of brine must always correspond to its freezing point at the ambient temperature, $S_b^{-1}(T) = T_f$, and thus brine salinity and consequently brine volume change with temperature. Warm sea ice, approaching $-2°C$, may have a large brine volume, 10%–20%, while in cold sea ice (below $-20°C$) the brine volume is less than 1%. To estimate brine volume, the bulk temperature, salinity, and density of sea ice need to be known. The relative brine volume is

$$\nu_b = \frac{\rho_i S_i}{\rho_b S_b(T)(1+k)} \tag{7.2}$$

where $\rho_b \approx 1 + 0.8 \times S_b$ is the density of brine; and k is the ratio of the mass of solid salts to the mass of salts in the brine. Equation (7.2) is a good approximation with $k = 0$ down to the eutectic point of sodium chloride. Corrections according to the amount of solid salt crystals can be directly made from the phase diagram. Phase equations for exact calculations are presented in Cox and Weeks (1983) for

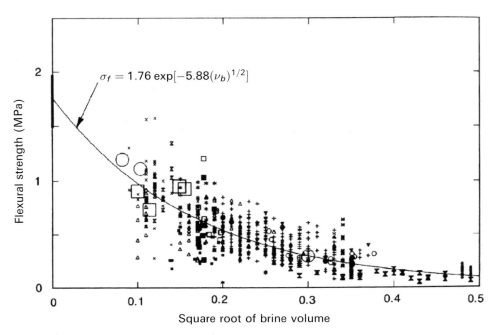

Figure 7.7. Flexural strength of sea ice as a function of brine content (ν_b). (Weeks, 1998, printed with permission from the Helsinki University Press.)

$T \leq -2°C$ and Leppäranta and Manninen (1988) for $-2°C \leq T \leq 0°C$ (see the Appendix).

As a result of salinity the properties of sea ice are different from those of freshwater ice. The difference is largest regarding electromagnetic properties and strength. Electromagnetic properties are normally not so relevant as such but they can be used indirectly to observe other ice properties. For example, microwave backscatter depends on the dielectric constant, and the salinity of ice consequently shows up in radar-mapping of sea ice. Due to the presence of brine pockets, sea ice is porous, and therefore the strength of ice is less than that of freshwater ice. Flexural strength[4] is a good reference since there is a lot of *in situ* data about it (Figure 7.7).

The thermodynamics of sea ice differs qualitatively from that of freshwater ice (Schwerdtfeger, 1963). Along with temperature changes, the brine volume changes according to associated phase changes at brine pocket boundaries. This raises an interesting point: sea ice has no definite melting point but always there is melting involved when sea ice warms in order to dilute the brine. This is accounted for by heat capacity, which consequently becomes strongly dependent on temperature, especially in warm sea ice. Seawater, on the other hand, has a definite freezing point (as discussed in Section 3.2). The thermal conductivity of brine is a little less than that

---

[4] Flexural strength is often used as a reference for floating ice because *in situ* tests can be easily performed with the ice sheet under natural temperature conditions.

of ice, and thus the thermal conductivity of sea ice is slightly smaller than that of freshwater ice.

The optical properties of sea ice are influenced by salinity, and this applies to brackish ice as well (Arst et al., 2006). Brine pockets absorb light and there is little scattering. Salt crystals scatter light but their number becomes significant only at very low temperatures (rarely observed in Baltic Sea ice). All in all, because of salt inclusions brackish ice is less transparent than freshwater ice. Arst et al. (2006) obtained the optical thickness of ice in Santala Bay, Gulf of Finland as about 1 m, while in the water beneath the ice it was 2 m–4 m.

7.2.3 Sea ice impurities

The impurities of sea ice (other than salts from seawater) consist of gas bubbles, sediments, and biota. The volume fraction of gas bubbles is $\nu_a \sim 1\%$. This influences the density of ice ρ_i as

$$\rho_i \approx (1 - \nu_a)\rho_{i0} \qquad (7.3)$$

where $\rho_{i0} = 917\,\text{kg/m}^3$ (at $0°C$) is the density of pure ice. In this way the total volume of gas bubbles can be estimated from density measurements. The size of gas bubbles is in the range 0.1 mm–10 mm; they originate from seawater or the sea bottom and in the snow ice layer air inclusions exist trapped by the parent snow cover. They are similar to those in freshwater ice and are good scatterers of electromagnetic radiation. In particular, as a result of their size they scatter all wavelengths of light equally resulting in the gray or white appearance of ice with a large amount of gas bubbles.

Non-living particles in the sea ice sheet are called *sea ice sediments*, but solid salt crystals originating from seawater are excluded (Leppäranta et al., 1998b). They originate from suspended particles in the water body, sea bottom, or atmospheric fallout. Congelation ice growth tends to reject impurities, but frazil ice crystals are effective in harvesting particles that are drifting freely in the water. Anchor ice may rise up due to its buoyancy and join the existing ice pack with bottom sediments still attached. When snow ice forms from the flooding of seawater, particles are trapped in the ice sheet. Finally, atmospheric fallout accumulates on the ice during the whole ice season. Sediments may influence the properties of ice, and they may also be taken by drifting sea ice to other locations; this mechanism, in particular, allows long-distance pollutant transport.

A characteristic of sea ice is that it has its own biota; this also applies to brackish ice (Ikävalko, 1997). Brine pockets serve as biological habitats. They contain nutrients, and with light penetration favorable conditions for primary production exist. The algae found in the Baltic Sea are also found in polar seas. They are captured inside the ice during the freeze-up process, and their growth in brine pockets is primarily light-dependent. The most active layer is the skeleton layer at the ice bottom, which may become colored brown-green by the algae. Phytoplankton absorb light and therefore influence the transparency of sea ice (Arst et al., 2006).

7.3 GROWTH AND DECAY OF ICE

7.3.1 Growth of ice

The growth of sea ice is a small-scale, vertical process. Growth velocity may be several centimeters per day for new ice, but when the ice thickness is more than 10 cm it is at most about 1 cm/day (Table 7.2). This corresponds to a heat loss of 30 W m^{-2} from the ice. The first ice layer is primary ice, with a thickness of the order of millimeters in calm water and some centimeters in much disturbed water. Then, as explained in Section 7.2, the ice grows at the bottom as congelation ice or frazil ice and at the top as superimposed ice.

The thermodynamic growth of ice is a classical geophysical problem. Latent heat released in freezing is conducted through the ice to the atmosphere. The thicker the ice the slower the conduction, and hence the rate of ice growth decreases when the ice becomes thicker. At the time of melting the ice receives heat from the Sun, the atmosphere, or water, but solar radiation only has the capacity to cause internal melting.

Thermal conduction in sea ice follows the Fourier law

$$\frac{\partial}{\partial t}(\rho_i c_i T) = \frac{\partial}{\partial z}\left(\kappa_i \frac{\partial T}{\partial z} - Q_{s,i}\right) \qquad (7.4a)$$

where T is the temperature of ice; c_i is the specific heat of ice; κ_i is the thermal conductivity of ice; and $Q_{s,i}$ is the solar radiation level in the ice. Boundary conditions stem from continuity of the heat flow and phase changes in the free boundaries:

$$\text{Top:} \qquad T \leq 0°C, \qquad \kappa_i \frac{\partial T}{\partial z} = Q_0 + Q_\pi \qquad (7.4b)$$

$$\text{Bottom:} \qquad T = T_f, \qquad \rho_i L \frac{dh}{dt} = \kappa_i \frac{\partial T}{\partial z} - Q_w \qquad (7.4c)$$

where Q_0 is net heat flux at the surface; Q_π is phase change at the top boundary; h is ice thickness; L latent heat of freezing; and Q_w is the heat flux from water. The heat flux Q_0 is as in open water except that the albedo is much larger for ice and snow; in addition, the surface temperature becomes lower pushing turbulent fluxes to a lower level. Approximate analytical solutions to the ice growth problem are available, and the full system can be numerically solved. If forcing data are known, resulting changes in thickness will be realistically given.

A simple, analytic approach to the ice growth problem is Zubov's (1945) model, which is an improvement over the classical Stefan's law (Stefan, 1891; Leppäranta, 1993). Zubov's basic assumptions are (i) thermal inertia is ignored, (ii) solar radiation is ignored, (iii) oceanic heat flux is ignored, and (iv) air temperature is known, $T_a = T_a(t)$. The heat exchange between the atmosphere and the ice is represented by the linear formula introduced in Equation (4.30). These assumptions are reasonable in the ice growth season; however, snow cover introduces difficulties since the thickness and quality of snow undergo change and necessitate a coupled snow and ice model for very good ice thickness results.

Table 7.2. Ice and snow thickness (mean ± standard deviation) in 1961–1990 in Kemi (65°41'N, 24°31'E) and Utö (59°47'N, 21°22'E). When the number of ice years is 30, this means ice has occurred every winter.

Kemi	Ice thickness (cm)	Snow thickness (cm)	Ice years
November 1	3.0 ± ×	0.0 ± ×	1
November 11	9.0 ± 6.3	0.0 ± 0.0	5
November 21	13.0 ± 8.3	1.3 ± 3.4	13
December 1	17.0 ± 8.9	3.7 ± 7.6	23
December 11	23.0 ± 8.8	4.2 ± 7.5	27
December 21	30.6 ± 8.8	5.9 ± 7.0	30
January 1	39.3 ± 8.2	8.7 ± 8.0	30
January 11	45.8 ± 9.2	12.7 ± 8.1	30
January 21	51.5 ± 9.6	16.1 ± 9.2	30
February 1	57.0 ± 10.6	19.5 ± 9.7	30
February 11	60.9 ± 10.5	21.9 ± 10.1	30
February 21	64.0 ± 10.5	24.8 ± 9.8	30
March 1	66.5 ± 11.2	25.9 ± 10.5	30
March 11	68.3 ± 11.0	26.2 ± 10.7	30
March 21	69.9 ± 11.1	26.2 ± 11.7	30
April 1	71.5 ± 10.1	23.6 ± 12.8	30
April 11	72.0 ± 10.6	16.3 ± 10.4	30
April 21	69.4 ± 13.1	7.2 ± 7.7	30
May 1	59.0 ± 18.0	4.4 ± 10.4	28
May 11	51.4 ± 13.1	2.9 ± 6.1	14
May 21	40.0 ± ×	0.0 ± 0.0	1

Utö	Ice thickness (cm)	Snow thickness (cm)	Ice years
January 1	× ± ×	× ± ×	1
January 11	11.5 ± 9.2	0.5 ± 0.7	4
January 21	22.7 ± 15.2	2.3 ± 1.5	6
February 1	22.4 ± 12.4	4.6 ± 2.9	8
February 11	23.4 ± 15.9	2.4 ± 2.6	14
February 21	30.1 ± 15.6	5.0 ± 5.4	15
March 1	28.4 ± 18.0	4.1 ± 4.9	17
March 11	30.4 ± 17.9	5.0 ± 5.7	18
March 21	37.6 ± 17.2	6.4 ± 4.7	16
April 1	34.8 ± 16.4	4.3 ± 5.9	15
April 11	33.2 ± 16.6	2.2 ± 2.3	8
April 21	31.2 ± 12.7	0.4 ± 0.9	5
May 1	28.0 ± ×	0.0 ± ×	1

× = no estimate.
Source: Finnish Institute of Marine Research, Kalliosaari and Seinä (1987) and Seinä and Kalliosaari (1991).

Zubov's model is based on transport of the heat released in freezing through the ice to the atmospheric surface layer. For bare ice, using the linear air–ice heat flux—Equation (4.30)—we have by the continuity of heat flow

$$\rho_i L \frac{dh}{dt} = \kappa_i \frac{T_f - T_0}{h} = -k_0 + k_1(T_0 - T_a) \geq 0 \qquad (7.5)$$

where T_0 is surface temperature; and k_0 and k_1 are parameters of the linear air–sea heat exchange law—see Equations (4.30) and (4.32). In growth conditions the terms must be positive, as indicated by the inequality on the right. The differential equation after elimination and solution is

$$\rho_i L \frac{dh}{dt} = \kappa_i \frac{T_f - T_a - k_0/k_1}{h + d} \qquad (7.6a)$$

$$h = \sqrt{(h_0^2 + 2h_0 d) + a(S + S_0) + d^2} - d \qquad (7.6b)$$

where h_0 is initial thickness; $a = 2\kappa_i/\rho_i L \approx 11\,\mathrm{cm}(°C\,\mathrm{da})^{-1}$; S is freezing degree-days; S_0 is the time integral of the ratio k_0/k_1; and $d = \kappa/k_1 \approx 10\,\mathrm{cm}$ is the insulation efficiency of the atmospheric surface layer. For example, 100 days at a temperature of $-10°C$ gives $S = 1,000°C\,\mathrm{da}^{-1}$. Then, if $k_0/k_1 = 2°C$, $d = 10\,\mathrm{cm}$, and $h_0 = 0$, we have $h = 105\,\mathrm{cm}$. The square root law accurately describes how ice insulates itself from a cold atmosphere during the growth process. Letting $S_0 \to 0$ and $d \to 0$, Stefan's law is obtained.

The main problem with Zubov's model is the lack of snow. A snow layer increases insulation but may also add to ice thickness via snow ice formation. The insulation effect can be examined by adding the heat flow through the snow, $\kappa_s(T_0 - T_s)/h_s$, to Equations (7.5). Here T_s is the surface temperature of snow, κ_s is the thermal conductivity of snow, and h_s is the thickness of snow. The solution can be found using Equation (7.6b) but with $h + d$ replaced by $h + (\kappa_i/\kappa_s)h_s + d$. Since the thickness and thermal conductivity of snow depend on time, an analytic solution is no longer possible. As a result of the effective insulation by snow, ice thickness becomes about half the thickness of bare ice under similar atmospheric forcing conditions (Leppäranta, 1993).

The growth of superimposed ice can be modeled along the same lines. But there are two major differences. Because the ice grows on top, the insulation effect is weaker. When the ice forms from slush, the latent heat release is less since the slush already contains ice crystals ($\sim 50\%$). The most important means of slush formation is flooding in the Baltic Sea. If snowfall is large enough relative to ice thickness—see Equation (7.1)—flooding may take place. In an extreme situation the rate of snowfall is just enough for the continuous growth of snow ice but not for congelation ice to form. In this case snow ice grows according to a modified Zubov's law: coefficient a in Equation (7.6b) is reduced by a factor of 0.7, and thus snow ice grows a little slower than bare ice (Leppäranta, 1993).

Empirical formulas are often based on the square root relationship $h = \sqrt{a^*(S - S_0) + d^{*2}} - d^*$, $\frac{1}{2}a < a^* < a$, where S_0 is the number of freezing degree-days used for cooling surface water and a^* and d^* are empirically corrected parameters a and d.

Frazil ice growth can be modeled in a simple way. With the water surface ice-free and assuming that temperature is at the freezing point, we have

$$\frac{dh_F}{dt} = -\frac{1}{\rho_i L} Q_0 \qquad (7.7)$$

where the "thickness" of frazil ice must be understood as the mean volume of frazil ice crystals per unit surface area. The heat flux Q_0 can be large when the atmospheric temperature is very low. For $Q_0 = -100\,\mathrm{W\,m^{-2}}$, the growth rate will be 2.8 cm per day.

7.3.2 Melting of ice

At the ice bottom, temperature is always at freezing point and melting only takes place when the heat flux from seawater is greater than the conductive heat flux into the ice. In spring the upper boundary and then the ice interior warm up. When the ice has become isothermal, there is no longer any conductive flux and the melt rate becomes more or less independent of ice thickness. Solar radiation has the leading role in the melting process as it turns the surface heat flux positive and directly melts interior ice. What makes the study of ice melting very difficult is that the key parameters—albedo and the light attenuation coefficient—undergo drastic variations during the melting season.

Melting of ice can be examined as a binary process:

$$\rho_i L \left(1 - \frac{n}{h}\right) \frac{dh}{dt} = -(Q_0 + Q_w) \leq 0 \qquad (7.8a)$$

$$\rho_i L \frac{dn}{dt} = (1-\alpha)\gamma[Q_s(0) - Q_s(h)] \geq 0 \qquad (7.8b)$$

where n is the accumulated internal melt (consequently n/h is the porosity of ice); and γ is the fraction of solar radiation penetrating the surface—see Equation (4.19). The first equation represents the thickness of ice (distance between top and bottom surfaces) and the second equation internal melting; net ice volume per unit area then becomes $h - n$. At the top surface, snow melts first followed by ice. Snow protects the ice cover due to its high albedo and small optical depth. Consequently, internal deterioration only starts up after snow melt.

Ice breaks up at time t_m, obtained from

$$t_m = \min\left\{ t_{m_1}, t_{m_2} \,\bigg|\, h(t_{m_1}) = 0, \frac{n(t_{m_2})}{h(t_{m_2})} = \nu^* \right\} \qquad (7.9)$$

where ν^* is the porosity through which ice loses its strength, $\nu^* \approx 0.4$. Time t_{m_1} is the theoretical time of ice melting at the boundaries, and t_{m_2} is the time at which ice breaks up as a result of reaching critical porosity ν^*. Thus, ice remains as a sheet to the very end and disappears when the top and bottom boundaries meet, or when ice becomes so porous in the interior that it cannot bear its own weight any longer. The way in which break-up takes place depends on the forcing conditions during the melting season.

It is clear that solar radiation plays a key role in the melting of ice. It melts the top and interior of the ice sheet, and the radiation penetrating the ice warms the water underneath, which may come back to the ice as oceanic heat flux. Albedo (α) varies widely during the melting season, starting at $\alpha \approx 0.9$ for dry snow it decreases to 0.5 for dry ice and down to 0.1–0.2 for wet ice. Studying the surface of melting sea ice, one gets a good idea of the high spatial variability of the albedo of melting ice. There is no simple way to quantify the changes in albedo during the melting season. Even advanced models take it as a step function with distinct values for dry snow, wet snow, dry ice, and wet ice.

The penetration of solar radiation can be described using Beer's law as in the case of liquid water:

$$Q(z) = (1 - \alpha)\gamma Q(0) \exp\left(-\int_0^z \kappa \, dz'\right) \qquad (7.10)$$

where Q is irradiance; and κ is the attenuation coefficient: $\kappa \sim 0.2 \, \text{cm}^{-1}$ for snow, $\kappa \sim 0.1 \, \text{cm}^{-1}$ for snow ice, and $\kappa \sim 0.02 \, \text{cm}^{-1}$ for ice. The ice sheet therefore consists of optically different layers. The attenuation distance can be taken as $3\kappa^{-1}$ or 15 cm in snow and 150 cm in ice.

Albedo and the attenuation coefficient in the melting season are difficult to estimate from routine meteorological data, and consequently such is the situation with the whole surface heat balance. Therefore, a simplified method, the degree-day formula, is often utilized in practice:

$$dh/dt \approx -A \max\{T_a, 0\} \qquad (7.11)$$

where $T_a = T_a(t)$ is air temperature; $A = A(t)$ is the degree-day coefficient[5]: $A \sim 0.1$–$1.0 \, \text{cm}/(\text{da} \, °\text{C})$. Ice and snow cover have their own, more accurate time-dependent degree-day coefficients. This approach is not directly connected to the physics of melting but it seems that the heat balance correlates well with the surface air temperature. For example, mean daily temperatures are 2°C in Utö, Northern Gotland Basin in April and 6°C in Kemi, northern Bay of Bothnia in May. Melting of ice in these months would mean $A \sim 0.3 \, \text{cm}/(\text{da} \, °\text{C})$.

7.3.3 Numerical modeling of the ice growth and melting cycle

The overall heat conduction and sea ice thickness problem—see Equations (7.4)—can be solved using a numerical model. The first attempt at this was the congelation ice model for Arctic sea ice (Maykut and Untersteiner, 1971). In the seasonal sea ice zone the formation of superimposed ice (in particular, snow ice) becomes important, and a

---

[5] In hydrology, the commonly used semi-empirical models that simulate the melting of snow and ice take melt rate as proportional to (positive) air temperature and the proportionality coefficient is called a "degree-day coefficient".

good sea ice model should always be coupled with a snow model. The first effort was made by Leppäranta (1983) and a more complete version was developed by Saloranta (2000). A snow ice model based on melting–freezing cycles was prepared by Bin (2002).

The model by Saloranta (2000) consists of the Maykut and Untersteiner (1971) model and a snow model that takes account of snow compaction, slush formation from flooding, melting or rain, and snow ice growth. The model consists of four layers: snow, slush, snow ice, and congelation ice. These layers interact: snow accumulation creates slush and snow ice depending on the total thickness of ice, while the growth and decay of congelation ice depends on snow and slush conditions. Slush and snow ice often take on a multiple-layer structure with snow ice and slush layers alternating. The snow layer needs its own model. The thickness of snow decreases for three different reasons: surface melting, compaction, and the formation of slush, which further transforms into snow ice.

The boundary conditions are written

Top: $\quad \kappa \dfrac{\partial T}{\partial z} = \rho L \dfrac{dh}{dt} + Q_0,$ (7.12a)

$$\dfrac{dh_s}{dt} = P - S - M - C \qquad (7.12b)$$

Ice/Snow interface: $\quad \kappa \dfrac{\partial T}{\partial z} = \kappa_s \dfrac{\partial T}{\partial z}, \quad \text{if } h_{sh} = 0$ (7.13a)

$$\rho_{si} L \nu \dfrac{dh_{si}}{dt} = \kappa_s \dfrac{\partial T}{\partial z} - \kappa \dfrac{\partial T}{\partial z}, \quad \text{if } h_{sh} > 0 \qquad (7.13b)$$

$$\dfrac{dh_{sh}}{dt} = S \qquad (7.13c)$$

Bottom: $\quad T = T_f,$ (7.14a)

$$\rho_i L \dfrac{dh_i}{dt} = \kappa \dfrac{\partial T}{\partial z}\bigg|_{\text{bottom}} - Q_w \qquad (7.14b)$$

where P is precipitation; S is slush formation rate; M is snow-melting rate; C is snow compaction rate; and h_{sh} is the thickness of the slush layer. The thickness of snow increases as a result of snowfall, given as water equivalent and changed to snow thickness using a fixed density of new snow. The threshold value between snow and no-snow conditions in the model is set to 3 cm. The thickness of snow decreases for three different reasons: surface melting, compaction, and transformation to slush. The density change in snow due to compaction can be formulated after Yen (1981). A depth-dependent snow density profile due to snow compaction is calculated. If negative freeboard conditions appear, the amount of new slush is calculated from the Archimedes principle. In the model a fixed snow overload is needed for the flooding event to begin; here it is taken as 3 mm of water equivalent. The thermal conductivity of snow is formulated after Yen (1981) as $\kappa_s = 2.22362 \times \rho_s^{1.885}$, where κ_s is in W/(m °C) and ρ_s is in g/cm^3. The ice growth modeling problem possesses a negative feedback to errors, which makes it relatively easy (i.e., too much ice in the model

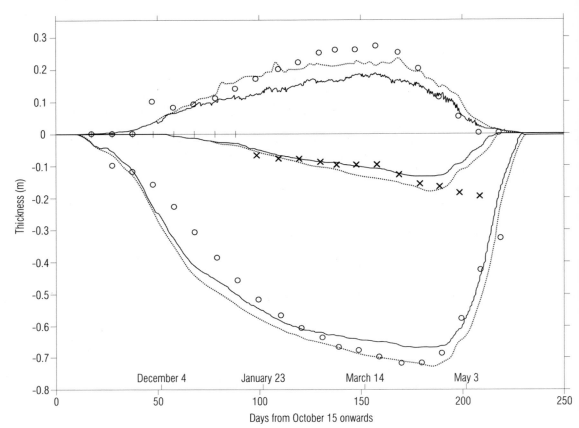

Figure 7.8. The mean thickness of congelation ice, snow ice, and snow at Kemi, 1979–1990. Observations are shown by ○ (snow thickness and total ice thickness) and × (snow-ice thickness); model calculations are shown by solid line (forced by SMHI gridded data) and dotted line (forced by Kemi Airport weather data). (Redrawn from Saloranta, 2000.)

reduces further growth and *vice versa*). In addition, the thicker the ice grows, the less is the thickness's sensitivity to atmospheric forcing. Indeed, the background Stefan's law states that squared ice thickness is proportional to freezing degree-days S, and consequently $\delta h_i \propto \delta S/h_i$. With snow accumulation, snow metamorphosis, slush formation, and snow ice growth, the sensitivity of ice thickness to the snow model also steps into the picture, and the model outcome may lead to large persistent deviations.

The model outcome for simulations in Kemi is shown in Figure 7.8. It can be seen that the model follows the evolution of the congelation ice and snow ice layers well, and the snow thickness is simulated quite well by the relatively simple snow model forced by precipitation. This figure shows mean evolution, but individual years come out well too. An independent validation of the model was made for landfast ice in the Sea of Okhotsk (Shirasawa *et al.*, 2005).

7.4 DYNAMICS AND MORPHOLOGY OF DRIFT ICE

7.4.1 Drift ice fields

The basins of the Baltic Sea are so large that no solid ice lid forms over them. In coastal and archipelago areas there is landfast ice, which is stable and smooth for most of the winter, supported by islands and grounded ice ridges on shoals. The landfast ice zone extends to depths of 5 m–15 m depending on the thickness of ice. Over large basins, wind fetch is long so that the resulting forcing is able to break up the ice cover and force it into motion, even at long distances, and deformation, and in consequence drift ice fields form (Figure 7.9).

The drift of sea ice can be considered a meso-scale or large-scale phenomenon the elements of which are ice floes. The drift ice landscape consists of leads, fields of ice floes, and deformed ice such as pressure ridges, rafted ice, and brash ice. In many places at the fast ice boundary heavy zones of ridged ice are found. Winds, currents, sea-level variations, thermal cracking, etc. keep the drift ice in its broken form allowing the dynamics to continue. The act of drifting makes sea ice an active player in the atmosphere–ocean interaction.

The extent of landfast ice increases with time. As the ice grows, the fast ice zone extends farther away from the coast with its boundary reaching outer islands and grounded pressure ridges on shoals. In an extreme situation, the whole basin may become fast ice and keep so for several weeks. In winter 1985 this happened in the Bay of Bothnia. This winter was very cold, ice grew to half-a-meter thickness in the whole basin, and the ice cover was immobile for two months. In that winter the record fast ice thickness was observed (as presented in Section 7.3). There was a drift ice base in the center of the basin for two weeks in February, and no motion was recorded

Figure 7.9. Drift ice at Nahkiainen lighthouse, Bay of Bothnia.

Table 7.3. Ice field classification based on ice compactness A (WMO, 1970).

Verbal	Numerical	Verbal	Numerical
Ice-free	$A = 0\%$	Very open drift ice	$0\% < A < 30\%$
Open drift ice	$30\% \leq A < 50\%$	Close drift ice	$50\% \leq A < 70\%$
Very close drift ice	$70\% \leq A < 90\%$	Compact drift ice	$90\% \leq A \leq 100\%$

beyond a position accuracy of 50 m (Leppäranta, 1987). Fast ice crosses basins first in narrow parts, resulting in "fast ice bridges". In the Bay of Bothnia fast ice formed across the Northern Quark in severe winters in the past. This was often utilized for on-ice traffic, even for car traffic, but since 1970 the northern harbors have been open for winter sea traffic and no fast ice is allowed to form.

Beyond the fast ice boundary there is drift ice.[6] A "drift ice landscape" consists of leads and ice floes with ridges, hummocks, and other variable morphological characteristics. As mentioned in Section 7.1 *ice types* are defined according to historical and practical shipping activities in ice-covered waters (WMO, 1970).

Open water regions are usually narrow linear formations, *leads*,[7] which form particularly at the lee side of the fast ice boundary. Narrow leads are also found in the interior of drift ice fields. Leads form at weak points in the drift ice field by means of mechanical processes and their structural arrangements can tell us a lot about the background process (Goldstein *et al.*, 2000). The opening and closing of leads are short-term phenomena in the Baltic Sea. *Ice concentration* or *ice compactness*, denoted by A, is normally given in percentages or tenths and further categorized into six standard classes (Table 7.3). Between the ice and open water surfaces there are strong contrasts that satellite remote-sensing methods can detect.

Definitions of the various ice types have been collected to form a sea ice nomenclature (Table 7.4) which is used on ice charts (Figure 7.3). Due to the drift of the ice, freshwater, latent heat, and ice impurities such as sediment particles are transported across long distances.

In the Baltic Sea the most important drift ice type is *ridged ice*, both from the viewpoint of drift ice physics and in practical applications (Figure 7.10). The floe accumulations at pressure ridges can be 5 m–15 m thick. The largest ridge observed was in the Bay of Bothnia and measured 31.5 m (Palosuo, 1975), 28 m constituted its underwater part, *keel*, and 3.5 m its freeboard, *sail*. The sail and keel are idealized as a triangle. They float according to the Archimedes law, and the ratio of sail height to keel depth is normally from 1:5 to 1:4 (Leppäranta and Hakala, 1992; Kankaanpää, 1998). Usually Baltic Sea ice fields contain ridged ice that makes up 10%–50% of the total volume. Ridges often form at the fast ice boundary but can be found all over the basins.

---

[6] Term used in a wide sense to include any sea ice other than landfast ice.
[7] Semi-persistent open water regions called *polynyas* do not form in the Baltic Sea.

Table 7.4. General and common Baltic Sea ice types (Armstrong *et al.*, 1966; WMO, 1970).

Sea ice	Any form of ice found at sea that has originated from the freezing of seawater.

New ice
  Frazil ice
  Nilas
	A general term for recently formed ice.
	Fine spicules or plates of ice in suspension in water.
	A thin elastic crust of ice, easily bending by waves and swell and rafting under pressure; matt surface and thickness up to 10 cm.

Young ice	Ice in the transition between new ice and first-year ice, 10–30 cm thick.
First-year ice	Ice of not more than one year's growth developing from young ice, thickness 30 cm to 2 m. Level when undeformed but where ridges and hummocks occur they are rough and sharply angular.
Fast ice	Sea ice that remains fast along the coast or over shoals. Also called *landfast ice*.
Grounded ice	Floating ice, which is aground in shoal water.
Ice field	Area of drift ice at least 10 km across.
Pancake ice	Pieces of new ice usually approximately circular, about 30 cm to 3 m across, and with raised rims due to the pieces striking against each other.
Ice floe	Any relatively flat piece of ice 20 m or more across.
Level ice	Sea ice which is unaffected by deformation; a substitute term is *undeformed ice*.
Deformed ice	A general term for ice that has been squeezed together and in places forced upwards and downwards; a substitute term is *pressure ice*.
Rafted ice	A form of pressure ice in which one floe overrides another. A type of rafting common in nilas whereby interlocking thrusts are formed—each floe thrusting "fingers" alternately over and under the other—is known as *finger rafting*.
Brash ice	Accumulations of ice made up of fragments not more than 2 m across, the wreckage of other forms of ice.
Hummocked ice	A form of pressure ice in which pieces of ice are piled haphazardly, one over another, to form an uneven surface.
Ridge	A ridge or wall of broken ice forced up by pressure; the upper (above water level) part is called *sail* and the lower part *keel*.
Fracture	Any break or rupture in ice resulting from deformation processes, length from meters to kilometers.
Crack	Any fracture that has not parted more than one meter.
Lead	Any fracture or passageway through sea ice which is navigable by surface vessels.

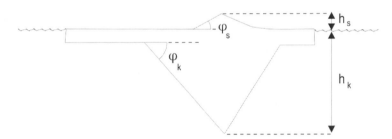

Figure 7.10. Mapping the topography of ridged ice in the Bay of Bothnia. Lower part shows a schematic cross-section of a ridge; h_s and h_k are sail height and keel depth, and φ_s and φ_k are sail and keel inclination angles.

Ice ridges are serious obstacles to winter shipping and are the main cause for concern about man-made marine structures in the Baltic Sea. Therefore, much research has been done on them. Palosuo (1975) reports their formation processes and presents structural mappings for more than ten large ridges. Leppäranta and Hakala (1992) made further structural investigations and also performed experiments on the cohesive strength of floe accumulations. Leppäranta et al. (1995) mapped the life history of one single ridge in the Bay of Bothnia, and Kankaanpää (1998) presented a large set of detailed structural information on Baltic Sea ridges.

Ridges may form in shallows as grounded ridges. Grounded ridges can also form when floating ridges drift into shallows. Such ridges act as support points helping landfast ice to expand farther out from the coast. Grounding ridges scour the bottom and need to be considered when lowering pipelines or cables across the sea.

On a scale of several kilometers or more, the amount of ridging is described by three parameters: cut-off size h_0, mean size h_r, and mean spatial density μ (the size is taken as the sail height or the keel depth). In the Baltic Sea the spatial statistics of ridges have been studied by Lewis et al. (1993). The cut-off size is the lower limit for floe accumulations defined as ridges, which needs to be known for observational systems, chosen as $h_0 = 40\,\text{cm}$ by Lewis et al. (1993). The size follows the exponential distribution, which has the probability density of

$$p(h; h_0, \lambda) = \lambda \exp[-\lambda(h - h_0)] \qquad (7.15)$$

where λ is the distribution shape parameter, and consequently $h_r = h_0 + \lambda^{-1}$. The distribution of spacings is usually considered logarithmic normal.[8] Mean spacing is related to ridge size and cut-off as

$$\mu(h'_0) = \mu(h_0) \exp[-\lambda(h'_0 - h_0)] \qquad (7.16)$$

where h_0 is the present and h'_0 is the new cut-off height.

Mean ridge sail height is 50 cm–70 cm in the Gulf of Bothnia when cut-off size is 40 cm (Lewis et al., 1993); thus $\lambda^{-1} = 10\,\text{cm}$–$30\,\text{cm}$, and for $\lambda^{-1} = 30\,\text{cm}$, 1% of ridges are higher than 1.8 m. Spatial ridge density was 1–10 ridges/km in Lewis et al. (1993). Then, with $\lambda^{-1} = 30\,\text{cm}$ and $\mu = 5\,\text{km}^{-1}$, ridges greater than 1.8 m are found on average once in every 20 km section. The total volume of ridged ice, per unit area, is given by

$$\tilde{h}_r = c\mu h_r^2 \qquad (7.17)$$

where c is a semi-empirical factor. When h_r represents the sail height, $c \approx 25$ (Lewis et al., 1993). Normally 10%–50% of the total ice volume is ridged ice, but in extreme cases a large ice field can be packed into an area as small as one-fifth of the original. On the coast where ridges are grounded and no longer freely floating, their sails can grow as high as 10 m–15 m. Such ridges are seen annually in coastal areas of the Bay of Bothnia.

Drift ice fields consist of ice floes of size 10^1–10^4 m, while deformed ice structures such as pressure ridges consist of ice blocks of diameter 0.1 m–10 m. The structure of ice fields is self-similar: geometrical relations are fixed although the thickness of ice and size of floes change. Any drift ice landscape photograph must therefore possess scale information so that actual size can be interpreted.

---

[8] A random variable X has a logarithmic normal distribution if log X has a normal or Gaussian distribution.

In sea ice dynamics, drift ice fields are described for their "mechanical state" using the following quantities:

Quantity	Definition
Ice type	Main types: level ice, pancake ice, rafted ice, brash, ridged ice
Ice compactness	Relative surface area of ice
Ice thickness	Distance from top to bottom
Size of ice floes	"Characteristic diameter", distribution

Ice types are based on visual observations and are used in operational ice charts (Figure 7.3). They are created by means of mechanical deformation processes. The thickness and compactness of ice are the principal properties in drift ice dynamics. They provide information about ice volume and strength.

Ice compactness provides information on how much open water exists and how mobile the ice field is. The exchange of heat, momentum, and matter between the ocean and atmosphere also depends on ice compactness. This quantity can be observed well by remote sensing, since it is a surface characteristic.

Ice thickness is the most important ice property in drift ice geophysics. The strength of ice and ice volume are proportional to the thickness of ice. The thickness may vary widely. Even in a small area there may be ridges thicker than 10 m and thin new ice. A severe practical problem is that spaceborne or airborne remote-sensing methods for ice thickness mapping are not yet good enough. In coastal and archipelago areas there are fixed observation sites for the thickness of landfast ice but nothing like that exists for drift ice.

Applicable remote-sensing methods for ice thickness mapping are upward-looking sonar (moored or on a submarine), the so-called airborne electromagnetic method used normally in ore prospecting (Multala et al., 1996), thermal infrared (Leppäranta and Lewis, 2007), and laser profiler (Lewis et al., 1993). The latter two methods have limitations such that the thermal infrared works best for thin ice and the laser profiler for ridged ice. Microwave methods have been examined to a very large extent but at their best they can produce a nominal classification of ice in some thickness classes.

For an arbitrary region, ice thickness can be described using the spatial distribution (Thorndike et al., 1975). The thickness distribution $\Pi = \Pi(h)$ is defined as

$$\Pi(h) = \frac{S(h)}{S(\infty)} \qquad (7.18)$$

where $S(h)$ is the surface area of the region where ice thickness is less than or equal to h. This distribution is analogous to the cumulative distribution in probability theory. An example of the thickness distribution in the Baltic Sea is given as a histogram in Figure 7.11. The upper tail of the thickness distribution covers ridged ice. A corre-

Figure 7.11. Ice thickness distribution in the Bay of Bothnia. The data are based on 3 km long AEM profiles in the Bay of Bothnia, March 17, 1993, showing sections from northern, central, and southern basins. The spatial resolution of the data is 90 m. The vertical axis shows frequency in m^{-1}. (Redrawn from Multala et al., 1996.)

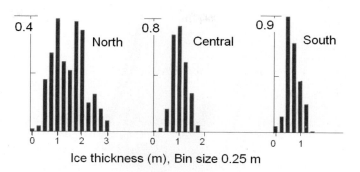

sponding distribution can be defined for floe size, whose width ranges from meters to kilometers.

The size and shape of ice floes seems to play a secondary role in the final outcome of sea ice dynamics. When an ice field is open and floes are small, diffusion is significant. In close ice, internal friction becomes large, floes group together, and diffusion weakens. The size and shape of floes are horizontal properties and therefore observable by aerial photography and satellite images. In midwinter, floes may be more difficult to distinguish, since they freeze together and receive snow cover. In spring, floes break into smaller pieces and their form becomes more rounded than their wintertime polygonal shape. Collisions between ice floes transfer momentum and consume mechanical energy but their significance is much less than the friction from ice floe accumulations.

Real-time information on ice conditions is available in the ice charts (Figure 7.3) published daily by ice services in the Baltic Sea (see *http://www.fimr.fi* or *http://www.smhi.se*). These ice charts present the ice extent, ice fields, ice types, ice compactness, and ice thickness. Ice drift can markedly change conditions in a few days, and therefore it is essential to update the charts daily. Ice information is based on coastal stations, ship reports, and satellite imagery (mainly NOAA, Terra/Aqua, and Radarsat).

7.4.2 Dynamics of drift ice

Drift ice can be considered as a granular, two-dimensional, non-linear, and compressible medium with ice floes serving as the "grains". Two-dimensionality means that an ice field moves horizontally on the sea surface while its velocity is constant across the vertical, and non-linearity refers to the ratio between stress and deformation. Since the sea surface is almost an equipotential surface, the compactness of ice can change very easily and can stay at any level. In open and close ice fields ($A < 70\%$) ice floes drift freely and independently, while compact ice fields ($A > 90\%$) are subject to large internal friction and can be stationary even

246 The ice of the Baltic Sea

Figure 7.12. Ship under ice pressure in the Baltic Sea. (Ramsay, 1947.)

under strong wind. The pressure built up in compact drift ice has created severe problems for winter shipping (Figure 7.12).

A drift ice field can be considered as a continuum, where the size of continuum particles is much larger than the floe size. The floe size limits the spatial resolution of the continuum approach. The state of drift ice is described by $J = \{J_1, J_2, \ldots\}$, a set of relevant ice properties (Leppäranta, 2005). In the Baltic Sea a common approach has been to take $J = \{A, h_u, h_d\}$, where h_u and h_d are the mean thicknesses of undeformed ice and deformed ice, respectively. The ice volume per unit area is then $\tilde{h} = h_u + h_d = (h_{uf} + h_{df})A$, where h_{uf} and h_{df} are the corresponding mean thicknesses of undeformed ice and deformed ice in the ice areas. The deformed ice thickness h_{df} is often taken as the mean volume of ridged ice per unit area—see Equation (7.17)—but to account for deformed ice smaller than ridges (below the cut-off size), an estimate is obtained by extrapolation: $h_{df} = c \exp(\lambda h_0) \mu (h_r - h_0)^2$ (Leppäranta, 1981b). More general is the thickness distribution approach, where the drift ice state consists of thickness classes $J = \{(A, 0), (p_1, h_1), (p_2, h_2), \ldots, (p_n, h_n)\}$, where the h_ks are the class centers and $p_0 (= A), p_1, \ldots, p_n$ give the spatial distribution of thickness classes. Usually, thickness classes are fixed but their spatial coverages change.

The rheology of a medium is a functional relationship between the internal stress field and the state of deformation. This relationship depends on the material properties of the medium. For example, the ideal gas law reads $p = \rho/(RT)$, where R

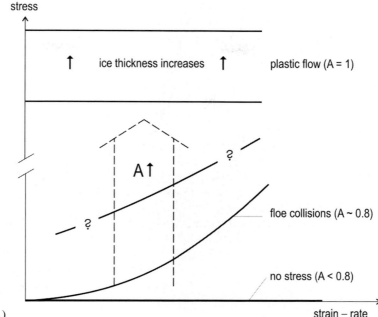

Figure 7.13. Schematic illustration of the rheology of drift ice, showing the dependence of internal stress on strain rate, ice compactness, and thickness. (Leppäranta, 2005.)

is the gas constant, and linear viscous (or Newtonian) fluids have the rheology $\sigma = -p\boldsymbol{I} + 2\mu[\dot{\varepsilon} - \frac{1}{3}\mathrm{tr}(\dot{\varepsilon})\boldsymbol{I}]$, where μ is viscosity, $\boldsymbol{I}$ is the unit tensor, and $\dot{\varepsilon}$ is the strain-rate tensor. Ideal gases are characterized by the gas constant and viscous fluids by their viscosity. Ideal gases are a very special case since there are no mechanical energy losses in deformation, while viscosity converts mechanical energy to heat (there is an insignificant amount in Baltic Sea dynamics, though). Differential internal stresses (i.e., taking stress differences across material particles) appear as internal forces in the equation of motion. Consequently, in linear viscous fluids the internal force is $\boldsymbol{\sigma} = -\nabla p + \mu\nabla^2 \boldsymbol{u}$, where the first term is the pressure gradient and the second term is the viscous internal friction.

Drift ice is a complicated medium for its rheology (Figure 7.13). The general form is

$$\boldsymbol{\sigma} = \boldsymbol{\sigma}(J, \varepsilon, \dot{\varepsilon}) \qquad (7.19)$$

where ε is the strain tensor. Mechanical behavior changes qualitatively with ice compactness. Compact drift ice is plastic ($A > 0.9$), very close drift ice is viscous ($0.7 < A < 0.9$), and close and open drift ice are stress-free ($A < 0.7$). The qualitative characteristics of drift ice behavior are

- stress ≈ 0, when $A < 0.7$ (little contact between ice floes);
- stress ≈ 0 under opening deformation (no resistance to tensile stresses);
- friction is less in shear than in compression;

- yield strength > 0, when $A \approx 1$ (compact ice field may be stationary);
- yield strength increases with ice thickness;
- no recoverable energy storage: and
- no memory.

In a plastic medium, a certain level of stress is needed for deformation, otherwise the medium stays as a rigid body. This stress level is called the *yield strength*. For example, with the Bay of Bothnia fully covered by a half-meter thick ice cover, a wind speed greater than 15 m/s is needed to start ice drift. Plastic flow is sufficiently stiff that over large areas spatial variations are small, but in transition zones intensive deformation may take place. Often there is a narrow ridging zone at the fast ice boundary. In viscous flow the deformation zones are smoother. For a stress-free ice field the free drift model can be employed, where ice floes move independently based on their size and shape. Finally, drift ice mechanics does not create recoverable energy storages, and consequently the internal force is purely frictional.

The most common rheology used for drift ice is Hibler's (1979) viscous–plastic rheology

$$\boldsymbol{\sigma} = (\zeta - \tfrac{1}{2}P)\boldsymbol{I} + 2\eta[\dot{\boldsymbol{\varepsilon}} - \tfrac{1}{2}\mathrm{tr}(\dot{\boldsymbol{\varepsilon}})\boldsymbol{I}] \quad (7.20\mathrm{a})$$

$$P = P^*\tilde{h}\exp[-C(1-A)] \quad (7.20\mathrm{b})$$

$$\zeta = P/2\max(\Delta,\Delta_0), \quad \Delta = \dot{\varepsilon}_I^2 + \dot{\varepsilon}_{II}^2 e^{-2}, \quad \eta = \zeta e^{-2} \quad (7.20\mathrm{c})$$

where ζ and η are bulk and shear viscosities; P is the unconfined pressure term; P^*, C, e, and Δ_0 are rheology parameters; $\dot{\varepsilon}_I$ and $\dot{\varepsilon}_{II}$ are strain-rate invariants equal to the sum and difference of the principal axis values; P^* is the compressive strength of ice of unit thickness; C is the strength reduction constant for opening (decrease of C^{-1} in ice compactness gives the e-folding scale for the ice strength); e is the ratio of compressive strength to shear strength; and Δ_0 is the maximum viscous creep rate. The following values are normal under Baltic Sea conditions (Zhang and Leppäranta, 1995; Haapala and Leppäranta, 1996; Leppäranta et al., 1998a): $P^* = 10\,\mathrm{kPa}$–$40\,\mathrm{kPa}$, $C = 20$, $e = 2$, and $\Delta_0 = 10^{-10}$–$10^{-8}\,\mathrm{s}^{-1}$.

The kinematics of drift ice is measured by means of drifters and remote sensing. Drifters move with the ice and are positioned using satellite or local geodetic methods (Figure 7.14). Remote-sensing methods are based on repeated recognition of ice features from sequential imagery (Figure 7.15). Drifters are limited by spatial resolution, while remote-sensing methods are limited in practice by temporal resolution.

The equation of the motion of drift ice is based on Newton's Second Law applied to continua. Momentum is changed due to internal and external forces. Coriolis acceleration needs to be taken into account as well. The equation of the motion of drift ice can be written in a general form (e.g., Leppäranta, 2005)

$$\rho_i\tilde{h}\left(\frac{\partial \boldsymbol{U}_i}{\partial t} + \boldsymbol{U}_i\cdot\nabla\boldsymbol{U}_i + f\mathbf{k}\times\boldsymbol{U}_i\right) = \nabla\cdot\boldsymbol{\sigma} + \boldsymbol{\tau}_a + \boldsymbol{\tau}_w - \rho_i\tilde{h}g\nabla\xi \quad (7.21)$$

where $\boldsymbol{U}_i$ is ice velocity; and $\boldsymbol{\tau}_a$ and $\boldsymbol{\tau}_w$ are tangential stresses due to wind and water on

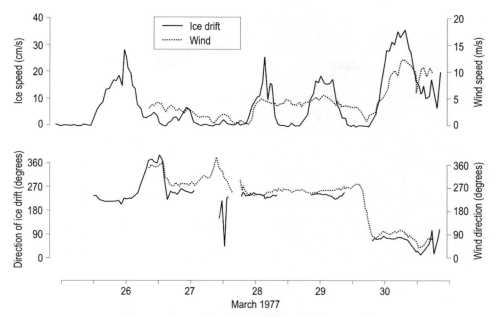

Figure 7.14. Ice drift and wind in the central Bay of Bothnia, March 1977. The ice speed scale is 2% of the wind speed scale. (Leppäranta, 1981a.)

the ice. The left-hand side contains the acceleration terms, and the right-hand side the internal friction and the three external forces, the last of which is the surface pressure gradient due to sea-surface tilt. This equation is somewhat similar to the vertically integrated circulation model (see Chapter 5). The essential difference is that the internal friction of drift ice is dictated by its plastic properties, while internal friction in ocean circulation is dictated by turbulence, and also in ice drift the internal friction belongs to the leading terms of the momentum balance.

The tangential stresses of winds and currents are given by the quadratic laws

$$\tau_a = \rho_a C_a U_a [\cos \theta_a U_a + \sin \theta_a \mathbf{k} \times U_a] \qquad (7.22a)$$

$$\tau_w = \rho_w C_w |U - U_i|[\cos \theta_w (U - U_i) + \sin \theta_w \mathbf{k} \times (U - U_i)] \qquad (7.22b)$$

where U_a is wind velocity; U is ocean current velocity; C_a and C_w air and water drag coefficients; and θ_a and θ_w boundary-layer turning angles in air and water. Table 7.5 presents the usual boundary layer parameters for ice drift in the Baltic Sea.

With a tilting sea-surface, there results a pressure gradient and a consequent flow. In deep areas this flow equals the surface geostrophic current U_g, and then $g\nabla\xi = -f\mathbf{k} \times U_g$. However, most of the Baltic Sea is too shallow for the geostrophic balance to be valid, and therefore it is preferable to estimate the surface pressure gradient directly from sea-level elevation.

250 The ice of the Baltic Sea [Ch. 7

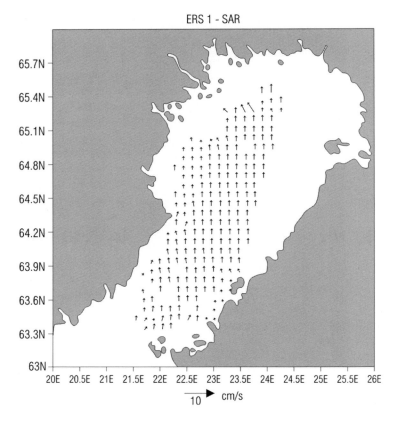

Figure 7.15. Ice field displacement March 5–8, 1994 obtained from ERS-1 SAR images. (Leppäranta et al., 1998a.)

The order of magnitude of terms in the momentum equation can be analyzed from the elementary scales $H = 50\,\text{cm}$, $U_i = 10\,\text{cm s}^{-1}$, $T = 1$ day, $L = 100\,\text{km}$, $P^* = 10^4\,\text{N m}^{-2}$, $U_a = 10\,\text{m s}^{-1}$, and $\nabla \xi = 10^{-6}$. We have:

Term	Scale	N/m²	Comments
Inertia	$\rho_i H U_i / T$	10^{-3}	$>10^{-2}$ under rapid changes
Advection	$\rho_i H U_i^2 / L$	10^{-4}	Long-term influence may be significant
Coriolis	$\rho_i H f U_i$	10^{-2}	Mostly $<10^{-1}$
Internal friction	$P^* H / L$	10^{-1}	$A > 0.9$ (0, if $A < 0.7$)
Wind stress	$\rho_a C_a U_a^2$	10^{-1}	Mostly significant
Water stress	$\rho_w C_w U_i^2$	10^{-1}	Mostly significant
Pressure gradient	$\rho_i H g \nabla \xi$	10^{-2}	Mostly $<10^{-2}$

Table 7.5. Atmospheric and oceanic boundary layer parameters for sea ice dynamics (Leppäranta, 1981b, 1990; Joffre, 1984).

Parameter	Notation	Value	Comments
Air density	ρ_a	1.3 kg/m^3	$T = 0°C$
Air drag coefficient	C_a	1.8×10^{-3}	Surface wind
Atmospheric turning angle	θ_a	$0°$	Surface wind
Water density	ρ_w	1,005 kg/m^3	$T = 0°C$, $S = 5‰$
Water drag coefficient	C_w	3.5×10^{-3}	Geostrophic flow
Oceanic turning angle	θ_w	$20°$	Geostrophic flow

In sea ice drift the ice conservation law must be satisfied. This law needs to be considered separately for each component of the ice state, influenced by advection, mechanical deformation, and thermodynamics. First, mean ice thickness must satisfy volume conservation, and ice compactness must satisfy area conservation:

$$\frac{\partial \tilde{h}}{\partial t} = -\boldsymbol{U}_i \cdot \nabla \tilde{h} - \tilde{h} \nabla \cdot \boldsymbol{U}_i + \phi_h \quad (7.23\text{a})$$

$$\frac{\partial A}{\partial t} = -\boldsymbol{U}_i \cdot \nabla A - A \nabla \cdot \boldsymbol{U}_i + \phi_A, \quad 0 \leq A \leq 1 \quad (7.23\text{b})$$

where ϕ_h and ϕ_A are thermal changes in ice thickness and concentration. These equations are well established and unquestionable. For any other ice state component, J_k, we have the general form

$$\frac{\partial J_k}{\partial t} = -\boldsymbol{U}_i \cdot \nabla J_k + \psi_k + \phi_k \quad (7.23\text{c})$$

where ψ_k is mechanical deformation and ϕ_k are thermal changes; the main open question is to formulate mechanical deformation functions. In general, the timescale of freezing and melting is much longer than the dynamics timescale, and often for short-term simulations the thermodynamics can be ignored.

If there is convergence in an ice field ($\nabla \cdot \boldsymbol{U}_i < 0$), compactness and mean thickness will increase. Compactness can reach 100%, and then mean ice thickness increases due to ridging. For example, if $\nabla \cdot \boldsymbol{U}_i = -0.01\,\text{h}^{-1}$, $A = 1$, and $H = 30$ cm, ridging increases mean thickness by 0.3 cm in an hour. A three-level ice state $J = \{A, h_u, h_d\}$, when compactness is treated the same way as in the two-level case, has been widely applied in the Baltic Sea. The thickness of ice is split into two parts, $\tilde{h} = h_u + h_d$: undeformed ice h_u growing due to melting and freezing and deformed ice h_d growing due to ridging and hummocking, which is only possible at $A = 1$ and $\nabla \cdot \boldsymbol{U}_i < 0$. As no mechanical "unridging" takes place, ridges only disappear through melting.

7.4.3 Analytical solutions of sea ice drift

In an enclosed or semi-enclosed basin, the drift ice cover is stationary as long as internal stresses are below the yield level. Otherwise, breakage of the ice cover takes place and then the ice will begin to drift. The breakage criterion is (Leppäranta, 2005)

$$\tau_a L > P^* h \tag{7.24}$$

where L is the length of the basin in the wind direction; and $P^* \approx 10\,\text{kPa}-100\,\text{kPa}$ is the compressive strength of ice. If the wind speed is 15 m/s, we have $\tau_a \simeq \frac{1}{2}$ Pa, and if $L \approx 100$ km the ice breaks when it is thinner than around 50 cm. In basins covered by compact ice, the ice may be stationary even under very windy conditions because of its internal friction. This is the case in severe winters in the Bay of Bothnia and in very severe winters in the Sea of Bothnia, Gulf of Finland, and Gulf of Riga (scales 100–250 km). In even smaller sea areas, such as Pärnu Bay (20 km long) in the Gulf of Riga, the ice is movable only when it is thin (i.e., less than about 10 cm).

In general, wind-driven ice less than 1 m thick has a velocity of 1%–3% of the wind speed and the direction of motion deviates 20°–30° to the right of the wind direction (in the northern hemisphere). Due to the dynamics of drift ice, ice conditions may change rapidly, and therefore ice charts are prepared for shipping on a daily basis by ice information services in the Baltic Sea. Opening and closing of leads have a major impact on the heat exchange between the atmosphere and the sea and consequently on the weather and climate. Ice drift is an important transport mechanism for latent heat, freshwater, and sediments trapped in the ice.

Internal friction has an effect on sea ice drift and understanding how the process works will ease solution of the problem. When internal friction is negligible, the situation is called *free drift*. This approach is valid when ice compactness is less than 0.7–0.8. The momentum equation is then easy to solve, but the ice conservation law can create problems, since there is no way to restrict the amount of ice that accumulates on the shoreline. When internal friction is included, this problem is avoided.

In the free drift approach, advective acceleration is ignored since it is very small in general (see the scale analysis above) as shown in the previous subsection, and when the timescale is longer than one hour, the inertial term can also be ignored. Consequently, the momentum equation reduces to an algebraic equation:

$$\rho_i \tilde{h} f \mathbf{k} \times \mathbf{U}_i = \boldsymbol{\tau}_a + \boldsymbol{\tau}_w - \rho_i \tilde{h} g \nabla \xi \tag{7.25}$$

In deepwater ($H \gg D$), combining Coriolis acceleration with pressure gradient and using the geostrophic current as the reference velocity in ice–ocean stress and for sea surface tilt, the solution can be written

$$\mathbf{U}_i = \mathbf{U}_i^a + \mathbf{U}_g \tag{7.26a}$$

where $\mathbf{U}_i^a$ is wind-driven ice drift, with a magnitude equal to 2.5% of the wind speed and a direction 20° to the right of the prevailing wind in the Baltic Sea (Figure 7.16). To a first approximation $U_i^a/U_a = [\rho_a C_a/(\rho_w C_w)]^{1/2}$ and the deviation angle equals

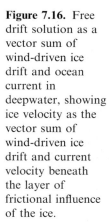
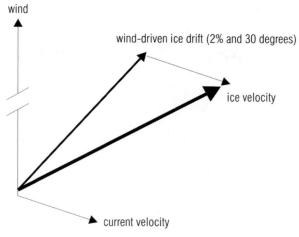

Figure 7.16. Free drift solution as a vector sum of wind-driven ice drift and ocean current in deepwater, showing ice velocity as the vector sum of wind-driven ice drift and current velocity beneath the layer of frictional influence of the ice.

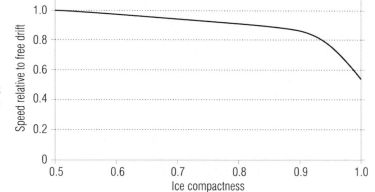

Figure 7.17. Ice drift speed relative to free drift speed as a function of ice compactness in the Baltic Sea. (Redrawn from Shirokov, 1977.)

θ_w for the surface wind. In shallow waters, we can take $U_g \approx 0$, and then

$$U_i = bU_i^{a\xi} \qquad (7.26b)$$

where $U_i^{a\xi}$ is the drift speed forced by wind stress and sea surface tilt; and $b = b(h/H)$ is a correction due to the small depth.

In a very close and compact ice field, internal friction is highly significant, and in the Baltic Sea this is clearly seen in the Bay of Bothnia, eastern Gulf of Finland, and the Gulf of Riga. The geometry of these basins strongly limits the drift of ice. Figures 7.17 and 7.18 give empirical information from the Baltic Sea about the dependence of the wind factor on ice compactness and on the tilt of the sea surface. Introducing rheology into the equation of motion results in a difficult non-linear differential equation. Analytical solutions are possible only for some one-dimensional and 1.5-dimensional cases but in general numerical models are needed.

254 The ice of the Baltic Sea [Ch. 7

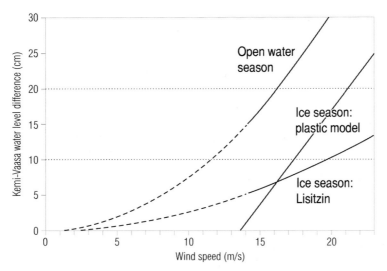

Figure 7.18. The sea-level difference between Kemi and Vaasa in open sea and ice conditions as a function of wind speed. (Redrawn from Lisitzin, 1957.)

Figure 7.19. A bay model for a compact ice field. Wind stress τ_a drives the ice (thickness h) to the shore is at $x = L$, and the ice edge is initially at $x = 0$.

A simple view of compact ice dynamics can be provided by a one-dimensional bay model (Figure 7.19). Wind drives the ice toward the end of the bay if it is strong enough—see Equation (7.24). Then ridging starts from the bay end, ice thickness increases, and the length of the ice field decreases. At time t_s the condition $\tau_a L(t_s) = P^* h(t_s)$ is achieved: ice motion comes to a stop as does the ridged ice field. Ice velocity is at maximum soon after the start-up and decreases during the process, with instantaneous levels as

$$U_i = \sqrt{\frac{\rho_a C_a}{\rho_w C_w} U_a^2 - \frac{P^*}{\rho_w C_w} \frac{dh}{dx}} \qquad (7.27)$$

where it has been assumed that the current velocity is zero. The argument of the square root must be positive, otherwise the solution is $U_i = 0$.

7.5 BASIN-SCALE SEA ICE MODELS

7.5.1 General

When considering the whole Baltic Sea or just one of its main basins, the evolution of sea ice is a three-dimensional problem. Ice grows and melts in the vertical direction, while the dynamics of drift ice transports and deforms ice fields on the sea surface. The growth and melting of ice influence its strength, and transport and deformation influence the heat exchange between ice and the atmosphere and between ice and the ocean. Thus, the thermodynamics and dynamics of sea ice constitute a coupled problem.

Basin-scale sea ice models consist of five elements:

(i) ice state $J = J(J_1, J_2, \ldots)$;
(ii) ice rheology $\boldsymbol{\sigma} = \boldsymbol{\sigma}(J, \varepsilon, \dot{\varepsilon})$;
(iii) equation of motion $\rho h \dot{\boldsymbol{u}} = \sum_k \boldsymbol{F}_k$;
(iv) ice conservation law $\dot{J} = \Psi + \Phi$; and
(v) thermal energy budget $Q_n = \sum_k Q_k, Q_w = Q_w(\cdots)$;

where Ψ and Φ formally represent the mechanical and thermal effects on the ice state. Ice models are forced by solar radiation, atmospheric forcing, and ice–ocean interaction. The heat balance freezes and melts ice, while winds and ocean currents influence ice drift. The salinity of ice is prescribed.

The modeler must choose an applicable ice state and rheology, and thereafter the model can be integrated to allow for the development of ice conditions. The free drift model has severe limitations in the small basins of the Baltic Sea. Plastic models are used, similar to those used in polar oceans and other sub-polar freezing basins such as the Gulf of St. Lawrence, Sea of Okhotsk, Bohai Sea, when the compressive strength of ice becomes one of the key tuning parameters.

Current sea ice models are essentially coupled ice–ocean models set up to simulate the exchange of heat, salts, and momentum both internally and forced by atmospheric fluxes. The oceanic part of the models is as described in Section 5.5. The ice–ocean system is strongly coupled with the atmosphere; therefore, full atmosphere–ice–ocean models would be needed for complete solution of the problem. In particular, the atmosphere is coupled with openings in the ice cover, surface temperature, and surface roughness.

7.5.2 Short-term models

Short-term sea ice modeling (1 hour–1 week) was the order of the day in the 1970s in the Baltic Sea. It was connected to ice forecasting for winter shipping. In the Baltic Sea the main harbors have been kept open all year since 1970 (Figure 7.20). Winter shipping is assisted by 20–25 icebreakers, and even then the transportation system has suffered from delays. This was the incentive for an extensive research program in Finland and Sweden, organized by a joint "Winter Navigation Research Board".

Figure 7.20. Convoy of ships assisted by an icebreaker in the Baltic Sea. Ice dynamics determines the length of the convoy that can follow an icebreaker.

The objective of short-term modeling is basic research into drift ice dynamics, including coupled ice–ocean–atmosphere modeling, simulations to examine the influence of ice dynamics on planned marine operations, and ice forecasting. One of the research aims was to develop an ice-forecasting system based on mathematical modeling, and the first real ice-forecasting season was winter 1977 (Leppäranta, 1981a). In the beginning the forecast period was 30 hours, limited by the period of available wind forecasts. The ice forecast, which was sent to icebreakers by facsimile telegraphy, contained ice speed, direction of ice drift, change in ice concentration, and qualitative pressure field. The system was continued for the following winters in parallel with model improvement and forecast period lengthening. The viscous–plastic three-level model was first used for operational ice forecasting in 1992 (Leppäranta and Zhang, 1992). This model was also integrated into the ice-forecasting system in Sweden (Omstedt et al., 1994). It has recently been examined in detail for the dynamics in different basins, in particular for scaling and the influence of coastal geometry and islands (Leppäranta and Wang, 2002; Wang et al., 2003). It has worked well down to bays 15 km in size with thin ice moving under a strong wind.

The approach in short-term modeling is often purely dynamic (i.e., thermodynamics is neglected). The initial field of the ice state must therefore be given. With the progress of satellite remote-sensing techniques the initial ice compactness became better mapped. From the outset a new initial thickness was constructed from a hindcast combined with new observations. However, the thickness of ice has continued to be a difficult quantity, with updates coming from occasional ship reports. Sea ice conditions may change in short timescales due to dynamics. Leads up to 20 km wide may open and close in one day, and heavy pressure may build up in the compression of compact ice. These processes have a strong influence on shipping, oil drilling, and other marine operations. Also changes in ice conditions such as the location of the ice edge are important for weather forecasting over a few days.

Oil spills are difficult problems in ice conditions. Good oil-combating[9] methods do not exist, and it is difficult to keep track of where the oil is going. The oil may penetrate into the ice sheet and drift with the ice or drift on the surface of openings and beneath sea ice. A simple modeling approach is an oil advection model with the ice drift and surface current with random dispersion superposed using, for example, a Monte Carlo method. An advanced, physical model treats oil as a viscous medium with density and viscosity dependent on the type of oil (e.g., Venkatesh *et al.*, 1990). They state that for ice compactness greater than 30% oil practically drifts with the ice, and that in slush or brash the thickness of oil film can be much larger than in open cold water. As the concentration increases, at the 80% level the oil is trapped between ice floes and above 95% the oil is forced beneath the ice.

The key areas of short-term modeling research are today ice thickness distribution and its evolution, and the use of satellite SARs for ice kinematics (Figure 7.21). The scaling problem and, in particular, the downscaling of stress from the geophysical to the local (engineering) scale is examined for its potential to combine scientific and engineering knowledge and to develop ice load calculation and forecasting methods. The physics of drift ice is well represented in short-term ice-forecasting models, in the sense that other questions are more critical for their further development. In particular, data assimilation has been little examined for sea ice models, and the user interface still leaves a lot to be desired.

7.5.3 Sea ice climate models

In long-term modeling the timescale is 1 month–100 years. The approach is dynamic–thermodynamic, and sometimes dynamics is neglected. The initial conditions are arbitrary on very long timescales but relevant to the ice state in monthly problems. The objectives are threefold: basic research into drift ice geophysics; ice climatology investigation; and coupled ice–ocean–atmosphere climate modeling. At the start only thermodynamic models were available for the times of freezing and ice break-up and for the evolution of ice thickness. But it quickly became clear that realistic ice dynamics is needed for ice transport and, in particular, for the opening and closing of leads. A large amount of heat is transmitted through leads from the ocean to the atmosphere. As a result of freezing and melting the ice has a major influence on the hydrographic structure of the ocean, especially since ice melts in a different region from where it was formed.

---

[9] The term "oil-combating" is used by environmental authorities and refers to discovering oil spills, forecasting their behavior, clean-up, and risk analysis.

258 The ice of the Baltic Sea [Ch. 7

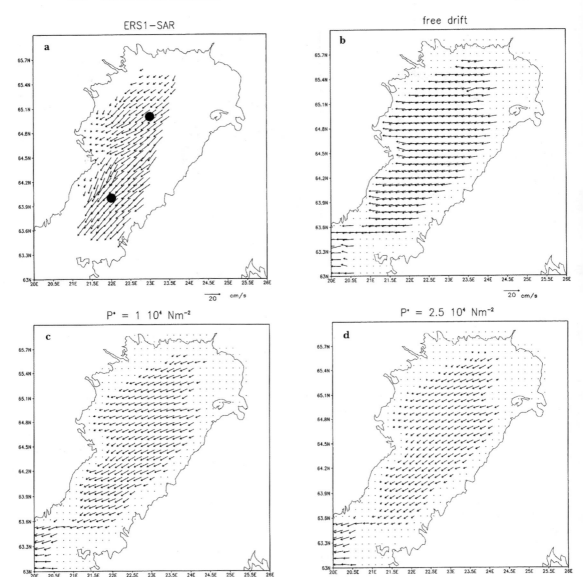

Figure 7.21. Velocity field calculated by the sea ice dynamics model and compared with observations. (Leppäranta et al., 1998a.)

The ice climate problem received increasing attention in the 1990s with the development of regional ice climate models (Haapala and Leppäranta, 1996). These ice climate models are forced by weather conditions. Model calibration is based on ice charts. The models are capable of simulating how the whole ice season from one

Sec. 7.5] Basin-scale sea ice models 259

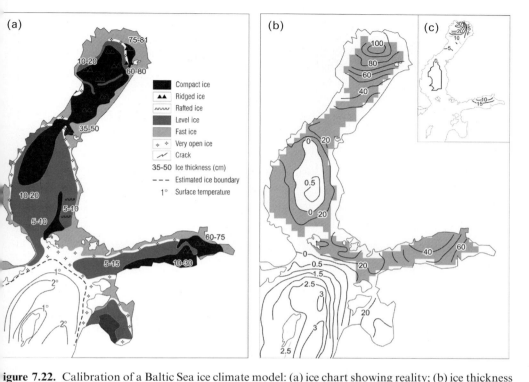

Figure 7.22. Calibration of a Baltic Sea ice climate model: (a) ice chart showing reality; (b) ice thickness (cm) and open water surface temperature from the model (°C) from the model; (c) mean thickness of deformed ice (cm). The model initial time was May 1, 1983, and the comparison here is with March 22, 1984. (Haapala and Leppäranta, 1996.)

summer to the next will develop. Calibrated models forced by atmospheric climate scenarios have then been used for ice season scenarios.

Calibration of the Baltic Sea ice climate model for a normal winter is shown in Figure 7.22 (Haapala and Leppäranta, 1996). The initial time was May 1. The figure compares modeled and observed ice conditions in March when the ice extent was at its largest. There are small discrepancies in that the surface temperature in the Gotland Sea is too high in the model, and in the Sea of Bothnia the model shows an open water region but the ice chart shows thin ice. Elsewhere the mean thickness came out quite well. Similar results were obtained for comparisons in mild and severe winters.

This chapter has presented Baltic Sea ice: ice formed from brackish water has been shown to be similar to sea ice in structure. The five sections covered topics across the hierarchy of scales: ice climatology; fine scale, structure; small scale, thermodynamics; and meso-scale to large scale, dynamics. Section 7.5 introduced full, basin-size sea ice models with vertically treated growth and melting processes and horizontally treated ice drift and mechanical deformation. Sea ice physics is a very specific topic usually less familiar to physical oceanographers, but necessary for physical oceanographers whose area of interest is ice-covered seas. Chapter 8 will look at coastal oceanography in the Baltic Sea, where all the classical oceanography and ice material presented in Chapters 3–7 will be utilized.

8

Coastal and local processes

Professor Eugenie Lisitzin (1905–1989) was an internationally well-known, distinguished oceanographer as a result of her research into sea-level changes and tidal motions. She was the head of the sea-level department at the Finnish Institute of Marine Research in 1955–1972 and she also was acting director of the institute. She was a cosmopolitan and could fluently speak nine languages. She was the "First Lady" in many respects. During her time (until 1961) it was not allowed to nominate a woman to a permanent position in the Finnish Institute of Marine Research but a dispensation was made for her. She was the first woman in Finland to obtain a Ph.D. in physics, in 1938, and she was also the first female member of the Finnish Society of Sciences and Letters. She published a book *Sea Level Changes* (1974) and was nominated as professor *honoris causa* in 1965.

The physical oceanography of the Baltic Sea, including the physics of sea ice, has been presented in the previous chapters. This chapter has a specific regional focus in considering the coastal zone, the area influenced directly by the presence of both land and open sea nearby. Due to the small size of the Baltic Sea, a significant part of it belongs to the coastal zone. In Section 8.1 the basic properties of the coastal zone are discussed. Sections 8.2 and 8.3 present the most important coastal oceanography field in the Baltic Sea: sea-level elevation with its variations and upwelling. Section 8.4 is about coastal ice—from landfast ice to the boundary of drift ice and landfast ice. Section 8.5 is about coastal weather.

8.1 COASTAL ZONE

8.1.1 How best to define the coastal zone?

Coastal zones play a specific role in the Baltic Sea, because their extent is relatively large compared with the entire area of the sea (Figure 8.1). The definition of a coastal area or a coastal zone is a complicated task in the Baltic Sea because there is no such clear topographic feature as the continental shelf in the World Ocean. Intensive physical processes take place in the coastal zone, and additionally water and material exchange between the coastal zone and the open sea is marked. Here we shall take into consideration several views on how best to solve this problem, each usable in certain situations. We mainly allow definitions that are based on physical criteria.

The coastal zone is, of course, the area between the land and the open sea. The width of this zone in the Baltic Sea is the key question. A simple approach is to define

Figure 8.1. Coastal landscape in the Baltic Sea from the Curonian Spit. (Courtesy of Dr. Inga Dailidienė.)

the coastal zone as an area where water depth is less than a fixed value, say 20 meters. This makes islands part of the coastal zone, except for a few in the central basins. Areas shallower than 20 meters cover about 25% of the total area of the Baltic Sea. In winter conditions, a natural boundary of the coastal zone is the fast ice edge, which normally lies near the 10-meter isobath (Leppäranta, 1981b).

The coastal zone can be defined as the area offshore that is subject to the influence of the shoreline. As an example, Lessin and Raudsepp (2007) concluded that the coastal zone is characterized by large freshwater fluxes, nutrients, and organic matter from the land and by light reaching the seabed due to the shallow depths. This classification reflects the fact that the coastal zone is highly productive and at risk of eutrophication.

A theoretical approach can be taken on the basis of ocean dynamics. According to the characteristic baroclinic Rossby radius of deformation, the width of the coastal zone is 3 km–10 km in the Baltic Sea proper (Fennel et al., 1991) while in the Gulf of Finland it is only 2 km–4 km (Alenius et al., 2003). This width must be taken for water outside archipelagic areas. The offshore extent of the wind-induced upwelling zone can also be employed, its width typically being 10 km–20 km and the same numbers represent the scale of the frictional influence of the coastline. A potential way to define the coastal zone is the area influenced by river plumes (Lessin and Raudsepp, 2007), but this is not feasible since the extent goes too far offshore. Using the zone of influence of wind-generated waves on the sea bottom as the definition would limit the coastal zone to depths less than about 20 meters.

In applied research it may be necessary to take an administrative point of view to define the coastal zone. The EU Water Framework Directive says that the coastal zone is: "surface water on the landward side of a line every point of which is at distance of one nautical mile on the seaward side from the nearest point of the baseline from which the breadth of territorial waters is measured" (EU Water Framework Directive, 2000). In practice, the width of the coastal zone is large according to this definition because the baseline mostly follows the outer archipelago.

8.1.2 Formation of fronts and river plumes

The most interesting parts of the thermohaline structure of the Baltic Sea are the boundaries between different watermasses (i.e., areas of large horizontal density gradients). The numerous fronts observed in the Baltic Sea can be formed either between meso-scale circulation patterns or in transition zones between sea areas; for example, at straits (Soomere et al., 2008a). Pavelson (2005) states that the diversity of watermasses follows from the estuarine-like character of the Baltic Sea, the circulation in and between the sub-basins, and vigorous modulation by coastal processes. Fronts can be found in the whole water body, but those formed in the uppermost layer are the most interesting and widely studied due to their implications for biochemical processes. Internal fronts, on the other hand, play a role in the horizontal mixing processes of deep layers.

There is a large variety of fronts in the Baltic Sea. According to Pavelson (1988) they can be divided up as follows: (1) quasi-permanent salinity fronts at the entrance

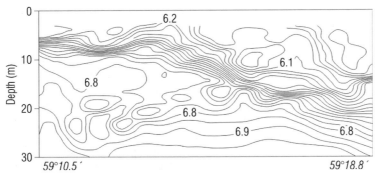

Figure 8.2. The quasi-permanent salinity front (in ‰) at the mouth of the Gulf of Finland. (Redrawn from Kononen et al., 1996.)

to the larger gulfs or fronts related to the general circulation pattern; (2) meso-scale salinity fronts, the most frequently found in the Baltic; (3) temperature fronts most likely formed by the interplay of watermasses of different thermal stratification (a rare phenomenon); (4) density-compensating fronts, sometimes with considerable cross-front temperature and salinity differences; and (5) wind-forced upwelling fronts in coastal zones with large temperature differences and gradients.

A good and well-known example of such dynamic features is the Type 1 quasi-permanent salinity (density) front at the entrance to the Gulf of Finland. Its existence has been confirmed by extensive field studies (see Kononen et al., 1996; Pavelson et al., 1997; Laanemets et al., 1997) and also by numerical modeling (Elken, 1994; Andrejev et al., 2004a). According to these authors the dynamical background of the front can be explained by the existence of cyclonic mean circulation in the Gulf of Finland (see Palmén, 1930; Andrejev et al., 2004a, see Figure 8.2). The saltier water of the northern Baltic Sea proper intrudes into the gulf along the Estonian coast whereas the seaward flow of fresher gulf water above the pycnocline occurs at the Finnish coast. The interface of these inflowing and outflowing waters characterized by a different salinity and/or temperature forms this quasi-permanent front at the entrance area of the Gulf of Finland. The front is typically oriented in a southwest–northeast direction (i.e., positioned approximately parallel to the bottom slope). The frontal area responds to wind forcing (Pavelson et al., 1997) so that under easterly winds the denser (saltier) watermass moves offshore. Thus, the front becomes sharp and strongly inclined to the sea surface. During westerly winds, the less dense Gulf of Finland watermass forms a surface layer over the denser watermass creating a secondary pycnocline approximately in the middle of the upper layer. Pavelson et al. (1997) also studied changes in potential energy (stratification conditions) in the upper layer. These changes are coupled with differential advection induced by along-front wind stress and wind-generated vertical mixing.

There are also other areas (Irbe Strait, Northern Quark, etc.) in the Baltic Sea with quasi-permanent or meso-scale salinity fronts (Type 1 or Type 2), the best known being the one in the Danish Straits (Bo Pedersen, 1993). This has already been discussed, albeit briefly, in Chapter 5 in connection with inflows to the Baltic and in relation to intrusions.

The horizontal temperature field also shows frontal structures (Type 3 or Type 5). Kahru et al. (1995) found from satellite images that surface temperature fronts occur predominantly in areas with a straight and uniformly sloping topography. Major frontal areas exist along the northwestern coast of the Gulf of Finland and near the eastern coast of the Sea of Bothnia. The main factors that produce these fronts are interaction between coastal upwelling and coastal jets, formation of eddies, differential heating and cooling, and water exchange between basins with different water characteristics (advection of cold/warm water). Temperature fronts are further discussed in Section 8.3 in connection with upwelling.

There are several large rivers flowing into the Baltic Sea whose plumes occasionally extend several tens of kilometers from the shore. Very little research, however, has been carried out into these plumes. River plumes are important for watermass formation processes and complicate the dynamics at river mouths. A flowing plume can be held trapped to the coast by a Kelvin wave mechanism for long distances (Feistel et al., 2008). The Oder's plume has been investigated in the southern Baltic Sea (Siegel et al., 1999). Figure 8.3 shows a Luleå River plume observed by ERS-1 SAR.

8.1.3 Specific features of archipelago areas

There are a few important archipelago areas in the Baltic Sea: the Archipelago Sea, Stockholm Archipelago, Vaasa Archipelago, Archipelago of the Northern Bay of Bothnia, and the Belt Sea. Circulation and hydrography in these areas are influenced by the morphology of island systems, and the short spatial scales require very high–resolution measurements. The steering effect of bathymetry and strong friction impose major difficulties for numerical modeling.

The Archipelago Sea is the largest archipelago. It has been investigated by Finnish oceanographers and its basics are well-known (Dr. Jari Hänninen, pers. commun.). The Archipelago Research Institute of Turku University is located nearby (Figure 8.4). The Archipelago Sea is between the Sea of Bothnia and the Gulf of Finland, within Finnish territorial waters. It is characterized by an enormous topographic complexity and includes some 25,000 islands. The average water depth is only 19 meters and the area is 8,893 km^2. The total drainage area is about 8,900 km^2 (of which lakes cover less than 2% and arable lands 28%). The salinity of surface water varies between 4‰–6‰, depending on the distance to the mainland and river mouths. In the Archipelago Sea the halocline is very weak or absent and the surface layer is subject to temperature stratification in summer with a thermocline at a depth of 10 m–20 m, and a deeper, denser, and colder layer extending down to the seabed. The emergence of the thermocline triggers micro-algae production in spring (HELCOM, 1993, 1996).

There are north–south channels with depths ranging from 30 m to 40 m through this sea area. They form a part of the water exchange system between the Sea of Bothnia and the Gotland Sea. The sill depth towards the Sea of Bothnia is about 18 meters. The main route goes through the Kihti Strait. The area is roughly divided into inner and outer archipelagos, with the outer archipelago consisting mainly of

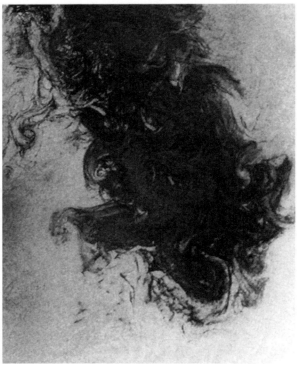

Figure 8.3. A river plume (Luleå River) at the Swedish coast of the Bay of Bothnia observed by ERS-1 SAR. The bottom photograph shows an enlargement of the tip of the plume. (Herlevi and Leppäranta, 1994.)

Sec. 8.1] Coastal zone 267

Figure 8.4. The island of Seili in the Archipelago Sea. This is the home of the Archipelago Research Institute of Turku University, Finland. (Courtesy of Dr. Ilppo Vuorinen.)

smaller, uninhabited islands. The archipelago covers a roughly triangular area with the towns of Mariehamn, Uusikaupunki, and Hanko at the corners.

The Stockholm Archipelago is the largest archipelago in Sweden, and one of the largest in the Baltic Sea. It stretches from Stockholm some 60 kilometers to the east, bordering the Åland Sea. It consists of approximately 24,000 islands and mainly follows the coastline of the Södermanland and Uppland provinces. The area is characterized by a mixture of different lengthscales in both horizontal and vertical directions. There are several basins that are a few kilometers wide, more than 100 m deep, and connected by straits whose widths and sill depths are clearly smaller. Surface water has a salinity of about 6‰ with somewhat higher salinities in the bottom layer. The most pronounced freshwater discharge is from Lake Mälaren with an outflow of approximately $165\,\mathrm{m}^3/\mathrm{s}$.

The area is traditionally partitioned into three parts: the inner, the middle, and the outer archipelago. These areas are forced by different mechanisms: in the inner archipelago the dominating exchange process is estuarine circulation, induced by marked freshwater discharge and vertical mixing. In the outer and middle archipelagos, density fluctuations due to Ekman pumping along the open Baltic Sea interface produce another type of baroclinic process, one that is clearly dominant.

Modeling such an area is a complicated task. It can be carried out, for example, by a cascade of models where local archipelago models get their open boundary conditions in the east from a large-scale, three-dimensional model (Engqvist and Andrejev, 2003).

8.2 SEA-LEVEL ELEVATION

8.2.1 Sea-level dynamics

Sea-level variability is influenced by meteorological, astronomical, and hydrological elements. Relatively rapid land uplift in some parts of the Baltic Sea has led to long-term apparent decrease in local sea-level. The concept of a "theoretical mean sea-level" represents an estimator of actual mean sea-level and provides a reference level to compare sea-levels in different sections and times. It includes the influence of land uplift and eustatic rise in global mean sea-level; therefore, it changes with time. Taking the measured sea-level relative to the theoretical mean sea-level tells whether the actual sea-level is high or low.

The frequency spectrum of the sea-level in the Baltic Sea shows periodical as well as irregular variations. In short periods of 1 to 24 hours sea-level variations are associated with long waves, currents, tides, and seiches. Witting (1911) concluded that tides in the Baltic Sea are mostly diurnal (components K_1 and O_1) with amplitudes of 0.5 cm–2 cm. Semi-diurnal tides (M_2 and S_2) have amplitudes of 0.5 cm–1.5 cm. Diurnal tides are generated within the Baltic Sea whereas semi-diurnal ones are partially externally forced. The amplitudes of tidal sea-level and currents are small in the Baltic Sea and thus accessing their signal from sea-level records is difficult.

Variations in wind and air pressure cause sea-level changes from periods of about one day to several weeks. The influence of wind is to pile up water, which is particularly so at the end of bays. An air pressure change of 1 mbar corresponds to about 1 cm in sea-level elevation (barometer effect). Air pressure variations thus cause sea-level variations of about ± 50 cm (Lisitzin, 1974; Carlsson, 1997).

Annual and semi-annual periods are next to be considered. Variations in the density of seawater cause small sea-level variations on timescales of several months. For example, if water density changes by 1 kg/m^3 this would lead to a change in sea-level of 5 cm. The permanent horizontal water density difference of slightly less than 10 kg/m^3 between the northern and southern Baltic Sea is reflected by the higher sea-level in the north than the south. On average, the sea-level declines 35 cm–40 cm from the Bay of Bothnia to the Skagerrak. The active water storage capacity of the Baltic Sea, taken as the difference between monthly maximum and minimum volumes, corresponds to about a 1 m thick water layer over the Baltic Sea. Landsort (off Stockholm) represents the mean sea-level of the Baltic and is also the nodal point of a uninodal seiche.

Mean sea-level varies according to the water balance as shown in Section 4.1. Large-scale processes causing variability in the NAO-index also strongly affect the

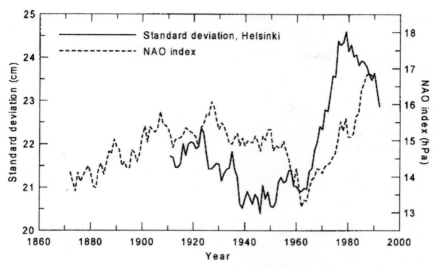

Figure 8.5. Detrended annual sea-level at Helsinki and the annual mean NAO-index, 15-year floating averages. (Redrawn from Johansson *et al.*, 2001.)

sea-level, this being especially true between December and May (Johansson *et al.*, 2001, see Figure 8.5). A positive NAO-index means strong westerly winds and consequently increased water transport to the Baltic Sea from the North Sea.

8.2.2 Long-term changes

Long-term changes to Baltic sea-level are due to land uplift which is a geological process, and to eustatic sea-level rise which is caused by climatological variations. The future climate change impact on sea-level is discussed in Chapter 10 and extremes in sea-level are introduced in Chapter 9. Ever since the Weichselian Glaciation the land is still rising some 8 mm–9 mm in the Bay of Bothnia whereas in the southernmost part of the sea the land rise is negligible (see Figure 3.1). The rate of uplift also varies in time. Ekman (1988) studied the longest continuous time-series on sea-level in the Baltic Sea[1]; this was in Stockholm, where measurements started in 1774. Land rise on average has been 5 mm/year and the rate has slowed somewhat during the last century. On the other hand, long-term changes also depend on the opposite effect of eustatic rise in the sea-level of oceans. This is today about 3 mm per year (IPCC, 2007) due to the melting of glaciers and the thermal expansion of seawater. The effect of tectonic processes is estimated to be very small.

The determination of land uplift is a complicated problem. When the water balance of the Baltic Sea is known it can be used to calculate more accurate estimates for land uplift. When the influence of yearly changes in water balance (δ) is subtracted from the observed sea-level, the result gives the linear effects of the land uplift (u_{ai})

---

[1] The St. Petersburg time-series started in 1703 but was not continuous during the early years.

and the global (eustatic) water-level rise (G). These terms can be combined as relative land uplift u_{ri}:

$$u_{ri} = u_{ia} - G + \delta \qquad (8.1)$$

Presently (*http://www.fimr.fi*) relative land uplift in the northern Bay of Bothnia in Kemi is 7.20 ± 0.27 mm/year whereas in the eastern Gulf of Finland the numbers are much smaller (in Hamina it is 1.62 ± 0.26 mm/year).

Land rise becomes visible at the Finnish coast of the Gulf of Finland and in the Gulf of Bothnia. Small bays become isolated from the sea, developing from weakly connected bays called *fladas* (Lindholm *et al.*, 1989) and eventually becoming lakes (called *kluuvi* in Finnish). A flada is formed when land gradually rises, water exchange in a bay gradually decreases, and finally is only connected to the main basin through a narrow strait.

8.2.3 Observed variability

General changes during the last century

The spatial variability of long-term sea-level changes is mainly caused by the postglacial rebound of the Fennoscandian Plate and eustatic sea-level rise. The latter dominates at the southern part of the Baltic Sea where the role of land uplift is negligible. In general, eustatic increase in sea-level was found to be accelerating in the last decades at many tide gauge stations (see Lisitzin, 1957, 1966a, b; Johansson *et al.*, 2004). This subsection in mostly based on the paper by Soomere *et al.* (2008a). There are several other studies devoted to eustatic sea-level rise (see Kalas, 1993; Stigge, 1993; Kont *et al.*, 1997; Johanssson *et al.*, 2001, 2004; Fenger *et al.*, 2001; Suursaar *et al.*, 2006a).

Johansson *et al.* (2001) examined conditions on the coast of Finland over the past 100 years based on time-series from 13 tide gauges. They provide evidence supporting a change in apparent land uplift along the northern coast of the Gulf of Finland around 1960. The first regression line for fitted mean sea-level, used in many applications in Finland as the "true" sea-level, was fitted to annual mean sea levels from the beginning of measurements up to 1960, the latter one showing a clearly different slope from then on. Suursaar *et al.* (2006a) studied sea-level changes on the Estonian coast (Gulf of Finland, Gulf of Riga) and found a faster increase in sea-level has taken place in recent decades. They found a eustatic sea-level rise varying between 9.5 cm and 15.4 cm depending on location between 1950 and 2002 (i.e., 1.8 mm–3.0 mm/year, which corresponds to our understanding of the present rate—IPCC, 2007).

In the southern part of the Baltic Sea extensive investigations have been made in Poland and Germany to understand the future of their coastal landforms (Zeidler, 1997; Pruszak and Zawadzka, 2005). In Lithuania, long-term sea-level shows a clear increase during the last 100 years due to eustatic sea-level rise and local effects (Dailidienė *et al.*, 2006). For example, in Klaipėda, long-term sea-level increased by about 13.9 cm between 1898 and 2002. Furthermore, it is remarkable that the increase was not linear during the study period: until World War II the increase was found to be negligible whereas from the 1950s the increase became faster, being since

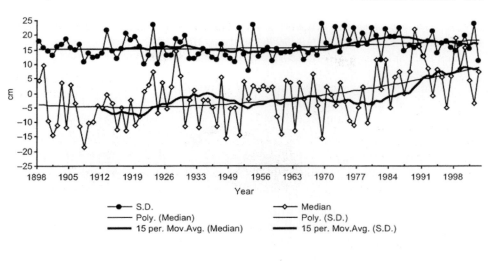

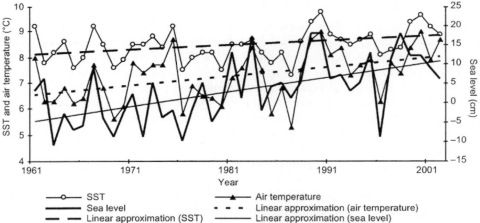

Figure 8.6. Annual mean sea-level, sea surface temperature, and air temperature along with linear trends in Klaipėda, Lithuania in 1961–2002. (Redrawn from Dailidienė et al., 2006.)

the 1970s about 3 mm/year (Figure 8.6). The reasons behind the local effects are increased water temperature and the thermal expansion of water due to the enhanced and frequent advection of warm and moist maritime airmasses during the cold season (October–November).

Periodic and aperiodic variability

Remarkable aperiodic inter-annual sea-level variability over 2 to 6 years has been documented by Sjöberg and Fan (1986) and Vermeer et al. (1988). It is best observed in the long record in Stockholm (Ekman, 1988) where it accounts for 10 cm–20 cm. Vermeer et al. (1988) suggested that its main cause is variation in wind forcing directed east–west over the Danish Straits pumping water into and out of the Baltic Sea. Samuelsson and Stigebrandt (1996) analysed this feature based on a box model in which each sub-basin constitutes one box and one sea-level series is used per box. They confirmed that this variability is forced mostly by sea-level variations in the Kattegat and by changes in freshwater flux to the Baltic Sea. About 50%–80% of annual sea-level variations are imported to the Baltic Sea from the North Sea. This argumentation is further supported by the statistical analysis by Lass and Matthäus (1996) concerning the seasonal cycle of winds and sea-level in the Baltic Sea.

The most dominant long-term component in sea-level variation is the annual cycle. Johansson et al. (2001) noticed that its range was about 20 cm at Helsinki and at Hamina during the first half of the 20th century; it increased rapidly and reached about 35 cm in the 1970s–1980s, and decreased again to around 25 cm at the end of the century. This is in accordance with the analysis of Ekman and Stigebrandt (1990) who found a statistically significant increase in the amplitude of annual variation in sea-level between 1825 and 1984. These changes only partially match the long-term changes in the wind regime described in Alexandersson et al. (1998). The average range in sea-level variation at the southern coast of the Gulf of Finland seems to be somewhat smaller. Its values, calculated from monthly mean values of the sea-level for a 5-year period (1978–1982), vary between 17 cm and 23 cm along the coast and between high water in October and low water in April (Raudsepp et al., 1999).

Seiches and tides are the only significant peaks in timescales from hours up to several months of sea-level variations in the Baltic Sea. The spectrum of daily mean values is also rather wide (Vermeer et al., 1988). The mid-range variability of sea-level in the Baltic Sea, with periods between 2 days and several years, has a significant spatial coherence (Samuelsson and Stigebrandt, 1996). Ekman (1996) analysed sea-level data from the 100-year period 1892–1991 from 56 sea-level stations. Sea-level variations from periods of 6 months up to about 10 years (including the *pole tide*[2] with a period of 14.3 months) have an almost identical geographical pattern and, most probably, a common origin. Maximum sea-level variation occurs towards the end of the Gulf of Finland and the Gulf of Bothnia and the minimum in the southwestern Kattegat. The sea-level of the entire Baltic Sea thus acts as a quarter-wavelength oscillator for these periods. The analysis of Samuelsson and Stigebrandt (1996) leads to the same result for shorter periods, from a few days up to one month. For periods longer than 1 month and up to 9 years, however, the sea-level regime was found to resemble a half-wavelength oscillator.

One major constituent in sea-level variability is due to short-term and long-term wind forcing. Andersson (2002) used historical sea-level time-series from Stockholm to demonstrate the key role played by winter climate. She concluded that there exists a statistical connection between the Stockholm sea-level and NAO winter values. Johansson et al. (2001, 2004) studied the correlation between sea-level variability and the NAO-index at the Finnish coastal station and find a high correlation (see below

---

[2] An ocean tide, theoretically 6 mm in amplitude, caused by the Chandler wobble of the Earth, has a period of 428 days. Earth's poles move in an irregular cycle 3 m–15 m in diameter during each oscillation.

for details). Additionally, the study by Suursaar et al. (2006a) confirmed the earlier results to be valid for the Estonian coast as well. The same kinds of results have been obtained for the southern Baltic. Janssen (2002) confirmed the connection between the wintertime NAO-index and sea-levels. At the Lithuanian coast, annual mean sea-level fluctuation is found to be correlated with the wintertime NAO-index (Dailidienė et al., 2006).

Trends

Ekman (1996) found a statistically significant (at the 99% level) increase in sea-level in December–January and a decrease in February–March at Stockholm when investigating long-term variability. The analogous trends were not significant for the Gulf of Finland except the decrease in February–March at Hanko (significance 97%). Johansson et al. (2001) reported that monthly means during April–October had decreased and those in November–March had increased, but both trends were not significant.

On the coast of Finland the timing of maximum monthly mean sea-level has shifted to a later timing during the last century, from the end of September to two weeks later in the 1980s and by a full month by 2000 (Johansson et al., 2001; cf. also Raudsepp et al., 1999). Thus, sea-level seems to be currently at its highest one month later than 50–70 years ago, but again this trend is not statistically significant. However, such shifts have been observed at other sites. For example, the month with the most frequent occurrence of maximum monthly mean in Stockholm over a range of years has shifted from July–August to December (Ekman, 1996).

The annual standard deviation in sea level (using a 4 h time interval), a simple measure of sea-level variability, correlates with the NAO-index at the 97%–99% level of significance. It has an apparent increasing trend, with a small minimum in the 1950s and with the most prominent increase in the 1960s–1970s, and shows large spatial variability. The largest values occur, as expected, in the innermost part of the Gulf of Finland (Johansson et al., 2001). The probability distributions of sea-level elevations over a long time period are asymmetric: high sea-levels are more probable than low sea-levels (Figure 8.7). The probabilities of extremely high sea-levels have also increased during the 20th century. The most pronounced variability occurs during the winter period (November to January) while the summer period (May to July) has clearly smaller variability. Probability distributions have different shapes during different seasons and in different sub-basins of the Baltic Sea (Johansson et al., 2001).

The overall trend in short-period variability of sea-level elevation (of the order of a few days) shows only a marginal increase during the 20th century. In the Gulf of Finland a roughly periodical behavior (∼30-year cycle) was found in short-term (8 days) variability of sea-level values in winter. A local minimum in the 1960s was followed by an increase which lasted until the 1980s, and then a decrease until the end of the century. The main cause of all this was the change in wind conditions in the Baltic Sea (Johansson et al., 2001) and possibly over the transition area between the North Sea and the Baltic Sea.

274   Coastal and local processes   [Ch. 8

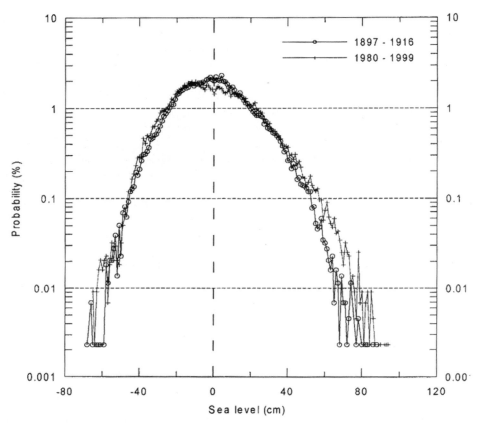

Figure 8.7. Sea-level probability distributions at Hanko for the 20-year periods of 1897–1916 and 1980–1999. The values have been referenced to the annual mean sea-level in question. The method excludes long-term interannual variations and makes the distribution representative of intra-annual sea-level variability. (Redrawn from Johansson et al., 2001.)

8.2.4 Modeling

The modeling of sea-level dynamics has been based on both two-dimensional and three-dimensional models. In practice, two-dimensional barotropic models have been shown good enough for short-term sea-level forecasting (see Häkkinen, 1980). The system of equations is obtained by vertical integration of the equation of motion and the continuity equation. Vertical friction then transfers the wind stress to the water body and motion is restricted by bottom friction. The linear equations are thus—see Equations (5.20)

$$\frac{\partial \tilde{U}}{\partial t} - f\mathbf{k} \times \tilde{U} = -g\nabla\xi + \frac{\tau_a}{\rho H} - r\frac{\tilde{U}^2}{H} \qquad (8.2)$$

$$\frac{\partial \xi}{\partial t} = -\nabla \cdot (H\tilde{U}) \qquad (8.3)$$

where $\tilde{U}$ is vertically integrated velocity; ξ is sea-level elevation; H is depth; τ_a is wind stress; and r is the bottom friction coefficient. In the shallow-water model—see Equations (6.5)—advection is correctly given because horizontal velocity is independent of depth. In a vertically integrated model with friction this is not true. Advection is ignored or approximated by $\tilde{U} \cdot \nabla \tilde{U}$.

A barotropic model is capable of producing realistic sea-level variations provided atmospheric forcing is realistic, and thus the forecasting timescale equals the timescale of weather forecasting (the dynamic memory of the system is short). The adequacy of boundary information in the Danish Straits is extremely important for correct simulations, and any bias there is reflected in the predicted sea-level after a few days (or even sooner, depending on the location of the site), a feature which is typical of most sea-level models in 2008. A solution to improve the situation is to extend the model to the North Sea, as was done, for example, in the HIROMB model (see Section 9.5), but this helps only to a limited extent since the main problem is to determine the flow through the Danish Straits. Today, three-dimensional modeling is commonly used in tandem with two-dimensional modeling. These modeling achievements are discussed in connection with operational modeling in Section 9.5.

8.3 UPWELLING

8.3.1 What is upwelling?

In general, upwelling is the vertical motion of watermasses caused by horizontal divergence in the surface layer of the ocean. As a result of mass conservation, dense, cooler, and usually nutrient-rich water raises up towards the ocean surface, replacing the diverging warmer, mostly nutrient-depleted surface water. Regions where upwelling is most pronounced in the World Ocean cover the coasts of Peru, Chile, Arabian Sea, West Africa, eastern New Zealand, and the California Coast (see Sverdrup, 1938; Defant, 1961; Philander and Yoon, 1982). Due to the importance of upwelling in the ocean, there are a large number of studies dealing with the topic. Detailed descriptions of upwelling are given in the textbook by Tomzcak and Godfrey (1994) and the review by Smith (1968). For the Baltic Sea a review of upwelling has recently been written by Lehmann and Myrberg (2008), and this paper is used as the main reference in this section.

Upwelling is an important process in the Baltic Sea as well; it is driven by wind forcing. Since the Baltic Sea is a semi-enclosed, relatively small basin, winds from virtually any direction blow parallel to some section of the coast and cause coastal upwelling. During the thermally stratified period, upwelling can lead to strong sea surface temperature drop, more than 10°C, changing drastically the thermal balance and stability conditions at the sea surface. Upwelling can additionally play a key role in replenishing the euphotic zone with the nutritional components necessary for biological productivity when the surface layer is depleted of nutrients.

8.3.2 Theory

In general, two classes of upwelling can be distinguished: open sea and coastal upwelling. The first class is of considerably larger scale and pertains to such vertical motions as those caused by the wind (Ekman pumping), by influences of the main oceanic thermocline, and by equatorial ocean currents. Coastal upwelling is more regionally limited than open ocean upwelling but its stronger vertical motion is associated with greater climatic and biological impact. Vertical motions in coastal upwelling are on the order of 10^{-5} m s^{-1} and in open ocean upwelling they are about 10^{-6} m s^{-1}, accounting for 1 m da^{-1} and 0.1 m da^{-1}, respectively (Dietrich, 1972). The reverse of upwelling is called "downwelling" which is correlated with surface convergence and divergence at a lower layer where descending water terminates. Downwelling causes not so clear signs at the sea surface and does not have such biological relevance as upwelling.

Ekman transport

The Ekman theory of ocean currents was described in Chapter 5. Here applications of the theory to upwelling will be briefly summarized. Due to the effect of the Earth's rotation and frictional forces, net transport due to wind stress is directed 90° to the right of the wind in the northern hemisphere. Thorade (1909) first applied Ekman's theory to describe an upwelling situation. He showed that coastal winds blowing parallel to the coast were sufficient to bring about offshore transport of surface water. The Ekman spiral is a theoretical consideration which has only been qualitatively verified by direct measurements. However, observed transports correspond well with the theoretical Ekman transport, which is independent of the eddy viscosity of water. In shallow areas close to the surf zone, current and transport are almost parallel to the wind.

For a continuously stratified ocean, the momentum exerted by wind stress rapidly penetrates downward by turbulent viscosity to create a well-mixed layer that has a certain depth. If a well-mixed surface layer already exists overlying a strong thermocline, the momentum generated by surface stress will be spread through the whole of the mixed layer in a time shorter than the inertial period, but the depth of the mixed layer will increase due to the vertical shear of inertial waves in a time longer than inertial. Vertical momentum transport due to wind-induced turbulence causes a more or less uniform current direction in the mixed layer (Pollard, 1970; Krauss, 1981). Once wind stress has been determined, the corresponding offshore transport can be calculated from Ekman transport relations (Smith, 1968). For two similar wind events the same transport will result, but whether upwelling will have a temperature signal in the sea surface depends additionally on bathymetry and thermal stratification.

The principal response of a stratified elongated basin to constant wind in the longitudinal direction of the basin can be described as follows (Krauss and Brügge, 1991):

(i) In the surface layer, Ekman transport runs perpendicular to the wind direction.

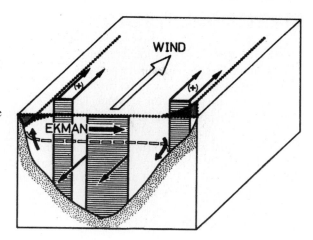

Figure 8.8. Principal response of an elongated basin to constant wind in the length direction of the basin. There are coastal jets at both sides of the basin in the wind direction and a return flow in the middle. Upwelling take place at the left coast, downwelling at the right coast. (Redrawn from Krauss and Brügge, 1991.)

(ii) This Ekman transport produces sea-level rise on the right-hand coast (viewed into the wind direction) and a fall on the left-hand coast. Furthermore, downwelling occurs on the right-hand coast and upwelling on the left-hand coast resulting in baroclinicity of the same sign at both coasts.

(iii) Consequently, coastal jets are produced along both coasts in the wind direction and a slow return flow compensates for this transport in the central area of the basin.

Figure 8.8 gives a scheme of this circulation which can be applied to the different subbasins of the Baltic Sea. Krauss and Brügge (1991) demonstrated that upwelling in the Baltic Sea should be regarded as a three-dimensional current system affecting not only the local coast but also the opposite coast and the interior of the basin (Fennel and Sturm, 1992; Fennel and Seifert, 1995). However, the vertical extension of Ekman compensation below the mixed layer is restricted in the Baltic Sea due to the existence of the halocline normally situated at a depth of 50 m–80 m.

Upwelling as a meso-scale feature is scaled by the baroclinic Rossby radius. As thermal stratification varies seasonally in response of solar heating and wind-induced mixing in the Baltic Sea, the baroclinic Rossby radius varies between 2 km and 10 km (Fennel *et al.*, 1991; Alenius *et al.*, 2003). Typical scales of upwelling in the Baltic Sea are

- vertical motion of 10^{-5} m s^{-1}–10^{-4} m s which is about 1 m da^{-1}–10 m da^{-1}
- horizontal scales of 10 km–20 km offshore and 100 km alongshore
- temperature change of 1°C–5°C da^{-1}
- temperature gradient of 1°C–5°C km^{-1}
- a lifetime of several days up to one month.

Abrupt changes in the alongshore component of wind stress, coastline irregularities, and topographical variations in the sea bottom generate internal waves that limit the amplitude of upwelling and give rise to a countercurrent under the thermocline (Gill and Clarke, 1974). These waves may sometimes resemble Kelvin waves (Crepon et al., 1984). The shape of coastline irregularity and the incident angle of the wind are the determining factors as to whether upwelling will be stabilized or destabilized by propagating waves. Continuous variations in the angle of incident wind on the coasts initiates Kelvin wave fronts, which lead to stabilization or destabilization of the upwelling (Crepon et al. 1984; Fennel and Seifert, 1995). Thus, upwelling areas are related to the shape of the coast and filaments will be generated at the same locations under similar upwelling conditions. Even remote areas, which are not affected by local upwelling directly, will be reached by propagating wave fronts. Observations show (Fennel and Seifert, 1995) that in the Southwestern Baltic Sea during upwelling-favorable winds, the intensity of coastal upwelling varies along-shore and even turns into downwelling in the western part. This is due to the generation of Kelvin waves.

8.3.3 Atmospheric forcing

Accurate descriptions of the wind, temperature, and humidity fields are essential for studying upwelling dynamics. From Ekman's theory, alongshore winds are most effective at generating upwelling in large basins, because the resulting surface layer motion created over a wide sea area is applied in a relatively narrow area of rapidly varying depth.

One measure of the characteristics of wind capable of producing upwelling is wind impulse I (Haapala, 1994),

$$I = \int_0^t \tau_a \, dt' = \int_0^t C_a \rho_a U_a^2 \, dt' \tag{8.4}$$

where ρ_a is air density; C_a is the drag coefficient; U_a is the wind speed at a 10 m height; and t is wind duration. The occurrence of upwelling depends on stratification and the strength of the wind impulse. During thermal stratification a $4{,}000 \, \text{kg m}^{-1} \, \text{s}^{-1}$–$9{,}000 \, \text{kg m}^{-1} \, \text{s}^{-1}$ wind impulse of about 60 h duration is needed to generate upwelling, and when the sea is thermally homogeneous the impulse required is $10{,}500 \, \text{kg m}^{-1} \, \text{s}^{-1}$–$14{,}000 \, \text{kg m}^{-1} \, \text{s}^{-1}$. This implies that under strongly stratified conditions wind stress has a direct effect only on the relatively thin water column over the thermocline. Even weak winds can lead to upwelling. If stratification is weak the influence of the wind penetrates much deeper, and more wind energy is needed to produce upwelling (Haapala, 1994).

In the Baltic area there exist different general weather conditions that are favorable for upwelling at various coastal areas. Bychkova et al. (1988) identified 22 typical areas in different parts of the Baltic Sea that were favorable for upwelling in relation to 11 different wind conditions (see Figure 8.9). For example, Wind Event I (northeasterly wind) is connected with Upwelling Regions 3, 5, 6, and 9 while for Wind Event VI (west-southwesterly winds) are connected with Upwelling Regions 2, 10–20, and 22 (see Bychkova and Viktorov, 1987; Bychkova et al., 1988 for details).

Sec. 8.3] Upwelling 279

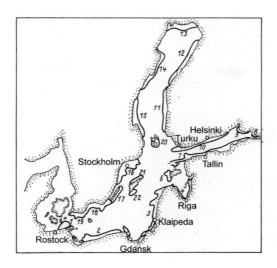

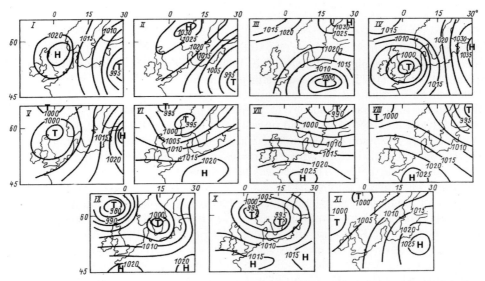

Figure 8.9. (Top) Common upwelling zones in the Baltic Sea (Bychkova et al., 1988). (Bottom) Eleven typical atmospheric conditions that are favorable for upwelling in different areas of the Baltic Sea. (Redrawn from Bychkova and Victorov, 1987.)

8.3.4 Early results based on *in situ* observations

Although initial results had been published before World War II, upwelling as a process itself remained poorly understood for a long while. The first documented scientific observation of the influence of upwelling on the properties of the surface layer in the Baltic Sea was carried out by Alexander von Humboldt in 1834 (Kortum and Lehmann, 1997). At the beginning of upwelling studies, measurements were to

Figure 8.10. Tvärminne Zoological Station of the University of Helsinki, located at the upwelling zone on the Finnish coast of the Gulf of Finland. (Courtesy of Tvärminne Zoological Station.)

some extent random in character; they were certainly not the results of well-prepared measurement campaigns. Only in the 1970s were more comprehensive results obtained. Walin (1972) detected upwelling at the Swedish east coast and found that temperature fluctuations extended only 5 km–10 km. He also proposed that these fluctuations may have a tendency to propagate along the coast as internal Kelvin waves. Svansson (1975) also found upwelling in the Hanö Bight, in agreement with Walin (1972). Svansson (1975) took up the question of the biological significance of upwelling in coastal regions where nutrients can be transported to the uppermost, euphotic layer of the sea. This fact was later found to be an important element in upwelling phenomena (see p. 282).

One of the main areas where upwelling has been observed in temperature measurements is the northern (Finnish) coast of the Gulf of Finland (Figure 8.10) (Hela, 1976; Niemi 1979; Kononen and Niemi, 1986; Haapala, 1994). All these papers confirm that upwelling is especially favored by southwesterly winds. In such cases temperature can drop by 10°C in 1–2 days during stratified periods. During such conditions, when the surface layer of water can be depleted of nutrients, upwelling plays a key role in replenishing the euphotic zone with the nutritional components necessary for biological productivity. Consequently, upwelling favors fishing in the area (Sjöblom, 1967). Niemi (1979) found that, in areas where upwelling brings phosphorus-rich deepwater to the surface, the nitrogen/phosphorus ratio becomes low which favors the blooming of nitrogen-fixing blue-green algae.

8.3.5 Analyses based on satellite measurements

The utilization of satellite measurements for upwelling studies started in the early 1980s and since then spaceborne measurements of various kinds (NOAA/AVHRR, etc.) have been utilized by numerous authors (see Siegel *et al.*, 1994; Kahru *et al.*, 1995; Lass *et al.*, 2003b; Kowalewski and Ostrowski, 2005). See Figure 8.11 in the color section. Among the most comprehensive studies is the one by Horstmann (1983) where the author studied upwelling at the southern coast of the Baltic Sea concluding that its genesis was easterly winds. Gidhagen (1987) made an analysis based on AVHRR data and concluded that upwelling at the Swedish coast takes place up to 10 km–20 km offshore and has a length of the order of 100 kilometers alongshore. According to Gidhagen (1987) water is lifted to the surface from depths of 20 m–40 m, which is somewhat deeper than previously estimated. He also found that in some areas upwelling takes place one-fourth to one-third of the time. Bychkova and Victorov (1987) found 14 upwelling cases around the Baltic Sea with different scales and lifetimes. In the southwestern Baltic Sea, at the German and Polish coasts, satellite observations of upwelling were analysed by Siegel *et al.* (1994).

8.3.6 Observed regional features

Upwelling in various parts of the Baltic Sea has some specific features based on topography, orientation, and shape of the coastline. A wind pattern favorable for the birth of upwelling depends on local features. At the Polish coast upwelling is favored by such meteorological conditions when a high-pressure system is located over northeastern Russia, accompanied by light or moderate easterly to southeasterly winds over the southern Baltic (Malicki and Wielbińska, 1992). Most often upwelling has been found to take place offshore the Hel Peninsula in the Gulf of Gdansk (see Matciak *et al.*, 2001). The potential maximum area of all upwelling at the Polish coast equals 10,000 km^2, which is as much as about 30% of the Polish economic zone (Kreżel *et al.*, 2005). At the Lithuanian and Latvian coasts upwellings are favored by northerly winds. At the east coast of the Gulf of Riga upwellings are observed in the presence of southeasterly winds.

In the Gulf of Finland upwelling takes place at the southern coast as a result of east-southeasterly winds; an excellent example of this is seen in Figure 8.12 (see color section). At the northern coast southwesterly winds in particular cause upwellings. Northerly winds may also cause upwelling there. Typically, upwelling takes place near the Hanko Peninsula (Haapala, 1994) or near the Porkkala Peninsula (Sjöblom, 1967). In the Gulf of Bothnia upwelling basically takes place at the Finnish coast as a result of northerly winds and at the Swedish coast as a result of south to southwesterly winds.

Upwellings play an important role in the formation of fronts (Kahru *et al.*, 1995). As an example, a temperature front can often be seen in satellite images at the Finnish side of the Sea of Bothnia. At the Swedish side of the Gulf of Bothnia upwelling can occur in some places one-fourth to one-third of the time. Such areas are Hornslandet (Sea of Bothnia) and the coastal area from Ratan to Bjuröklubb (Bay of Bothnia). At the Swedish coast in the Baltic Sea proper the most well-known upwelling region is the Hanö Bight (as well as the Trelleborg–Ystad region) where the lengthscale is about 60 km and the width 5 km–10 km (Gidhagen, 1987). The steep west coast of Gotland is also a well-known upwelling area as a result of northerly winds (Shaffer,

1979). The most pronounced upwelling area at the German coast is that stretching from the west coast of Hiddensee island in a north-northwesterly direction (see Siegel *et al.*, 1994; Lass *et al.*, 2003b). However, this observed upwelling is not driven by local Ekman offshore transport at the west coast of Hiddensee island. The lowering of the sea-level in the Kattegat caused by easterly winds triggers an adjustment process in the Belt Sea resulting in a pressure step trapped at the Darss Sill. Currents geostrophically adjusted to this step are fed by upwelled water from the west coast of Hiddensee island maintaining the mass balance of the dynamical system. Characteristic scales of the upwelling feature at the west coast of Hiddensee island are its 20 km alongshore basis, 60 km offshore length, duration of 5 days, and temperature drop of about 4°C (Lass *et al.*, 2003b).

8.3.7 Ecosystem implications

The effects of upwelling on late-summer cyanobacteria growth has been studied in detail by Vahtera *et al.* (2005). Usually, phytoplankton growth in the Baltic Sea is nitrogen-limited, an exception being the filamentous cyanobacteria, which fixes atmospheric nitrogen. Cyanobacteria growth is thus phosphorus-limited; it is also limited by temperature. So, the effects of upwelling on cyanobacteria bloom is not straightforward due to the decrease of temperature in upwelling regions and due to potential changes in DIN:DIP (dissolved inorganic nitrogen:dissolved inorganic phosphorus) ratios. According to Laanemets *et al.* (2004) nutriclines, at least in the Gulf of Finland, lie in the thermocline, the phosphacline being shallower than the nitracline. Nutricline, phosphacline, and nitracline are zones of sharp changes in corresponding substances (nutrients, phosphorus, nitrates), analogous to the thermocline and halocline. Thus, upwelling leads to phosphorus enrichment and low DIN:DIP ratios in the euphotic layer. It is possible that filamentous nitrogen-fixing cyanobacteria would benefit from phosphorus enrichment (cf. Niemi, 1979).

However, the direct biological effect of upwelling at the same time as decreasing surface temperature leads to a decline in filamentous cyanobacteria biomass. According to Vahtera *et al.* (2005), *Nodularia spumigena* is most severely affected due to its strong buoyancy and vertical displacement near the surface. The lifetime of a typical upwelling remnant is too short for populations outside the upwelled water to benefit directly from the nutrient input. Owing to the low DIN:DIP ratio of upwelled water, the nitrogen-fixing *Aphanizomenon flos-aquae* populations that live at the top of the thermocline are the most likely to have been in a good position to exploit the additional phosphorus in good light conditions.

However, utilization of the lowered DIN:DIP ratio does not lead in a straightforward way to enhanced cyanobacteria blooming. There is a clear lag (2–3 weeks) between upwelling and biomass increase (Vahtera *et al.*, 2005). Thus, upwelling may enhance cyanobacteria blooming only after a certain "delay time".

8.3.8 Modeling

Due to the small values of the baroclinic Rossby radius and the limited offshore extent of the upwelling area at a steep coastal slope with a width no greater than

10 km–20 km, very high–resolution models are needed to describe the dynamics of upwelling and related processes properly. In spite of these requirements, several model studies have been carried out; however, they were not able to fully resolve the upwelling process (Fennel and Seifert, 1995; Jankowski, 2002; Lehmann et al., 2002; Myrberg and Andrejev, 2003; Myrberg et al., 2008). Recently, Zhurbas et al. (2007) have used a very high–resolution model for the Gulf of Finland and found interesting meso-scale features of upwelling (filaments and squirts—Figure 8.13, see color section). Some models were used to statistically describe upwellings in an attempt to determine their location and corresponding frequency of occurrence (Myrberg and Andrejev, 2003; Kowalewski and Ostrowski, 2005). The annual average frequency of upwelling appeared to be higher than 30% in some parts of the Baltic Sea coast.

The model results presented so far demonstrate the applicability of numerical models for further deepening our understanding of upwelling and related statistical properties. With increasing computer power, numerical models can have horizontal resolution, which allows the full range of meso-scale dynamics for long-term simulations to be studied.

8.4 ICE CONDITIONS

8.4.1 Structure of the coastal ice zone

The ice cover close to land forms a *coastal ice zone*, which extends from the coastline to the drift ice off outer islands. This zone consists of two parts: landfast ice and the landfast ice edge. The extent depends on the thickness of the ice and bottom topography. Landfast ice, or briefly fast ice, is stationary, while the edge may be dynamically active.

At the beginning of the ice season freezing starts at the coast due to shallow depths and the proximity of cold airmasses. The plasticity of ice means that ice can resist loads up to its yield level, which increases with the thickness of ice, and remain as fast ice. A compact sea ice field is stable as long as (see Chapter 7)

$$P^*h > \tau_a L \qquad (8.5)$$

where $P^* \sim 50$ kPa is the ice strength; h is ice thickness; τ_a is wind stress; and L is fetch. In the inner archipelago this condition is satisfied when ice thickness has reached 10 cm or so, and the cover becomes undeformed, smooth fast ice. The landscape is similar to the winter landscape in northern lake districts. However, there are regions with no islands, and then the drift ice may even ride up on the shore under exceptional conditions.

As the ice grows thicker, the ice cover becomes stable in larger areas and the fast ice cover expands laterally. Data analyses from Finnish coastal areas show this growth very well (Figure 8.14). In the outer archipelago the ice may move in the early season and form pressure ridges, making the topography of the outer part of the coastal zone rougher. In straits with strong-flow velocities the surface may be ice-free

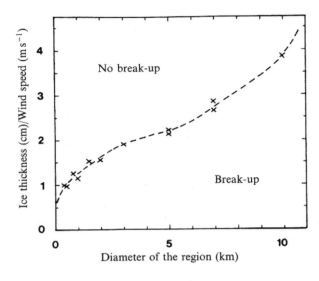

Figure 8.14. The stability of the ice cover in the archipelago as a function of ice thickness, fetch, and wind speed. (Leppäranta, 1981b, based on Palosuo, 1963.)

for part of the winter (as in rivers), and this could lead to frazil ice generation, and in shallow places to anchor ice[3] formation as well. However, these phenomena have not been reported in observational studies.

The edge of fast ice is in contact with drift ice when drift ice is pressing against the coast. This leads to high forces and consequently frequent pressure ridge formation. The sizes of pressure ridges depend on wind speed, fetch, and ice thickness. In the outer archipelago the growth of ridges is fetch-limited, but beyond the outer archipelago basin-wide fetches are available; in fact, the largest ridges form at the edge of the coastal ice zone. Occasionally, deep ridges ground and become additional support points for the landfast ice zone to extend farther offshore. Grounded ridges can grow high, often 5 m–10 m above sea-level (Figure 8.15). The largest recorded was 15 m off Hailuoto island in the northern Bay of Bothnia (Alestalo et al., 1986). Grounded ridges are a common coastal landscape feature on eastern sides of Baltic Sea basins, since strong winds usually blow from the west. Floating ridges may also drift into the shallow coastal area and become grounded.

In the grounding process, the keels of ridges influence bottom material. The keels *scour* the sea bottom, penetrating into bottom sediments. This has to be taken into consideration when constructing undersea cables or oil and gas pipelines in the coastal zone. Ridge keels have been known to damage shipwrecks and thus destroy marine archeological objects.

A fundamental problem in the fast ice zone is the bearing capacity of ice. This problem can be approached using the theory of plate on elastic foundation from civil engineering. For a point load, ice is forced down while it is supported by water

---

[3] Anchor ice accumulates on the sea bottom attached to the bottom sediments. Due to its buoyancy, anchor ice may rise up and come to the surface to join the ice cover.

Figure 8.15. Grounded ridge at the fast ice edge, eastern side of the Bay of Bothnia.

pressure (Figure 8.16). The supporting pressure is $\rho g w$, where w is the deflection, and thus a liquid water body acts mathematically in the same way as an elastic foundation. The equation for deflection is (Michel, 1978)

$$\nabla^4 w = \rho g w \tag{8.6}$$

The load acts in the origin and can be accounted for by boundary conditions. Bending of the ice causes stresses that eventually break the ice (Michel, 1978). This equation can be analytically solved using Bessel functions. First-order approximation is obtained from the first term of the series solution; the ice can bear mass M if

$$M < ah^2 \tag{8.7}$$

where $a \approx 4\,\mathrm{kg\,cm^{-2}}$ for Baltic Sea ice. The deflection takes the form of an exponentially damped wave with wavelength $2\pi\lambda$, where λ is the characteristic length of ice on a water foundation:

$$\lambda = \sqrt[4]{\frac{Eh^3}{\rho g 12(1 - \nu^{-2})}} \tag{8.8}$$

where $\nu \approx \frac{1}{3}$ is Poisson's ratio. Here $E = 5\,\mathrm{GN\,m^{-2}}$ is the elastic modulus of ice. The radius of influence of a point load is 2λ and consequently this is the safe distance between loads if ice thickness is close to the critical level. For moving loads there is an

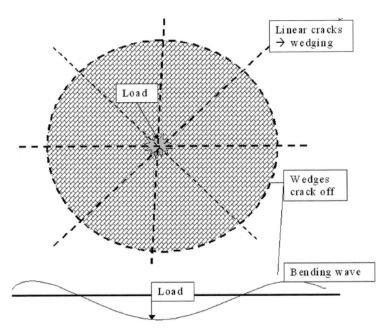

Figure 8.16. Schematic diagram on the bearing capacity of ice. Load causes elastic bending. Linear cracks form wedges, which break off. Then the ice loses its bearing capacity.

additional danger when the speed of the load is close to the speed of a shallow-water wave beneath the ice, $c = \sqrt{gH}$. Due to resonance in the bending of the ice, the bearing capacity in practice becomes half that of static loads.

In spring, melting of the coastal ice zone begins from the land side, due to the shallow depth of the sea, where temperature quickly increases as a result of solar radiation and the advection of warm air from the land. The melting time is proportional to the thickness of ice, and therefore the outer edge of the coastal ice zone with its heavy ridges is the last to melt.

8.4.2 Fast ice edge

When ice gets thicker in the fast ice zone, it becomes more stable and grounded ridges increase in number for additional support. Ridges then contact landfast ice and the coastal ice zone extends farther out from the coast. The boundary conditions between drift ice and the coastal ice zone are the following: if drift ice moves out from the coastal zone, it has a free boundary (no normal stress); and if drift ice moves toward the boundary, the velocity of ice is zero at the boundary and ridging takes place. Under normal stress at the boundary, pressure ridges form, whereas if drift ice is in motion along the coastal ice boundary long shear ridges form.

The motion of drift ice at the boundary can be examined using a simple 1.5-dimensional model (Leppäranta and Hibler, 1985). Mathematically, the y-axis is aligned longitudinally, and the situation is assumed invariant in the y-coordinate

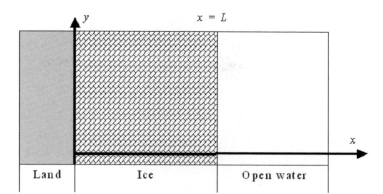

Figure 8.17.
Boundary zone configuration.

or $\partial/\partial y \equiv 0$ (Figure 8.17). For the ice state, a two-level description $J = \{A, h\}$ is employed.

The y-invariant equations of the dynamics of ice are directly obtained from the full form: the equation of motion from Equation (7.21), ice rheology from Equation (7.20), and the ice conservation law from Equations (7.23). The resulting model is (Leppäranta and Hibler, 1985)

$$\rho h \left(\frac{\partial U_i}{\partial t} + u_i \frac{\partial}{\partial x} U_i - f\mathbf{k} \times U_i \right) = \frac{\partial}{\partial x}(\sigma_{xx}\mathbf{i} + \sigma_{xy}\mathbf{j}) + \boldsymbol{\tau}_a + \boldsymbol{\tau}_w - \rho h g \nabla \xi \quad (8.9a)$$

$$\boldsymbol{\sigma} = \boldsymbol{\sigma}(h, A, \dot{\boldsymbol{\varepsilon}}), \quad \dot{\varepsilon}_{xx} = \frac{\partial u}{\partial x}, \quad \dot{\varepsilon}_{xy} = \dot{\varepsilon}_{yx} = \frac{1}{2}\frac{\partial v}{\partial x}, \quad \dot{\varepsilon}_{yy} = 0 \quad (8.9b)$$

$$\frac{\partial \{A, h\}}{\partial t} + \frac{\partial u_i \{A, h\}}{\partial x} = 0 \quad (0 \le A \le 1) \quad (8.9c)$$

Here the boundary conditions can be taken as a lack of onshore motion at the fast ice edge and zero stress a long distance from the coast. The free drift velocity U_{iF} is first evaluated. If the onshore component is $u_{iF} > 0$, then ice will drift out of the fast ice zone with free boundaries and in a free drift state. Otherwise, ice will stay in contact with fast ice and internal stress will spread the friction from the coast deeper into the drift ice zone. The steady state condition requires $u_i \equiv 0$ since no further ice state redistribution is possible. The ice conservation law is automatically satisfied, and the momentum equation and rheology are left for the longitudinal component of ice velocity and for the ice stress:

$$\frac{d\sigma_{xx}}{dx} + \tau_{ax} + \rho C_w \sin \theta_w |v - v_i|(v - v_i) + \rho_i hf v_i - \rho g h \beta_x = 0 \quad (8.10a)$$

$$\frac{d\sigma_{xy}}{dx} + \tau_{ay} - \rho C_w \cos \theta_w |v - v_i|(v - v_i) = 0 \quad (8.10b)$$

Only the shear strain rate can be non-zero. In this case, for general conditions $|\sigma_{xy}/\sigma_{xx}| = \gamma = \text{constant}$. Then, the ice stress derivatives can be eliminated from

Equations (8.10), and an algebraic equation is obtained for the longitudinal ice velocity v_i. Sea surface tilt β_x is small and we can think of it added into the on-ice wind stress component.

When ice drifts north, $v_i \geq 0$ and then $\sigma_{xy} > 0$. An acceptable solution is

$$v_i - v = \sqrt{\frac{\tau_{ay} + \gamma \tau_{ax}}{C_N} + \left(\frac{\gamma \rho h f}{2C_N}\right)^2} - \frac{\gamma \rho h f}{2C_N} \tag{8.11}$$

where $C_N = \rho_w C_w (\cos \theta_w - \gamma \sin \theta_w)$. In northward onshore forcing $\tau_{ay} > 0$ and $\tau_{ax} < 0$, and thus coastal friction cuts the portion $\beta \tau_{ax}$ from the momentum input. Setting $v \equiv 0$ we obtain the condition $\tau_{ay} + \gamma \tau_{ax} > 0$. This means that the northward component τ_{ay} must be larger than $\gamma \tau_{ax}$ or the direction of the wind stress must be less than the $\arctan(\gamma)$ on-ice angle, or otherwise a stationary ice field results. For example, if $\gamma = \frac{1}{2}$, then $\arctan(\gamma) = 27°$. The deviation angle between wind and ice drift is simply equal to the on-ice angle of wind velocity. It is noteworthy that even though sea ice drift is highly non-linear in the presence of internal friction, a steady zonal flow allows superposition of wind-driven and ocean current–driven components. The southward flow is treated in similar fashion. The general solution is illustrated in Figure 8.18.

The characteristics of shear flow become clearer if we ignore Coriolis acceleration and boundary layer angle effects. The northward flow solution is

$$v_i - v = \sqrt{\frac{\rho_w C_w}{\rho_a C_a} (\sin \vartheta + \gamma \cos \vartheta)|\boldsymbol{U}_a|} \tag{8.12}$$

where the wind direction is in the range $90° \leq \vartheta < 180° - \arctan(\gamma)$. Thus, $v_i - v \propto |\boldsymbol{U}_a|$, but the proportionality factor is affected by wind direction, strength parameter γ, and drag coefficients.

During the time of zonal sea ice flow, the ice state is adjusted to allow steady state conditions to be satisfied. With the present two-level ice state, the ice strength

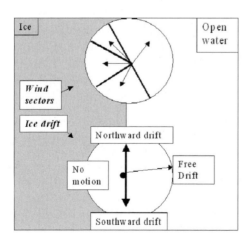

Figure 8.18. Steady state solution of wind-driven zonal flow, northern hemisphere. The wind sectors are shown in the upper circle, and the resulting ice velocity is shown in the lower circle: free drift, northward or southward boundary flow, or stationary ice depend on the ice strength. (Based on Leppäranta and Hibler, 1985.)

depends on ice compactness and thickness. At the fast ice boundary ridging takes place if forcing is strong enough. The geometric profile depends on the functional relationships between ice strength and ice thickness.

When there is open water at the fast ice edge, upwelling and downwelling may take place in the same way as on the coast in the open-water season. As soon as drift ice moves along the fast ice boundary, there is stress transfer through drift ice to the water, while the water beneath fast ice receives no surface stress. If fast ice is to the left of the direction of motion, upwelling results; in the opposite case the result is downwelling. The situation necessary for the start of upwelling or downwelling is well understood, but no cases have been reported from the Baltic Sea; however, elsewhere there are such reports (Gammelsrød et al., 1975).

8.5 COASTAL WEATHER

Coastal areas have specific weather conditions where the presence of open sea and land areas is felt. Here, we discuss some specific key elements of coastal weather, but we do not intend to cover them thoroughly. Coastal weather is the result of a mixture of different factors. The different characteristics of atmospheric stability, coupled with temperature, wind, and moisture distributions between the land and the sea may cause specific weather features at coastal areas (convective precipitation, fog, etc.). This is furthermore important in the Baltic Sea where many narrow bays exist. The sparsity of meteorological observations in the coastal area and in the open sea has made it difficult to study in detail the specific features of coastal weather.

8.5.1 The land–sea breeze

The wind conditions in coastal areas are specific due to the asymmetry of the air–sea interaction and surface roughness conditions between land and the open sea. A well-known feature at any coastal area is the sea breeze, this being true of all coastal areas of the Baltic Sea during the summertime, too. A sea breeze (or onshore breeze) is formed when land surface temperature increases in the daytime and becomes higher than air temperature while the sea surface temperature and air temperature over the sea are fairly stable. This creates a pressure minimum over the land and forces higher pressure, cooler air from the sea to move inland.

The sea is warmed by the Sun to a greater depth than the land due to its greater thermal diffusivity. The sea therefore has a greater capacity for absorbing heat than does the land and so the surface of the sea warms up more slowly than the land's surface. As the temperature at the surface of the land rises, the land heats the air above it. Warm air is less dense and so it rises. This rising air over the land lowers sea-level pressure by about 0.2% (about 2 mbar). The cooler air above the sea, now with relatively higher sea-level pressure, flows towards the land into lower pressure, creating a cooler breeze near the coast. The strength of the sea breeze is often directly proportional to the temperature difference between the land and the sea. If the wind speed in the general wind field is larger than $4 \, \text{m s}^{-1}$ and opposing the direction of a

290 Coastal and local processes [Ch. 8]

possible sea breeze, the sea breeze is not likely to develop. Sea breezes occur most often in early- and mid-summer during daylight hours when there is a large difference between the temperature of the air over the land and the temperature of the air over the still-cold sea. Sea breeze dynamics over the Gulf of Finland region has been studied by Savijärvi *et al.* (2005) and found to have specific characteristics.

A sea breeze front is a weather front created by a sea breeze, also known as a convergence zone. When cold air from the sea meets the warmer air from the land, a boundary like a shallow cold front is created. When it grows in power, this front creates cumulus clouds, and if the air is humid and unstable cumulonimbus clouds are created; the front can sometimes trigger thunderstorms. At the front, warm air continues to flow upward and cold air continually moves in to replace it, and so the front moves progressively inland. Its speed depends on whether it is assisted or hampered by the prevailing wind and the strength of the thermal contrast between land and sea. At night, a sea breeze usually vanishes (Figure 8.19).

At night, land cools down quicker, which forces daytime sea breezes to die. The pressure over water will be lower than that over land, setting up a land breeze as long as the general surface wind pattern is not strong enough to oppose it. If there is sufficient moisture and instability available, the land breeze can cause showers or even

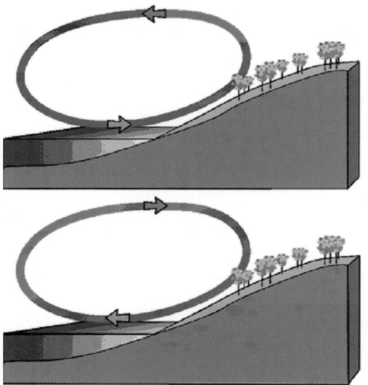

Figure 8.19. Schematic diagram of sea breeze (top) and land breeze (bottom).

thunderstorms over the water. Overnight thunderstorm development offshore can be a good predictor of what will happen on land the following day, as long as there are no expected changes to the weather pattern over the following 12–24 hours. A land breeze will die once the land warms up again the next morning.

8.5.2 Specific air–sea interactions

There are few good examples of the specific weather phenomena that develop over the Baltic Sea due to asymmetry between the thermal balance and moisture balance over the land and sea. During autumn and early winter, when the sea surface temperature is higher than atmospheric temperature, a so-called "thermal low" can develop due to thermal imbalance between sea and land areas. This situation takes place when the overall vertical stability of large-scale circulation is favorable for cyclone development. In such a situation a thermal low deepens over the sea based on the release of warm and moist air to the atmosphere.

Favorable conditions are met when winds blow from the cold land towards the warm sea. In such a case there is asymmetry in vertical heat exchange: over the open sea the atmosphere gains heat and moisture whereas over the land the opposite is the case; that is, horizontal Laplacian or diabatic[4] heating is non-zero at the coastal area. Such heating is important for vertical motion, in addition to thermal advection, and vorticity advection (see Holton, 1979; Myrberg et al., 1990). Energy gained from the sea surface will furthermore weaken atmospheric stability, and deepening of the low can be rapid when convective processes and baroclinic instability interact. When the low is advected over the land it will rapidly weaken due to the existence of horizontally homogeneous vertical heat fluxes. Such a thermal low often develops in the Bay of Bothnia during autumn, coupled with strong wind and heavy precipitation. Such low-pressure systems also have been observed in other parts of the Baltic Sea as well.

Another example in which the land–sea interaction plays a key role is the formation of convective snow bands (e.g., over the open Gulf of Finland and at the Swedish east coast). Such features are observed in a cold airmass with neutral atmospheric stability when the geostrophic wind is blowing along the latitudinal axis of the gulf. Convective showers can be very intense (Figure 8.20). The main causes for the formation of such snow bands over gulfs are related to differences in the thermal and evaporational characteristics of land and open sea surfaces. The geometrical configuration of the coast is also expected to play a crucial role. Due to the small size of the Baltic Sea both the coast of departure and the coast of arrival play important roles.

At the coast of departure the open gulf triggers the formation of snow bands. The reason for their genesis is the thermal difference between the land and sea: land breezes form at one or both coasts which results in convergence over the gulf. This in turn forces convection, and organized snow bands form moving downwind (Figure 8.20). The warm and moist surface may lead to intensive growth of the snow bands

---

[4] $\nabla^2 Q_d$, where Q_d is diabatic heat flux.

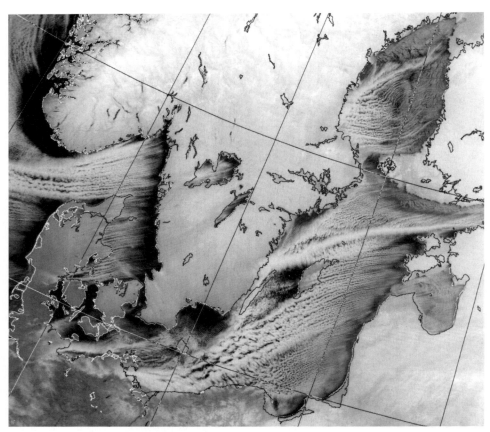

Figure 8.20. NOAA satellite image of snow bands over the Baltic Sea. The width of the image is about 1,200 km. (Dr. Nils Gustafsson, pers. commun.)

when they move for a long enough time over the sea. When snow bands approach the coast of arrival, clear intensification of the snow bands may occur depending on the shape of the coastline. A land breeze circulation may form at the coast of arrival directed against the mean flow and a new convergence zone may be established. An example of this is when snow bands drift to the west to the Swedish coast from the Gulf of Finland with a prevailing easterly mean flow. Over the cold land the snow bands quickly disappear due to a lack of moisture and heat supply (Andersson and Nilsson, 1990; Andersson and Gustafsson, 1994).

A heavy snowstorm caused by such a mechanism took place in December 1998 in Gävle, Sweden when 1.4 meters of snow fell in a few days (Andersson and Michelson, 1999). Martial law was declared in Gävle as a result. The forecasting of such an event is difficult and requires the use of coupled atmosphere–sea models.

Especially during late summer there are situations when unexpected (i.e., not forecast) fogs appear at the coastal area. The role of the air–sea interaction is very important in such situations. What is interesting is that the sea can be either warmer or colder than the air above. In a warm and moist airmass the triggering mechanism might be coastal upwelling which causes an abrupt drop in sea surface temperature. Thus, the relative humidity of the lowest air layer will rapidly increase and finally the airmass temperature will equal the dew point temperature. If the wind is weak, a fog can form rapidly. In such a situation, coupled air–sea models are needed in which sea surface temperature is realistically forecast. Such unpredicted fogs have caused problems for navigation.

Another specific feature is sea smoke which typically occurs during late autumn or early winter. In such a situation sea surface temperature is clearly higher than atmospheric temperature. When evaporated water vapor from the sea surface moves upwards, it quickly reaches the dew point temperature and a fog called "sea smoke" forms. The mechanism is no different from that of fog, but condensation in the upward air motion leads to a smoke-like visual impression. Sea fog is typically observed when cold air dominates near the ice boundary or at cracks in the ice.

This chapter has dealt with the physics of the coastal zone, the area influenced directly by the nearby presence of land and the open ocean. Due to the small size of the Baltic Sea, a significant part of it belongs to this zone. The treatment included fronts, archipelago areas, sea-level elevation, upwelling, coastal ice conditions, and coastal weather. This chapter closes our discussion of the physics of the Baltic Sea. Chapter 9 introduces the interface between physics and ecology and human life—a theme already touched on in previous chapters.

9

Environmental questions

Stig Fonselius (1921–2003) was a chemical oceanographer who contributed widely to the development of physical oceanography in the Baltic Sea. He was born in Tampere, Finland. After World War II he studied chemistry in the University of Helsinki and in the 1950s he moved to Sweden and completed his Ph.D. in Gothenburg [Göteborg] in 1969. He worked most of his years in the Hydrographical Laboratory of the Fishery Board of Sweden in Göteborg, and was laboratory chief during 1979–1987. He published a large number of scientific articles about the Baltic Sea and an excellent monograph on the general oceanography of the Baltic Sea and Kattegat–Skagerrak region (Fonselius, 1995).

Physical oceanography and sea ice science are closely connected to ecology and human life in the Baltic Sea area, and these connections need to be understood by physicists themselves. Principally, this relates to the physics of the Baltic Sea which provides the background conditions for ecology in terms of light conditions, temperature, salinity and stratification, current dynamics including transport and diffusion, and the ice season. The life of Baltic Sea peoples has always been influenced by shipping conditions and fishery, and presently eutrophication, pollution, and the construction of infrastructure is an additional severe load to "our sea" (so-called by local people) by humankind. Section 9.1 treats light conditions, and Section 9.2 is about oxygen conditions in deep basins. Human impact is discussed in Section 9.3, and extreme situations for key physical quantities are presented in Section 9.4. Section 9.5 introduces operational oceanography, which has come to the fore in recent years and will have a key role in the future protection of the Baltic Sea.

9.1 LIGHT CONDITIONS

9.1.1 Basic concepts

Solar radiation is a key forcing factor in the physical oceanography of the Baltic Sea (as discussed in Section 4.3). Distributed in the surface layer, it is the principal heat source for seawater. In addition, solar radiation is a key forcing factor in marine biology providing light quanta for photosynthesis, and the behavior of light signals in natural waters can be used as an indicator of their ecological state and as a tracing method of watermasses (Jerlov, 1976; Dera, 1992; Arst, 2003). It is absorbed and scattered in the water body by the water itself and by *optically active substances* (OAS), which include colored dissolved organic matter (CDOM), also known as yellow substance, suspended matter, and chlorophyll *a*. In distilled water the transmission distance of sunlight is about 100 m, and in natural waters it is less due to the influence of OAS.

The spectrum of solar radiation covers the wavelength band 0.2 μm–3.0 μm, from the ultraviolet to the near-infrared, with maximum intensity at 0.48 μm (Figure 9.1). The optical band (or light) is usually taken as 0.38 μm–0.76 μm. In water only the optical band and small sections of the ultraviolet and near-infrared bands of the solar radiation spectrum can be transferred to any significant distances. In the optical classification of Jerlov (1976) the Baltic Sea as well as all coastal seas belongs to "Case III ocean waters", where the concentrations of all OAS components are significant.

Ice cover and in particular snow on ice has much higher reflectance than liquid water, and the transmission of radiation is affected by scattering due to gas bubbles in addition to OAS.

Radiation concepts are illustrated in Figure 9.2. The level of radiation attenuates with distance from the source due to *absorption* and *scattering*. Absorption of radiance L obeys Beer's law (Dera, 1992)

$$\frac{dL}{dz} = -aL \qquad (9.1)$$

where a is the absorption coefficient and the z-axis is positive downward. Scattering has angular dependency given by the scattering function (Dera, 1992), and often the principal object of interest is calculation of total scattering or backscattering. The former gives the attenuation of light due to scattering (by spreading the beam, scattering also causes attenuation), while the latter is used in remote sensing (backscattered signals bring information about the water properties back to the observer).

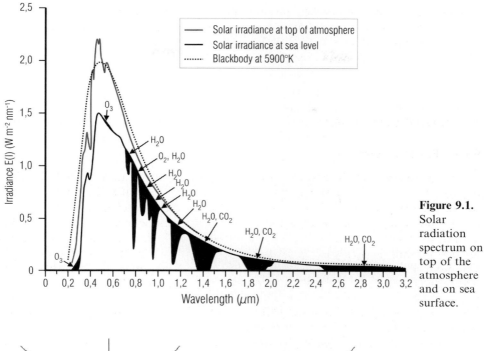

Figure 9.1. Solar radiation spectrum on top of the atmosphere and on sea surface.

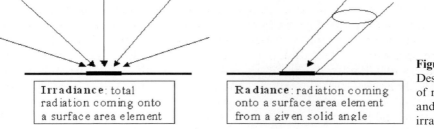

Figure 9.2. Description of radiance and irradiance.

Attenuation due to scattering is described by a law similar to Equation (9.1) with scattering coefficient b replacing the absorption coefficient, and backscattering itself is given as $b_b = p_b b$, where p_b is the probability of backscattering.

In optical investigations, two important assumptions are made. First, scattering and absorption are independent, and hence the total attenuation law is also similar to Equation (9.1) with a replaced by the attenuation coefficient $c = a + b$. Second, the influences of OAS are independent and can be superposed together. Thus

$$c = c_w + c_y + c_{chl} + c_{sm} \qquad (9.2)$$

where c_w, c_y, c_{chl}, and c_{sm} are the attenuation coefficients due to pure water, CDOM, chlorophyll a. and suspended matter, respectively.

CDOM strongly absorbs short wavelengths,[1] modeled as

$$a_y(\lambda) = a_y(\lambda_0) \exp[-\gamma(\lambda - \lambda_0)] \tag{9.3}$$

where λ_0 is a reference wavelength (often 380 nm); and $\gamma \approx 0.017\,\text{nm}^{-1}$ is a damping constant in the wavelength space with γ^{-1} corresponding to the width of the absorption band at $\lambda \geq \lambda_0$ (Arst, 2003). Chlorophyll a has two absorption bands (430–440 nm and 660–690 nm) and it too causes scattering.

The absorption and scattering of suspended particles depend on the size, shape, and optical properties of the particles. When the particle size is less than 1 μm, short waves scatter more strongly, while in the case of larger particles the wavelength dependence is weak. In pure water, scattering takes place due to fluctuations in water density. Finally, using independence assumptions the absorption and scattering coefficients can be expressed as

$$a = a_w + a_y + a_{chl} + a_{sn}, \quad b = b_w + b_{chl} + b_{sm} \tag{9.4}$$

Absorption and scattering are *inherent optical properties* (i.e., they are independent of the directional distribution of incoming light). A small part of the radiation is scattered back from the water body to the atmosphere, and the task of optical remote sensing is to interpret the backscattered signal for OAS.

Many observations of solar radiation have been based on irradiance sounding as a function of depth. The downwelling and upwelling planar irradiances E_d and E_u and scalar irradiance E_0 are, respectively

$$E_d(z;\lambda) = \int_0^{2\pi} d\phi \int_0^{\pi/2} L(z,\theta,\varphi;\lambda) \cos\theta \sin\theta\, d\theta \tag{9.5a}$$

$$E_u(z;\lambda) = -\int_0^{2\pi} d\varphi \int_{\pi/2}^{\pi} L(z,\theta,\varphi;\lambda) \cos\theta \sin\theta\, d\theta \tag{9.5b}$$

$$E_0(z;\lambda) = \int_0^{2\pi} d\varphi \int_0^{\pi} L(z,\theta,\varphi;\lambda) \sin\theta\, d\theta \tag{9.5c}$$

where $L(z,\theta,\varphi;\lambda)$ is the radiance; and θ and φ are the zenith and azimuth angles (see Figure 9.2). Planar irradiance accounts for the radiance reaching an infinitesimal element on a horizontal plane from the upper (downwelling) and lower (upwelling) hemisphere, and scalar irradiance accounts for radiation falling on an infinitesimal sphere. Integrated over wavelength bands, band irradiances are obtained. An important band is the optical band $\Omega = [380\,\text{nm}, 760\,\text{nm}]$, which approximately overlaps with the PAR (photosynthetically active radiation) band, denoted below by a subscript Ω in optical quantities.

The irradiances and optical characteristics derived from them provide *apparent optical properties*, which depend on the directional distribution of radiation, as clearly

---

[1] Absorption of light at short wavelengths removes them from the backscatter to the observer, and the water color shifts towards the yellow band—hence the substitute term *yellow substance* for CDOM.

shown by definitions (9.5). An important task in the optics of natural waters is to establish the relations between apparent and inherent optical properties (see Kirk, 1994; Arst, 2003).

In exact terms photosynthesis takes light quanta (photons) rather than energy, and therefore the number of photons is of interest in marine biology. Consequently, the concept of quantum irradiance is introduced:

$$q_d(z) = \int_\Omega \frac{\lambda}{hc_0} E_d(z;\lambda)\, d\lambda \qquad (9.6)$$

where $h = 6.6255 \times 10^{-34}$ J s is Planck's constant; and $c_0 = 2.9979 \times 10^8$ m/s is the speed of light in a vacuum. The unit of quantum irradiance is quanta/(m^2 s), normally changed into μmol/(m^2 s), 1 μmol $= 6.022 \times 10^{17}$ quanta.[2] Transformation between irradiance power and irradiance quanta can be made from $hc_0 q(z) = \lambda^* E_{d\Omega}(z)/\Delta\lambda$, where λ^* is the irradiance-weighted mean PAR wavelength and $\Delta\lambda$ is the PAR bandwidth.

Two very important optical characteristics derived from the irradiances are *reflectance r* and the *diffuse attenuation coefficient*[3] K_d:

$$r(z,\lambda) = \frac{E_u(z;\lambda)}{E_d(z;\lambda)} \qquad (9.7a)$$

$$\frac{dE_d(z;\lambda)}{dz} = -K_d(z,\lambda) E_d(z;\lambda) \qquad (9.7b)$$

This latter equation integrates into

$$E_d(z;\lambda) = E_d(0^+;\lambda) \exp\left[-\int_0^z K_d(z',\lambda)\, dz'\right] = E_d(0^+;\lambda) \exp(-\tilde{K}_d z) \qquad (9.8)$$

where 0^+ refers to the level just beneath the water surface and a tilde on a symbol stands for vertical averaging.

For band reflectances and attenuation coefficients, corresponding band irradiances are used. Albedo is, by definition, reflectance with irradiances taken over the whole wavelength space at the surface, $\alpha = E_u^*(0^-)/E_d^*(0^-)$, where the notation $f(0^-)$ means just above the sea surface or $\lim_{z \to 0^-} f(z)$. In the water body, only the PAR band penetrates and irradiance measurements provide the PAR attenuation coefficient, obtained using spectrally integrated PAR irradiance $E_{d\Omega}(z)$. The result is the total attenuation coefficient $K_{d\Omega}(z)$. When $K_{d\Omega} \approx$ constant, the Secchi depth[4] is about twice the attenuation length, $z_D \approx 2(K_{d\Omega})^{-1}$. A common definition of euphotic depth, z_{ed} (Arst, 2003) is the depth at which irradiance has fallen to 1% of the level at the surface; with $K_{d\Omega} \approx$ constant we then have $z_{ed} = \log 100 (K_{d\Omega})^{-1} \approx 4.6(K_{d\Omega})^{-1}$.

---

[2] The number of photons in one mole is equal to Avogadro's number, 6.022×10^{23}.
[3] Also called "extinction coefficient" (Warren, 1982).
[4] The depth at which a white disk 30 cm in diameter is seen from the surface by the human eye.

9.1.2 Optical characteristics of Baltic Sea water

Our knowledge of the optical properties of Baltic Sea watermasses is to a large degree based on irradiance data. A long time-series, over 100 years, exists for the Secchi depth (Sandén and Håkansson, 1996; Laamanen et al., 2004). A limited amount of information about inherent optical properties is available, determined from water samples and *in situ* soundings. Most of the research has been performed in the Southwestern Baltic Sea (Jerlov, 1976; Dera, 1992; Højerslev et al., 1996), and more recently the number of field programs have also grown in the Gulf of Finland (Arst et al., 1990; Sipelgas et al., 2004).

The present level of Secchi depth is 5 m–10 m in the Baltic Sea, depending on the time and place (Laamanen et al., 2004). Secchi depths in coastal areas are lower than those in central basins, and are lower during runoff peak, which brings large amounts of suspended matter into the basin. Since Secchi depth equals about twice the e-folding scale of light attenuation it can be linked with physical quantities, and because of the large database it serves well as the first reference for the optical state of water. Mean Secchi depth has experienced a remarkable decrease (Figure 9.3) over the last 100 years, which can be accounted for by anthropogenic influence. This change is obviously due to eutrophication, which has led to increased levels of CDOM and chlorophyll, both in the Baltic Sea and in the lakes of the drainage basin.

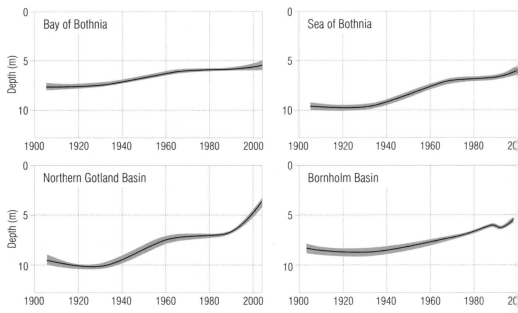

Figure 9.3. Secchi depth in the Baltic Sea 1900–2000. (Laamanen et al., 2004; http://www.helcom environment2/ifs/ifs2004)

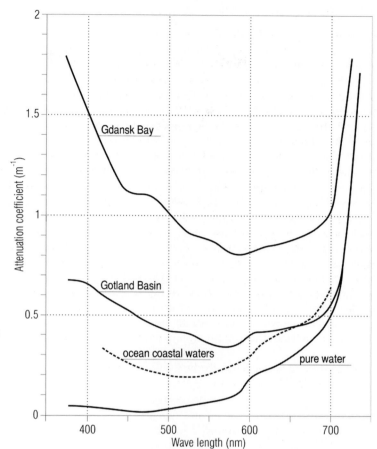

Figure 9.4. Attenuation spectra of light in Gdansk Bay and the Gotland Sea compared with general coastal seawater and pure water. (Redrawn from Dera, 1992.)

Baltic Sea watermasses are characterized by high levels of CDOM and suspended matter from nearby land areas which strongly influence light transfer, as is the case with Jerlov's Type III waters (Jerlov, 1976). In such optically multi-component waters the inversion of optical measurements for water properties is not necessarily unique. The large amount of CDOM in the Baltic Sea turns the color brownish (Figure 9.4).

Field measurements of the optical properties of Gulf of Finland waters are available for coastal sites. Sipelgas et al. (2004) performed a field campaign offshore the Estonian northwestern coast. They obtained a regression formula between the beam absorption coefficient and CDOM concentration C_{cdom} as $a(412\,\mathrm{nm}) = 0.4288 C_{cdom} + 0.1603$, where a is in m^{-1} and C_{cdom} in mg/L. The range of $a(412\,\mathrm{nm})$ was from 0.5 m^{-1} to 2.0 m^{-1}, and thus was dominated by C_{cdom}. They further obtained a good connection between the concentration of suspended matter C_{sm} and the beam-scattering coefficient. The scattering coefficient was only weakly dependent on wavelength, $b_{sm} \propto \lambda^n$ with $n \leq -0.5$—the closer to zero the more turbid the water. Sipelgas *et al.* (2004) combined their results into the mathematical expression of the beam attenuation coefficient (in m^{-1}) over the optical band $c_\Omega - c_w = 0.3C_{sm} + 0.35C_{chl} + 0.86C_{cdom}$, where c_w is the attenuation coefficient of distilled water, C_{chl} (mg/m^3) is the chlorophyll concentration, and C_{sm} and C_{cdom} are given in mg/L. Thus, all three optically active substance groups are important. They reported that for variability of the attenuation coefficient, CDOM was the most important component in their data. In general, it is also clear that chlorophyll has large seasonal variability which is reflected in optical properties.

Ice and snow cover have a major influence on light transfer. The albedo of snow and ice is very large mainly depending on the wetness of the surface and the thickness of ice and snow (see Section 4.3). Some winter studies on optics have been made in the coastal ice zone in Santala Bay (59°55′N, 23°03′E) on the Hanko peninsula in the Gulf of Finland in winters 1998–2002.

Reflectance measurements were made with ice surface conditions varying from melting wet ice to refrozen snow patches (Leppäranta *et al.*, 2003). PAR-band reflectance ranged from 0.28 to 0.76, being lowest for ice covered by a thin water layer and highest for snow patches. Fresh snow reflectances were typically 0.8–0.9 and constant across the PAR band (as reported earlier for Santala Bay by Rasmus *et al.*, 2002). In the early melting season the reflectance of snow became lower and had a weak maximum at 550 nm with the overall level above 0.4. The main optically active impurities in sea ice are gas bubbles and liquid brine pockets that, due to their size, scatter all wavelengths of light equally well, and chlorophyll with its selective absorption.

The ice and water beneath the ice were examined by Leppäranta *et al.* (2003) and Arst *et al.* (2006). The diffuse attenuation coefficient of ice was 1.5 m^{-1}–5.5 m^{-1} depending principally on the proportion of snow ice, while in the water the diffuse attenuation coefficient was 0.5 m^{-1}–2.2 m^{-1}. In 2001 alone the coefficient was above 1 m^{-1} in water and was greater in ice. Thus, the ice and snow cover restricts the transfer of sunlight into the water body. The optical depth of snow is about 0.2 m, and therefore snow causes strong attenuation, but a significant amount of light is transferred to the water body in the Baltic Sea as soon as ice has become bare (i.e., no slow lying on top).

The results (Arst *et al.*, 2006) further show that the apparent optical properties in the water beneath the ice are different from those in ice-free conditions because incoming light is much more diffuse under the ice cover. Scalar irradiance decreases with increasing depth more rapidly than downwelling planar irradiance. Thus, the diffuse attenuation coefficient for scalar irradiance in under-ice water exceeds that for planar irradiance. These data are of interest, because in the surface layer of ice-free water bodies an opposite result has been obtained.

In the water body in 2000 the diffuse attenuation coefficient over the PAR range was 0.5 m^{-1}–0.9 m^{-1} for planar irradiance and 0.7 m^{-1}–1.0 m^{-1} for scalar irradiance. The planar values would mean a Secchi depth of 1 m–2 m. The water showed the following characteristics: suspended matter 1.3 mg/L–3.9 mg/L, chlorophyll 1.6 mg/m^3–1.9 mg/m^3, beam attenuation coefficient at 380 nm wavelength 1.2 m^{-1}–3.9 m^{-1}, and beam attenuation coefficient at PAR band 0.6 m^{-1}–1.4 m^{-1}.

9.1.3 Optical remote sensing

Optical spectral measurements by satellites and airborne systems give information on OAS in the surface layer with a thickness approximately equal to the attenuation length. The remotely sensed signal, or surface reflectance, contains the surface reflection and backscatter from the surface layer, which constitutes the essential information on OAS. So-called "bio-optical" models have been fine-tuned for inversion of the remote-sensing signal (Reinart and Kutser, 2006; Zhang, 2005). Figure 9.5 (see color section) shows an example of a chlorophyll chart based on an interpretation of satellite imagery.

The optical complexity of Baltic Sea waters (and, in general, Jerlov's Type III waters), in particular the mixed influence of CDOM and suspended matter, causes limitations to the identification of optically active substances from remote-sensing data. Therefore, methods have been successful mainly in identifying algae blooms due to their strong signal. Although this method shows future promise for identifying different types of water in the basin, it is not clear how to separate the signal of climatic change from the effect of local anthropogenic loads. Optical methods may well be able to identify different types of water in the basin, including river plumes. An effort was made by Sipelgas *et al.* (2006) to examine suspended particle transport in Pakri Bay, west of Tallinn, using MODIS images together with a hydrodynamic model. MODIS has been quite widely used in this region, and Reinart and Kutser (2006) concluded that MERIS provides results of the same quality.

Mapping of harmful algal blooms is presently one of the most important applications in marine optics, especially in the Baltic Sea (Zhang, 2005; Kutser *et al.*, 2006). To evaluate the background of long-term changes, it is not clear how to separate the signal of climatic change from the effect of local anthropogenic loads. Chlorophyll absorption peaks can be utilized to estimate primary production in surface water. In particular, in connection with algal blooms the signal is so strong that it can even be detected from panchromatic images. The use of the MODIS satellite to monitor dredging blooms was examined by Kutser *et al.* (2007). The band 620 nm–670 nm was found suitable for a suspended matter detection algorithm.

9.2 OXYGEN BUDGET OF BOTTOM WATER

Anoxic conditions are frequently met in strongly stratified brackish water basins such as the Black Sea and the Baltic Sea where ventilation in the lower layer is weak. In the Black Sea the renewal of bottom water through the narrow Bosporus Strait is extremely slow with a renewal time of 3,000 years, and consequently all deepwaters are permanently anoxic. In general, when oxygen concentration is below 2 mL/L, conditions are considered unsuitable for life.

In the Baltic Sea the strong and persistent halocline in the Gotland Sea (see Section 3.3) inhibits vertical convection, due to cooling in fall and winter, from reaching the lower layer. Consequently, ventilation of deepwater may take place in these areas only by horizontal advection and thus also oxygen conditions can improve only through inflows. This delicate situation is dictated by natural conditions—water and salt budgets—in the Baltic Sea, but eutrophication due to environmental loads increases oxygen consumption and becomes a major risk to the ecological state of bottom water. Major Baltic Inflows improve the oxygen conditions for some years, but on the other hand the related strengthening of stratification weakens vertical mixing and the input of oxygen from the air to deepwater is strongly reduced.

The Gotland Sea is strongly stratified and deep with a relatively slow water renewal rate in the bottom layer. Anoxic bottom water with aperiodically oscillating areal extent is found dictated by Major Baltic Inflows and the consumption of oxygen (Figure 9.6). Because of recent eutrophication the anoxic area has become larger and extended in some years even to the Gulf of Finland. In the Gulf of Bothnia, water renewal is good and the nutrient load is not as high as in the Gulf of Finland, and thus there are no oxygen problems, at least up to the present. In other basins, normally there are no oxygen problems.

Irregular inflows of oxygen-rich and highly saline water through the Danish Straits are the reason that oxygen concentrations have such a high natural variability in the Baltic Sea. These ocean-originated watermasses fill the deep basins of the Baltic. Inflowing new water has such a large density that it remains at the bottom for a long time (decades), while its density gradually reduces by mixing and entrainment with surrounding water. During these long periods with no inflows, called *stagnation periods*, all oxygen will be used to oxidize dead organic material. In such a situation hydrogen sulfide is formed. Usually, this is observed only at the deepest measurement stations (Fonselius and Valderrama, 2003). According to Fonselius (1969), hydrogen sulfide is expressed as "negative oxygen".

Fonselius and Valderrama (2003) investigated changes in deepwater oxygen concentration during the last 100 years using the inflow index (Q_{87}) by Franck *et al.* (1987), discussed in Section 5.4. The time-series is shown in Figure 5.21. After the relatively strong inflow in 1914, salinity and oxygen concentration decreased in the Gotland Deep (Station BY15). Stagnant water was removed by the inflow of 1951, the strongest ever observed. After that the oxygen level once again went down in the Gotland Deep and hydrogen sulfide was observed. Stagnation lasted about 10 years after which a moderate inflow took place in 1961. The period 1961–1977 was rich in inflows. However, afterwards no major inflows were observed for a long time, and in the 1980s hydrogen sulfide was often observed in the Gotland Deep. This stagnation ended only in 1993 with a Major Baltic Inflow and oxygen concentration was somewhat improved, but the inflow brought about no long-term improvement. In 1998 hydrogen sulfide was again found in the deepwater of the Baltic Sea proper and in the early 2000s its concentration was higher than ever. The series of inflows in 2002–2003 slightly improved the oxygen concentration to about 3 mL/L. Since then the situation has again become worse due to the absence of major inflows (as of 2008).

The extent of bottom areas where hydrogren sulfide is found can be up to one-third of the total area of the Baltic Sea. However, according to Elken (1996) and Elken and Matthäus (2008) the Gotland Deep regularly receives saline water

Sec. 9.2] Oxygen budget of bottom water 305

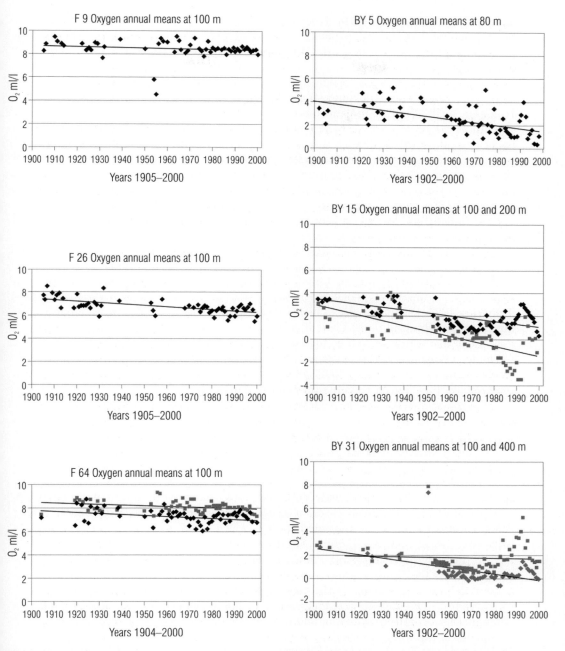

Figure 9.6. (Left-hand column) Long-term oxygen variation in deepwater (100 m, ■) and bottom water (◆) at Stations F 9, F 26, and F 64. (Right-hand column) Long-term oxygen and hydrogen sulfide variation in deepwater (100 m, ■) and bottom water (◆) at Stations BY 5, BY 15, and BY 31. Hydrogen sulfide is expressed as "negative oxygen". (Redrawn from Fonselius and Valderrama, 2003.)

interleaving at a depth of 80 m–130 m with a maximum northward flow across the basin reaching 2,500 m^3 s^{-1} around a depth of 100 meters. This ventilation process explains why hydrogen sulfide does not appear at depths above 140 m–150 m, even during long stagnation periods.

A few years after a major inflow, the area of anoxic bottom water once again starts to extend. This is a quite straightforward procedure in the Gotland Deep. However, in the Gulf of Finland the situation is much more complicated. Oxygen conditions can become worse while a major inflow takes place because old and anoxic water from the Gotland Deep are pushed farther into the gulf. During long stagnation periods the halocline can disappear from the Gulf of Finland, and thus oxygen conditions improve due to the input of oxygen from the air to the entire watermass. In addition, at times when the halocline is expected to be stable in the gulf, counter-estuarine transport can weaken the stratification of watermasses at the entrance to the Gulf of Finland and lead to improvement of oxygen conditions. Also strong winds in autumn and winter can produce strong mixing, erode the halocline, and consequently improve oxygen conditions (see Chapter 5 for details).

Figure 9.7 (see color section) gives two examples of how oxygen conditions have developed in the Gotland Basin–Gulf of Finland system. During the last 20 years the situation has dramatically worsened and the extent of the anoxic area has increased. The years 1993–1994 show the influence of the major inflow of 1993 which slightly improved oxygen conditions. Thereafter, the situation has gradually worsened due to the lack of major inflows and increased oxygen consumption. Thus, the influence of the major inflow in 2003 on oxygen conditions was not marked and long-lasting; the oxygen situation in the years 2006–2007 shows the anoxic area to have extended in the Gotland Basin, even reaching the Gulf of Finland.

9.3 HUMAN IMPACT

9.3.1 Major constructions

In the Baltic Sea region there are 85 million people and several major coastal cities and harbors. This has brought about major construction projects with regard to general infrastructure, and the interaction between building projects and the sea itself has always been a big issue. The most famous of these projects have been the St. Petersburg (Leningrad) dam, oil terminals, nuclear power stations and bridges across the Danish Straits, and presently (2008) there is a plan to construct a gas pipeline from Russia, at the eastern Gulf of Finland, across the Baltic Sea to Germany. The debate continues around the pipeline plan on the possible consequences of bottom dredging and potential environmental catastrophes.

Sea-level variations have been a major problem in St. Petersburg from the time of its foundation (1703). Due to topographic constraints in the eastern Gulf of Finland, the sea-level can be as much as four meters above the mean in the city waterfront, a maximum of 421 cm occurred in 1824. St. Petersburg has been constructed on lowland marsh, and exceptionally high water means flooding in the streets and basements of houses. Major damage due to very high sea-levels has been the consequence. To protect the city, construction of a dam[5] commenced in the 1970s under the old Soviet Union and has continued up to this day.

A few model-based investigations into the influence of the dam on the Gulf of Finland have been carried out (Dr.Sci. Vladimir Ryabchenko, pers. commun.; Rukhovets, 1982; Voltzinger et al., 1990). Simplified shallow-water equations have been used to ascertain the pattern of stationary currents and pollutant distribution in Neva Bay (Rukhovets, 1982). Comparison between simulations with a dam and without one showed that water transport across the dam differs considerably from natural conditions. Stagnant zones in Neva Bay became more extensive with a dam in place. The disposal of waste water into the Neva Bay can lead to significantly increased pollution in these stagnant zones. According to Voltzinger et al. (1990) the dam has no influence on the general pattern of water transport but marked changes do occur in the neighborhood of the dam.

Environmental considerations have been integral to the construction of bridges across the Öresund and the Belts (Figure 9.8), and have been decisive regarding the alignment and determination of the design of the construction. As far as the water flow is concerned, bridge construction is expected to comply with the so-called "zero solution" (i.e., no influence on the water exchange through a strait). This has necessitated dredging corresponding parts of the straits, so that the water flow cross-section has been increased. Such excavation compensates for the blocking effect caused by bridge pylons and approach ramps. The general conclusion concerning the water flow is that it is now almost at the level that existed before the bridges were built.

The bridges across the Danish Straits have generated increased road traffic, which in itself has meant increased air pollution. However, there has been significant savings in the energy consumption of east–west traffic by switching from ferries to the bridge. Train ferries and car ferries consume a lot of energy for propulsion, and high-speed ferries consume even more and cause coastal erosion. Air transport is very expensive in terms of energy consumption. Domestic air travel in Denmark over the Great Belt was greatly reduced after the opening of the bridge, with former air travelers now using trains and private cars.

The construction of the Great Belt and Öresund bridges represents a marked turning point in the history of humankind and the Baltic Sea. Bridges are fixed constructions, and their height defines the maximum height (the Great Belt bridge at 65 m, the Öresund bridge at 57 m) for ships and any sea traffic to enter the Baltic Sea (e.g., no high oil-drilling platforms can be taken in or out of the Baltic Sea). In this way the Baltic Sea has ceased to be part of the "free ocean".

9.3.2 Influence on coastal erosion

The ongoing massive increase in sea traffic in the Baltic Sea has many environmental effects, coastal erosion being one of the major ones. For example, in the Gulf of

---

[5] A flood protection barrier, briefly called "dam" here.

Figure 9.8. The Öresund bridge, seen from the east looking toward Copenhagen. (Courtesy of Bridgephoto, Denmark; *http://bridgephoto.dk/search/start.php*)

Finland, between Helsinki and Tallinn, there has been a strong increase in fast-ferry[6] traffic during the last 10 years. This has created serious concerns about its potential influence on the marine environment. Another critical area is the Archipelago Sea, where ferry routes pass through narrow straits (Figure 9.9, see color section).

The Helsinki–Tallinn seaway experiences high-speed vessels crossing the gulf nearly 70 times daily during the high season. They frequently sail in the transcritical velocity range (i.e., within $\pm 15\%$ of the maximum phase speed $\sqrt{gH}$ of shallow-water waves). Wave systems excited at these speeds may contain solitonic waves that are fundamentally different from waves from conventional ships (Soomere, 2006).

In Tallinn's harbor the influence of fast ferries on bottom sediments has been studied (Erm and Soomere, 2004, 2006). Optical methods have been used to identify and quantify the influence of wakes from fast ferries on bottom sediments and the properties of watermasses. Bottom erosion was evident in the range of water depths from about 2 m to 15 m. It usually lasts about 6–15 minutes but is limited to a water

---

[6] High-speed car-carrying catamarans and monohull vessels.

layer with a thickness of about 1 m near the seabed. Rough estimates suggest that about 10,000 kg of sediments per meter of affected sections of the coastline may be disturbed by ship wakes annually (Erm and Soomere, 2006).

The part played by ship waves in total wave activity may be markedly high in the vicinity of certain open-sea fairways (Soomere, 2005; Soomere et al. 2003a, b). The properties of ship waves were measured in areas with a depth of 5 m–7 m in different parts of Tallinn Bay and at distances of 2 km–8 km from the fairway (Soomere and Rannat, 2003) at sites that apparently were the most vulnerable with respect to ship waves (Kask et al. 2003). The waves from conventional passenger ships, cargo ships, and from hydrofoils have typical heights of 20 cm–30 cm and periods of 3–4.5 s, and are almost indistinguishable from the natural background (Soomere and Rannat, 2003).

Leading wake waves from large high-speed car-carrying catamarans and monohull vessels frequently have heights about 1 m and periods of 10–15 s; such waves seldom occur in natural conditions in Tallinn Bay. The wakes typically consist of two or three groups with periods of 9–15 s (the longer waves arriving first), 7–9 s, and about 3 s, respectively. The highest waves are usually the leading ones; yet the third group may produce the highest waves in remote areas (about 70 cm as far as 8 km–10 km from the ship lane, Soomere and Rannat 2003).

9.3.3 Oils spills and sea traffic

Oil spills are a major environmental risk in the Baltic Sea. The basin is shallow and has a variable bottom topography, and marine sea routes are narrow in places. Coasts and islands are nearby everywhere, and an oil spill would quickly reach them. In the large archipelago areas, oil spills would be held captive in the bays of islands and the consequences would be dramatic. Especially in winter, oil combating in ice conditions poses severe difficulties, as there is as yet no suitable infrastructure for observation, prediction of drift and dispersion, and collecting the oil. Biota and human recreational activities would be harmed by oil spills, and critical questions may arise if an oil spill approaches the water take-up sites of nuclear power plants.

The prediction of oil spill drift and dispersion is an applied problem in physical oceanography. Oil spill movement is forced by shear stresses from wind and water to the oil; therefore, an oil spill model can be linked to a three-dimensional circulation model. This linking is straightforward since there is no feedback from oil movement to current dynamics. The drift and dispersion of oil spills can be modeled as a sum of translation and diffusion or treating the oil with the physical properties taken exactly for the oil type in question. The latter method, together with an excellent Lagrangian advection scheme, is used by Ovsienko et al. (1999).

The Gulf of Finland is potentially a problem area. Due to increasing tanker traffic and the opening of new big Russian oil terminals, the risk level is even increasing. In this narrow and shallow basin with its large archipelago areas serious oil spill accidents would cause damage to the ecological and natural environment. Such a major oil spill occurred in winter 1987 in the middle of the Gulf of Finland.

The most recent oil accident happened in the Gulf of Finland in March 2006. The accident is described in Wang *et al.* (2008). On the night of March 5, the *Runner 4*, a Dominica-flagged cargo ship, collided with the merchant ship *Svjatoi Apostol Andrey* and sank in Estonian territorial waters. The ship was carrying a cargo of aluminum and had 102 tons of heavy fuel oil, 35 tons of light fuel oil, and 600 liters of lubricant oil in its tanks when it sank 70 m to the bottom. Initially, the oil was either beneath the ice cover or mixed with the broken ice. Clean-up operations started when the wind pushed the ice floes away and the spill was observed in open-sea areas. The ice conditions were severe as a result of ridged ice. For example, in some areas the clean-up vessel *Seili* was unable to pass through the ridged ice zones. It was also observed that drifting oil tended to accumulate on ice edges.

At noon of March 15, combined observation of the oil spills was made using aerial surveillance and clean-up vessels (HELCOM, 2006). The five main patches of oil leaked from *Runner 4* were identified; they were mainly in the routes of vessels sailing in an east–west direction (Figure 9.10). Oil Spills No. 1 and No. 2 were close to the site of the wrecked *Runner 4*. Oil Spill No. 1 was identified in open water as being rainbow-colored (thickness of rainbow-colored oil was 0.3 μm–5.0 μm, quantity 0.30 m^3/km^2–5.0 m^3/km^2), whereas all the other patches were seen as brown oil mixed with ice. The size of Oil Spill No. 1 through No. 4 was around 0.01 km^2 each, while that of No. 5 was about 0.09 km^2. There were also reports of oil close to the shore in Parispeä. The clean-up vessel *Hylje* was following and cleaning up the oil spills. It left Helsinki on the night of March 14–15, and encountered Oil Spill No. 3 on the afternoon of March 15. On March 19, 2006, the wind pushed the oil-polluted field of ice southward to the Estonian shore, and the large clean-up vessels were unable to work near the shoreline due to shallow water. As a result, the three Finnish clean-up vessels returned home. Nevertheless, they had managed to collect nearly 15 m^3 of oil from the sea during their oil-combating operations.

9.3.4 Eutrophication

The most serious environmental problem in the Baltic Sea is eutrophication as a result of the large amount of nutrients entering the sea from different sources. According to HELCOM (2000), nitrogen and phosphorus enter the Baltic either waterborne or airborne. In addition, there are direct discharges from ships. In 2000, total input equaled 100,970 tonnes of nitrogen and 34,500 tonnes of phosphorus. About 75% of nitrogen entered the Baltic waterborne and 25% airborne. Agriculture and managed forestry contributed 60% of waterborne nitrogen to the sea. About 28% of nitrogen entered the sea from natural background sources and 13% came from point sources. Agriculture together with managed forestry contributed nearly 50% of the waterborne phosphorus whereas point sources and natural background sources each contributed about 25%. The amount of airborne phosphorus was just a few percent (HELCOM, 2000).

However, eutrophication in the Baltic Sea is not simply about high nutrient loads getting into the sea as a result of strong human impact and related problems in the marine environment. Therefore, because of its specific physics, eutrophication-related

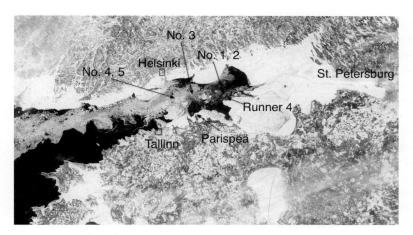

Figure 9.10. (Top) The operation area of the clean-up vessel *Hylje* during March 15-18, 2006 shown on a MODIS image of NASA's Terra/Aqua satellite. (Courtesy of MODIS Rapid Response Project at NASA/GSFC). This ship was following and combating oil spills since the afternoon of March 15. (Bottom) Oil-combating ship in operation. (Photograph by Jouko Pirttijärvi, printed with permission from the Finnish Environment Institute.) (Wang *et al.*, 2008.)

questions are briefly discussed in this book. The long residence time of Baltic Sea water—up to 40 years (Meier, 2007)—makes the sea very vulnerable to eutrophication. Another factor which promotes the effects of eutrophication is the strong haline stratification and the consequent anoxic conditions during stagnation periods in the Baltic Sea. As described in Section 5.4, it looks as if Major Baltic Inflows have become rarer.

The high level of nutrient concentrations leads to increased algal growth and to production of excess organic matter. This in turn leads to increased oxygen consumption by heterotrophic organisms and finally to deepwater oxygen depletion, especially in deep basins that have strong vertical stratification. Under oxygen-depleted conditions, benthic organisms die. Oxygen depletion leads furthermore to the release of redox-sensitive phosphorus from sediments, a process known as internal loading. This has many consequences and feedbacks in the Baltic Sea ecosystem. In particular, a vicious circle exists: in anoxic bottom areas phosphorus is released from the bottom leading to increased biological production and then increased sedimentation of organic matter increases oxygen consumption in the bottom. In conclusion the special nature of Baltic Sea physics has a strong impact on how the ecosystem functions.

There has recently been discussion about the feasibility of ventilating deepwater by human actions. One option is to pump a lot of oxygen to the bottom to prevent the formation of anoxic areas (as has been done in some lakes with relatively good results). However, ways and means of applying this method to the Baltic Sea are at present (2008) highly limited. It looks like this kind of "watermass engineering" is very expensive and time-consuming; the scale of the problem is very much larger than in Finnish lakes, there is no guarantee that oxygen conditions would improve, and even if some progress were obtained, the original problem—eutrophication—remains. However, all schemes should be carefully examined in the attempt to improve the state of the Baltic Sea.

Another option is to dredge the Danish Straits to enhance water exchange between the Baltic Sea and the North Sea and thus increase the amount of oxygen-rich waters entering the Baltic. This idea has its pros and cons. The pro side is really more intense water exchange. However, the Danish Straits are already eutrophied and the water with high oxygen concentrations that is expected to flow in may not necessarily be so "clean". Stratification would most probably become stronger due to more saline bottom water, and the absolute level of salinity would increase in the Baltic Sea with possibly harmful consequences to Baltic Sea brackish water biota. On the other hand, mixing might be more intense due to stronger tidal effects than currently. These problems should be studied using modeling tools to ascertain the total effect of these actions on the state of the sea.

9.3.5 Winter traffic

Ice conditions in the Baltic Sea have had a major impact on human life around its shores. This was especially so before the advent of steamships when the shipping season finished as soon as the ice formed, and consequently northern parts of the Baltic Sea became isolated.[7] In winter 1877, there was the first ever regular winter ship route, plied by the S/S *Express*, between east and west, from Hanko, Finland to Stockholm, Sweden. The first icebreaker was the *Murtaja*, which began operations in 1890 in Finland. Gradual progress has led since 1970 to all the main harbors of the Baltic Sea being kept open by 20–25 icebreakers. Even so, ice conditions still have a major impact on shipping: ice-strengthened vessels are needed, and delayed schedules result from such difficult ice conditions as heavily ridged ice or ice under pressure.

Another question related to winter is the forcing of ice on structures. This applies

---

[7] This was called the "ice barrier" in the past.

Figure 9.11. Ice road between the island of Hailuoto and mainland at Oulu, eastern coast of the Bay of Bothnia. (Photograph by Mr. Kyösti Marjoniemi, Oulu, Finland, printed by permission.)

mainly to navigational constructions such as lighthouses and buoys that need to keep their position through the ice season; drift ice in particular has caused serious damage. Even lighthouses have been damaged as a result of ice pressure. Ice pressure is a serious risk for wind farms constructed in sea areas.

In the northern Baltic Sea, the ice cover has not only been a barrier, ice roads have been constructed when the ice has become thick enough (Figure 9.11). These roads are well known from the distant past, when even lodging houses were built beside the ice roads for travelers (Olaus Magnus Gothus, 1539). A very convenent way of travel and transport in archipelago areas has thus been available in winter. The bearing capacity of floating ice is ah^2, where $a \approx 4\,\mathrm{kg/cm^2}$ is an empirical coefficient and h is ice thickness. Thus, an ice thickness of 25 cm is enough to bear a normal automobile. In severe winters there used to be a road across the Gulf of Bothnia in the Northern Quark, active as recently as 1966. In the Bay of Bothnia there still is an ice road from the mainland of Finland to the island of Hailuoto. The length of this ice road is 10 km. Several other ice roads are kept active in the coastal zone in the Baltic Sea. However, weak ice seasons in fall and spring create difficulties for archipelago traffic, with airboats being the feasible solution.

Special methods have been developed for winter fishing, and seal hunting in the spring has been mastered. As recreation has come more to the fore, so ice has been increasingly used as an excellent platform for skiing, skating, ice fishing, and outdoor life in general.

9.4 EXTREME SITUATIONS

9.4.1 Meaning of extreme cases

The physical characteristics of the Baltic Sea have an influence on environmental issues. Extreme cases bring particular risks such as those in Table 9.1, which need to be prepared for.

People have lived on the shores and islands of the Baltic Sea since the arrival of the ancestors, and the sea has been an important source of nutrition and a travel route. Table 9.2 illustrates some extremes of the physical characteristics of the Baltic Sea and how they impact human activities. The most critical extreme cases are due to sea-level elevation and ice, which will be discussed in more detail below. In fact, these two topics have been at the core of research and monitoring since the start-up of oceanography in the Baltic Sea.

9.4.2 Sea-level

Large anomalies in sea-level elevation in the Baltic Sea are caused by strong winds, air pressure differences, progressive waves, and seiches; this is especially so when they combine. On the sea shore these anomalies are found to be greatest at the end of bays: St. Petersburg in the Gulf of Finland, Kemi in the Gulf of Bothnia, and Pärnu in the Gulf of Riga. Very low levels can damage boats in docks, but most damage results when the sea-level becomes exceptionally high.

The maximum values of Baltic Sea levels are reached when there is a combination of several factors. First, the average sea-level needs to be above the normal (by about

Table 9.1. Extremes of the physical characteristics of the Baltic Sea and their environmental impact.

Characteristic	Low	High	Comments
Temperature	Freezing	25°C	High is connected to harmful algae blooms
Salinity[a]	Fresh	25‰	Species control due to salinity level
Currents	0	1 m/s	Transport, bottom erosion
Sea-level[a]	−1 m	1–4 m	Shoreline conditions
Wave height	0	14 m	Shoreline conditions, bottom erosion
Ice extent	12.5%	100%	Seals, spring bloom, mixing of riverine loads

[a] Extremes depend strongly on location.

Table 9.2. Extremes of the physical characteristics of the Baltic Sea and their impact on human activities.

Characteristic	Low	High	Comments
Temperature	Freezing	25°C	Recreation
Density[a] (above 1,000 kg/m^3)	0	15 kg/m^3	Displacement tonnage of ships
Currents	0	1 m/s	Transport, bottom erosion
Sea-level[a]	−1 m	1–4 m	Floods, harbors
Wave height	0	14 m	Navigation safety
Ice extent	12.5%	100%	Navigation time and safety

[a] Extremes depend strongly on location.

one meter). Such a situation becomes possible when long-lasting winds, as a result of a cyclone in the North Atlantic, persist for several weeks. The resulting westerly winds could cause the mean sea-level in the Baltic Sea to increase by as much as one meter as a result of the inflow of water of about 500 km^3. This needs to be followed by a rapid storm surge produced by a deep cyclone over the Baltic Sea for the development of very high sea-level elevation. The storm surge and the depression come from the west, with the storm surge principally in the form of a forced progressive wave. On its way to shallow areas in the east (Gulf of Riga and Gulf of Finland) this wave is reinforced by a decrease in the basin cross-section as well as by local winds. Thus, the sea-level may increase (e.g., from $\sim \frac{1}{2}$ m at the entrance to the Gulf of Finland up to 3 m–4 m at its eastern end, St. Petersburg). Although maximum sea-level rise is due to forced progressive wave development, the role of seiches can be pronounced when external forcing ceases, which has an especially important role in the Gulf of Finland.

The maximum sea-level height ever recorded was in 1824 in St. Petersburg when the sea-level was 421 cm above the zero level. The critical sea-level for flooding is 1.6 m in St. Petersburg, which has been exceeded 280 times since the start of measurements in 1703. Typically, sea-levels in other parts of the Baltic are not as high as in the eastern extremity of the Gulf of Finland, but values some 2 meters above mean sea-level have been recorded in many sites. The deviation of sea-level elevation from the mean is smaller for minimums than for maximums in any site. This is most likely due to northerly winds, which cause the minimums, to be weaker and wind events to be shorter than corresponding westerly winds.

9.4.3 Extremes of ice seasons

Baltic Sea ice is located at the edge of the seasonal sea ice zone, where regional climate variations show up drastically in the ice extent. During very severe winters the entire

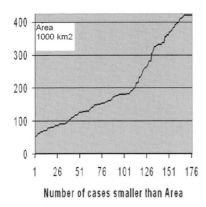

Figure 9.12. Cumulative distribution of the maximum annual ice extent in the Baltic Sea (1830–2005). (The data are from Jurva, 1952; Palosuo, 1953; Seinä and Palosuo, 1993 and from the FIMR website: *http://www.fimr.fi/*)

Baltic Sea is ice-covered, while in very mild winters the maximum annual ice extent is 12.5%–15% of the surface area of the Baltic Sea, covering the Bay of Bothnia and the eastern part of the Gulf of Finland (see Figure 3.24). Figure 9.12 shows the cumulative distribution of the ice extent based on data through the years 1830–2005.[8] The distribution has a folding point at about 180×10^3 km^2, which is very close to the average value.

Extreme ice conditions or lack of them depend of course on whether the ice is present or not in a given winter. Over large areas of the Baltic Sea the probability of ice occurrence is less than 50%, but the appearance of ice makes life difficult both for wildlife and for human activities especially when nature and people are not well prepared. A similar conclusion also holds for exceptionally early freezing, freezing dates of one month before the average date are not rare (Figure 9.13). In contrast, early or late melting is not so critical, late melting being more a nuisance than a major problem. The melting date, like the freezing date, can deviate about one month from the average (Figure 9.13). Since melting is governed by solar radiation, the later it takes place the faster the rate will be.

When the sea has frozen, the characteristics of the ice season depend on ice extent and ice thickness. Most concern centers on winter shipping conditions. In the past, at the time of sailboats and before the advent of railroads, people were largely dependent on marine transportation. When the Baltic Sea started to freeze from the north and east, harbors were closed and communities became isolated until the following summer. At times lighthouse staff could be imprisoned in their lighthouse when the sea froze unexpectedly, and ships were stuck in the ice or even destroyed by the ice pressure. Even now, the ice season brings extra costs for society such as keeping the icebreaker fleet operational and delays in ship schedules. Ice difficulties are largely influenced by drift ice dynamics, and therefore navigation conditions are not directly related to the severity of the ice season. Thick, stable or nearly stable 50 cm ice is not

---

[8] The ice extent series is known as "Jurva's time-series" after its originator. It starts in 1720 but Jurva indicated that there are significant uncertainties in the data until 1830. His notes about the data were destroyed when Helsinki was bombed during World War II.

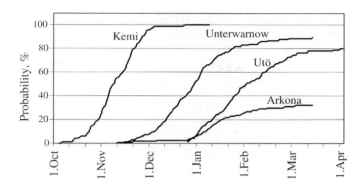

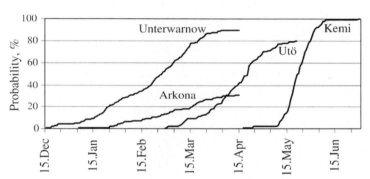

Figure 9.13. The cumulative distributions of freezing and melting dates. Kemi—northern coast of Bay of Bothnia; Utö—Archipelago Sea; Unterwarnow—Warnow river estuary at Rostock; Arkona—Arkona Basin. (Jevrejeva et al., 2002.)

difficult for icebreakers to go through, and dynamic ice with a thickness half of that can easily move and form strong pressure patterns. This is because the strength of an ice sheet depends on its thickness. The yield strength of thick-ice fields may be non-reachable, and therefore in the middle of winter the ice covering large basins such as the Bay of Bothnia and the Gulf of Riga may be stationary.

In the northern part of the Baltic Sea, ice roads are maintained for cars in coastal archipelago areas when the ice cover has become stable and grown in thickness to at least 25 cm. Increasingly much on-ice traffic concerns snowmobiles, and when the ice is thin there are major risks of ice breakage. Fatalities have resulted.

In the Baltic Sea there are two species of seals (ringed seal and gray seal). They both give birth on spring ice. The puppies grow fat from their mothers' milk and live on the ice before they start swimming; drift ice is a safe growing environment for them. In very warm winters the ice conditions may be too weak for seal nests.[9] In spring the first bloom in the surface layer is connected to melting ice and its influence on hydrography, and therefore the nature of the ice seasons is reflected in the spring ecological conditions in the Baltic Sea.

---

[9] In the Bay of Bothnia, seal hunters call the thick, heavy ice floes where seals are often found "seal ice".

9.5 OPERATIONAL OCEANOGRAPHY

9.5.1 General

Operational oceanography has a relatively long history in the Baltic Sea area. It commenced in the 1800s when winter shipping needed (near) real-time information on ice conditions. Consequently, ice-monitoring methods and ice-charting systems were developed. In the 1920s the Baltic Sea countries agreed the Baltic Ice Code which brought about ice information transfer using telegrams and standardized ice chart symbols and nomenclature. Two doctoral theses were prepared in the last century about ice season and charting methods (Jurva, 1937a; Palosuo, 1953), and ice forecasting based on numerical models commenced in 1977.

Apart from the oceanographic study of ice conditions, the development of operational oceanography in the Baltic Sea had to wait until the 1970s. Then, surface temperature charting in the fall was commenced by ice services to help predict the start of the ice season. Although sea-level forecasts could be made on a daily basis, development of the service was slow. However, operative activity slowly gained in momentum and a major step was taken in 1991 when the Global Ocean Observing System (GOOS) was created. This is based on routine collection, interpretation, and presentation of data from the ocean and atmosphere resulting in reliable description of the actual conditions of the sea including its living resources. An important element is the provision of prognoses for future conditions in the sea. Operational oceanography also concerns the establishment of a marine database from which time-series and statistical analyses can be obtained for descriptions of trends and changes in the marine environment (see Gorringe and Håkansson, 2005).

In operational oceanography there is now ongoing active co-operation under EuroGOOS, which is an association of agencies founded in 1994 to support the goals of GOOS and in particular the development of operational oceanography in European sea areas and adjacent oceans. In the Baltic Sea, BOOS (Baltic Operational Oceanographic System) is an association of institutes from all Baltic Sea countries taking national responsibility for operational collection of marine observations, forecasts, and dissemination of information and services including warnings. Real-time data collection by means of weather and wave buoys commenced at the same time as operational wave forecasting.

There is a wide range of users for these services. The most common are forecasting services needing information concerning harmful algae blooms, oil spills, sea ice, sea level, and warnings of, for example, extreme wave heights and icing on ships. Forecasts are also needed for planning fishery operations, ship routing, and monitoring purposes. Data and results are also important information sources for marine consultant work. Further development of operational systems is of great value to future oceanography in the Baltic Sea.

9.5.2 Observational systems

One of the key cornerstones in operational oceanography is the existence of a network of observations connected to a common data exchange system in near real-time.

There is ongoing collaboration with the EDIOS project (European Directory of the Ocean-observing System) in utilizing a proper observation network system (*http://www.edios.org/*).

A number of routine oceanographic observation stations exist in the Baltic Sea (see Gorringe and Håkansson, 2005 for details). Some stations measure in real-time while others have a delay. One of the key real-time observables is sea-level along the Baltic Sea coast. There are 130 real-time and 53 near real-time observation sites. Sea surface temperature can be observed using a number of different methods. Data collection can be in real-time from fixed stations or collected from ships in connection with monitoring programs and by so-called "ships of opportunity" (SOOP). Satellite thermal images provide basin-wide temperature charts and form the key element in basin-wide surface temperature mapping, but surface data from the sea still constitute the critical calibration information since satellites measure the temperature of just a very thin surface film.

Salinity measurements are usually carried out at the same time as temperature measurements in field surveys. Satellites are expected to help in this matter in the very near future. For temperature and salinity cross-sections relatively few real-time or nearly real-time observations are available. One profiling system at least should be working in all sub-basins of the Baltic Sea at any given time. Wind-driven surface wave measurements are still rare and only a few observational sites work operationally. One of the major drawbacks in operational observations is undoubtedly the availability of current data (or lack of them). These are measured in real-time only at a few locations on German and Swedish coasts. Oxygen is observed only at five operational sites, all on the German coast. Otherwise, oxygen is measured during monitoring cruises and other field campaigns.

SOOP refers to ships with marine measurement systems that are connected to information centers. A popular method of sampling water ("flow through water sampling") is applied at a few transects in the Baltic Sea. While the ship is moving, salinity, temperature, and chlorophyll *a* are registered. In the laboratory many other parameters can be analysed as well. The HELCOM COMBINE monitoring program, carried out by many institutes around the Baltic Sea, covers 571 stations, where a large number of seawater properties (physical, chemical, and biological) are frequently observed. There are also other national coastal monitoring programs with varying sampling frequency of temperature, salinity, nutrients, and biological parameters.

Ice observation systems in the Baltic Sea are well developed. Ice mapping is based on satellite imagery and surface observations from icebreakers and other ships and from coastal stations. The end-product is an ice chart with standard symbols and nomenclature, which can be downloaded in near real-time from ice service websites (*http://www.smhi.se*, *http://www.fimr.fi*). The operational ice-mapping system also includes transmission of digital ice charts from ice service centers to ships.

Remote-sensing satellites form an important part of observation systems (Table 9.3). There are several satellite-receiving stations around the Baltic Sea. The main areas of application are ice charting, sea surface temperature charting, and detection of harmful algae blooms. In addition, the consequences of environmental accidents

Table 9.3. Satellite systems used in operational oceanography in the Baltic Sea.

Satellite	Window	Objects
NOAA	Optical–infrared	Ice, surface temperature, algae blooms
MODIS	Optical–infrared	Ice, surface temperature, algae blooms
Radarsat	C-band radar	Ice

such as oil spills and pollutant inflows can be monitored using remote-sensing techniques.

9.5.3 Modeling tools

A large number of oceanographic models have been developed for the Baltic Sea (waves, sea-level, circulation, drift ice, oil spills, ecology, etc). Most are used for research purposes only.

Some operational models are used by a number of institutions. Of such models the following are discussed here: the wind-generated wave model WAM (the Wave Prediction Model) (WAMDI Group, 1988), the ocean circulation models HIROMB (High Resolution Operational Model for the Baltic Sea) (Funkquist, 2001) and BSHcmod (the model code was originally developed at *Bundesamt für Seeschifffahrt und Hydrographie*, BSH in Hamburg; Kleine, 1994), and sea ice–forecasting models in FIMR and SMHI Ice Services. In addition, there are now many more pre-operational models, which can be quickly run by real-time forecast services according to particular needs. These semi-operational models are used to generate hindcasts, nowcasts, and forecasts.

WAM model

The Wave Prediction Model (WAM) is a third-generation wave model developed by the Wave Model Development and Implementation (WAMDI) group (WAMDI Group, 1988). It is forced by the surface wind field and resolves the energy balance equation (Komen *et al.*, 1994) with no previous assumptions on the spectral shape of the wave field. The model output is a two-dimensional wave spectrum; it also provides derived wave properties such as significant wave height, mean wave direction and frequency, swell wave height, and mean direction.

The WAM model has been used by several teams in the Baltic Sea for research and operational prediction. Wave forecast providers in this area include the Danish Meteorological Institute, SMHI, and *Deutsche Wetterdienst*. In the Finnish Institute of Marine Research operative forecasts are produced four times a day based on the WAM model forced by the HIRLAM weather-forecasting model of the Finnish Meteorological Institute (Figure 9.14, see color section).

HIROMB

HIROMB is a Baltic Sea–North Sea circulation model at SMHI (Swedish Meteorological and Hydrological Institute). It is a three-dimensional baroclinic ice–ocean model and is run at three different horizontal resolutions: 12 nautical miles over the North Sea and 12, 3, and 1 nautical miles over the Baltic Sea. In the vertical direction HIROMB uses up to 24 layers, ranging from 4 m thick at the surface and increasing to 60 m at the deepest parts.

The model is mainly forced by the atmospheric circulation model HIRLAM. For freshwater inflow, daily data from the river runoff model HBV (Bergström, 1992) are used. At the open ocean boundary a storm surge model (NOAMOD) is used for sea-levels together with tides, climatological salinity, and temperature. The HIROMB model is used for forecasting currents, salinity, temperature, sea ice, and sea-level in connection with the meteorological model HIRLAM. Both these models are required to provide forcing data for the oil spill model (Sea Track Web), which calculates the drift and fate of oil spills from ships. The system itself is operating at SMHI but many users do their own oil spill forecasts (*http://www.smhi.se*).

BSHcmod

The DMI (Danish Meteorological Institute) operates a three-dimensional ocean model (BSHcmod) for the North Sea–Baltic Sea (Kleine, 1994) for short-term forecasting of the physical state of Danish and nearby waters. The model code has undergone extensive revision in DMI implementation, in co-operation with two model groups.

BSHcmod is forced with output from DMI's numerical weather prediction model, DMI HIRLAM. The meteorological parameters used are forecasts for wind, mean sea-level, atmospheric pressure, air temperature, humidity, and cloud cover. The sea model is further forced by real-time freshwater fluxes from 79 rivers. It is run twice a day and provides forecasts for the sea 60 hours ahead. The model gives information about sea-level variations, sea temperature, salinity, currents, and ice coverage (*http://www.dmi.dk*).

Sea ice model

Operative sea ice modeling dates to 1977 in the Baltic Sea based on a viscous model (Leppäranta, 1981a). Later, viscous–plastic models were employed (Leppäranta and Zhang, 1992; Omstedt et al., 1994; Haapala, 2000) and are now used in many ice services in the Baltic Sea region. These models provide 1–5 day forecasts of ice drift and consequent pressure and change in ice compactness and thickness and they are designed for the winter shipping service. Due to the dynamics, ice conditions may markedly change over synoptic timescales which is very important for shipping to know. The direction of motion with respect to the fast ice boundary is critical—to know whether there is pressure or a lead opening at the boundary. Recently, efforts have been made to develop navigation condition indexes for compact ice reports to ships but this has yet to be resolved (see Kõuts et al., 2007).

The model physics is fairly well-known for short-term forecasting. The main tasks for future research include improving knowledge of the initial conditions for the ice thickness field, developing data assimilation schemes to improve the quality of forecasts, and working toward a proper index for shipping conditions.

9.5.4 Case study: Storm Gudrun, 2005

Operational modeling systems have been used successfully in some extreme cases. This is very important because strong storms are known to bring unusually high sea-levels and waves that cause a lot of damage. A well-known example is from 2005. Storm Gudrun was one of the strongest storms in the last four decades in the Nordic area. An extratropical low depression reached the power of a hurricane, developed above the North Atlantic, and traveled over Ireland, Scotland, Scandinavia, and Finland on January 7–9, 2005. This hurricane, called *Gudrun* in Nordic countries, caused massive forest damage and disruption of power and phone lines. The storm killed 17 people (Suursaar *et al.*, 2006b). In the Baltic Sea this storm approached hurricane level (Soomere et. al., 2008b) and the storm surge reached record sea-levels. It set a new maximum at many measurements sites in the Gulf of Finland and Gulf of Riga. As a result of long-lasting strong cyclonic activity, the Baltic Sea level was already very high (+70 cm) before the storm (Suursaar *et al.*, 2006b).

The consequences of the storm were at their most dramatic in Estonia, causing lot of damage. A large part of the city of Pärnu was flooded (Figure 9.15), and a new sea-level record was also set (+275 cm). New records were also set at Tallinn (+152 cm) and at Toila (+160 cm). At the southern coast of Estonia the sea-level was close to historical maximums, and in all four Fininish stations around the Gulf of Finland new maximums were reached (Turku +130 cm, Hanko +132 cm, Helsinki +151 cm, and in Hamina +197 cm). However, the sea-level thankfully reached the relatively modest value of 230 cm in St. Petersburg.

An important fact is that these main features were captured very well by the sea-level models used, for example, in Finland, Sweden, Estonia, and Denmark. The authorities were informed in plenty of time and really serious damage was avoided. Storm Gudrun was reported widely in the media and set the standard for others to follow on the importance of forecasting such events in the present world (Professor Kimmo Kahma, Finnish Institute of Marine Research, pers. commun.). In Estonia, Academician Tarmo Soomere (Tallinn Technical University) warned the authorities in time so that the worst damage could be avoided. As a result of this he was nominated "person of the year" in 2005 in Estonia by the leading newspaper *Postimees*.

A special feature of Storm Gudrun was the extreme wave conditions in the northeastern part of the Baltic Sea (Soomere *et al.*, 2008b). The maximum recorded significant wave height in the central part of the northern Baltic Sea proper was 7.2 m, which is close to the all-time record (Kahma *et al.*, 2003). The dominating wave period was about 12 s in the northern Gotland Basin, probably reaching 13–14 s in the eastern sector. This longwave system penetrated the Gulf of Finland with periods up to 12 s. The wave height exceeded 5 m between Helsinki and Tallinn, close to the

Figure 9.15. Flooding in Pärnu during Storm Gudrun in January 2005. (Courtesy of Ms. Küllike Rooväli.)

historical maximum of 5.2 m. However, wave model results (e.g., Finnish Institute of Marine Research, German Weather Forecast System, Danish Meteorological Institute) suggested that the highest wave exceeded 10 m off the coasts of Saaremaa and Latvia, which were located far from the wave buoys. These models captured well the basic features of wave conditions about 48–54 hours ahead and accurately reproduced conditions in 24–36 hour forecasts. Actual measured wave fields were found to fit in the values predicted by different models (Soomere *et al.*, 2008b).

This chapter has discussed the links between the physical oceanography of the Baltic Sea and ecological and human life conditions. The topics included light conditions, oxygen conditions in deepwater, anthropogenic load, extreme cases, and operational oceanography. The link is mostly uni-directional, with physics providing the background conditions for ecology and people. However, there is a reverse link in that water quality influences the physical properties of seawater, mainly light absorption and scattering, and the building of major infrastructure influences the boundary conditions of physics. The next and last chapter (Chapter 10) summarizes the content of the book and briefly discusses the future of the Baltic Sea—influenced by anthropogenic load and climate change—and takes a look at future research in physical oceanography and sea ice in the Baltic Sea.

10

Future of the Baltic Sea

Professor Kazimierz Demel (1889–1978) studied at the universities in Lvov and Geneva. In 1916–1917, he served in the Russian Army and then the Polish Army. He participated in the Polish–Russian War in 1920 as well as in the Third Silesian Uprising in 1921. In 1923 he became the first permanent scientific employee of the Marine Fisheries Laboratory in Hel, which over time became the Marine Station, and moved to Gdynia in 1938. In that same year, Kazimierz Demel earned his doctor of philosophy degree from the Jagiellonian University in Kraków. He lived in Warsaw during the Nazi occupation of Poland, and participated in the clandestine education system. In 1949, the Marine Fisheries Laboratory joined forces with the Sea Fisheries Institute in Gdynia, and Professor Demel became the Head of the Department of Oceanography. Later, he was acting vice-director of scientific matters for a period. In 1950, he earned the Degree of Habilitation from the Jagiellonian University in Kraków. Professor Kazimierz Demel was the first in Poland to lecture on marine biology at the university level. For many years he served as an expert on the International Council for the Exploration of the Sea. Retirement in 1960 did not mean the end of his scientific work; Professor Demel remained active until his last days. The fruit of Professor Demel's career, some 250 popular and scientific publications, embody extensive knowledge of the sea and its resources. (Photograph: Courtesy of Sea Fisheries Institute in Gdynia, Poland.)

10.1 BALTIC SEA—UNIQUE BASIN

The Baltic Sea is a small and shallow brackish water basin, or rather a series of basins connected to the main Atlantic Ocean via the Danish Straits. The variable coastal geomorphology and the wide archipelago areas give the Baltic Sea its characteristic appearance (Figure 10.1). Mean salinity is about 7‰—one-fifth of the salinity of normal ocean water—and haline stratification is strong. This elongated sea lies between maritime temperate and continental sub-Arctic climate zones. In winter it is partly ice-covered and during the most severe winters it is completely frozen over. The Baltic Sea is young, has undergone several brackish and freshwater phases since the Weichselian glaciation, and from about 2,000 years ago salinity has been close to the present level. Land uplift is slowly changing the Baltic Sea landscape. Here it is possible to observe how land rises from the sea and how life on land gradually takes over.

The Baltic Sea is bounded by nine coastal countries, and there are about 85 million people living in its drainage basin. Its marine environment is highly vulnerable, which has been known for several decades. This has activated close co-operation to monitor the state of the sea as well as to strengthen its protection, even though Baltic Sea countries have different historical, political, and cultural backgrounds. The Baltic Sea Protection Agreement was signed in 1974, and as a consequence the Baltic Sea Commission or Helsinki Commission (HELCOM) was founded. Despite a number of different agreements the state of the Baltic Sea has become worse, loadings

Figure 10.1. The Baltic Sea landscape. (Courtesy of M.Sc. Riku Lumiaro.)

Figure 10.2. The Baltic Sea herring is not only a very important source of nutrition but also a low-cost seafood for all people. Increased levels of dioxins and PCBs have led to strong limitations in its utilization. These herrings are about 15 cm long.

and risks have increased, and the protection of "our sea" (as people in the area call it) has become even more important. Recently, in 2007, all nine Baltic Sea countries signed the HELCOM Baltic Sea Action Plan, which is an ambitious program to restore the good ecological status of the Baltic marine environment by 2021.

Increased anthropogenic load has led to eutrophication, which has further increased the anoxic bottom area and worsened the state of the Baltic Sea. Eutrophication is undoubtedly the most serious environmental problem in the Baltic Sea. Fish populations are accumulating harmful substances, and overfishing threatens the fish stock—in particular, the Baltic herring, cod, and salmon, which have been traditionally very important sources of nutrition (Figure 10.2). Ship traffic is also increasing (in particular, oil transport), which means higher risks to the vulnerable Baltic Sea environment. The impact of climate change is an additional potential threat on the centennial timescale.

In this book we have examined the present brackish Baltic Sea. Its mean depth is 54 meters, and its fundamental feature is the permanence of its salinity stratification. This limits vertical convection, and consequently the ventilation of deepwater masses is weak over large areas resulting in oxygen depletion. The salinity field shows a continuous decrease from oceanic levels at the North Sea boundary to freshwater in river mouths. The permanent circulation is weak, and sea ice is found for 5–7 months annually in the basin. The renewal time of the entire watermass is around 40 years. The book has presented—as it is titled—the physical oceanography of the Baltic Sea including the physics of sea ice. It is targeted at the research community and as a course textbook for Ph.D. students. It is anticipated that the book would also be very useful to environmental monitoring units, marine engineers, and decision-makers, who work for both exploitation and protection of the Baltic Sea and to people with a deep interest in getting to know the Baltic Sea. Oceanography has been central to this book from the outset, so that researchers from other fields such as physics, environmental sciences, ecology, or geography can use the book as a handbook or as self-study material.

There are ten text chapters, an appendix containing a summary of the properties of brackish seawater and some useful constants and formulas, a list of study questions, a reference list, and an index. Chapter 1 was an introduction to our topic, and Chapter 2 presented a brief history of the research, geography, and climate of the region. The classical oceanography of the Baltic Sea was presented next: topography and hydrography in Chapter 3; water, salt, and heat budgets in Chapter 4; and circulation and waves (Chapters 5 and 6, respectively). The theoretical background was also provided, introducing brackish water, the air–sea interaction, mixing of watermasses, ocean dynamics, and mathematical modeling methods. The physical system was described using observations (Figure 10.3). Chapter 7 presented the ice of the Baltic Sea, a unique brackish ice, which forms every winter and is a major factor in physics and ecology as well as in everyday life in this region. Then followed coastal processes with upwelling, sea-level elevation, weather and ice, and the interface between physics and ecology and physics and human activities in the Baltic Sea (Chapter 8). Chapter 9 considered environmental questions. The present chapter is the last text chapter and reaches conclusions about the Baltic Sea and its future.

The future of the Baltic Sea is beset by two classes of risk. Climate change scenarios (IPCC, 2007) would increase atmospheric temperature in the Baltic Sea region during the next 100 years, and this would lead to a warmer Baltic Sea with further consequences for the annual cycle of hydrography, circulation, ice season, and ecological state. The second, and more immediate problem, is eutrophication and pollution in the Baltic Sea. This latter problem we have to address now, irrespective of whatever happens with the climate.

10.2 CLIMATE CHANGE IMPLICATIONS

10.2.1 Climate change scenarios

The climate of the Baltic Sea region together with climate scenarios and climate impact on the Baltic Sea up to 2100 has recently been reviewed thoroughly by the BACC (Baltex Assessment of Climate Change) team (BACC Author Group, 2008), and all existing information was summarized in a HELCOM report (HELCOM, 2007) a year before. These publications are used here as the basic sources for our short summary below.

The present climatic and oceanographic conditions in the Baltic Sea have been given in previous chapters. In this chapter the present (year 2008) understanding of future climate change for the next 100 years is introduced, and its impact on Baltic Sea physics will be outlined. The first question concerning climate change is its proper definition. It has been defined to be (see BACC Author Group, 2008) *any change in climate over time whether due to natural variability or as a result of human activity* (Pielke, 2004). In the sense of this definition, climate has been changing continuously thus far. The more delicate question is whether climate change during the next 100 years includes a significant component due to anthropogenic influences. A "zero scenario" means that climate variability continues with the same characteristics as

Figure 10.3. (Top) The Finnish research vessel *Aranda* is a Super 1A Ice Class ship where research cruises have been carried out in Baltic Sea, Arctic, and Antarctic waters. (Photograph: Courtesy of Mr. Henry Söderman.) (Bottom) In spite of the rapid development of numerical modeling, satellite remote sensing, and automated observation systems, ship expeditions will be a critically important part of oceanographic research and technical development is moving along apace in research vessel technology. The Finnish R/V *Aranda* has a dynamic positioning system which makes it possible to keep the ship in an exact location (Sea Captain Riitta Kyynäräinen is keeping watch on the bridge). (Photograph: Courtesy of Mr. Tero Purokoski.)

before, as hypothesised by several scientists. However, the majority of scientists seem to lean toward "warming scenarios", the most authoritarian of which is the International Panel for Climate Change (IPCC).

Forecasts of future climate change are made using general circulation models (GCMs). These models describe the physics of the atmosphere and ocean in detail, and are applied to simulate future climate. However, there are many uncertainties. Present and past climate can be hindcast by GCMs, and consequently one may investigate how accurate simulations can be. In fact, these hindcasts give very good results. However, in order to gain accurate results of present climate, the increased concentration of greenhouse gases should be taken into consideration in model simulations, otherwise the models would give a cooling trend for the last 50 years. There is strong evidence that global climate change based on anthropogenic influences is really happening and warming our planet.

The forecast of future climate consists of many factors. First, models should have a proper description of the internal dynamics of atmospheric circulation. Second, the external forcing factors of future climate conditions should be well specified. This includes many problematic and hardly known factors (e.g., long-term change in solar radiation and volcanic activity). Third, anthropogenic factors in the future need to be evaluated. The IPCC has constructed future emission scenarios, which naturally depend on anticipated human activity in the future. Therefore, different kinds of scenarios for greenhouse gases and aerosol precursors have been developed. IPCC future scenarios are built to describe four different possibilities for future development ("storylines"). The so-called A1 and A2 storylines describe a future in which people strive for personal gain and pay less attention to the environment. In such a situation, emissions would further increase. The other two storylines (B1, B2) are based on sustainable development in which emissions would at least not increase from the present level. According to these storylines, the IPCC has constructed a number of different emission scenarios and these are the basis for present estimations of climate change.

The BACC's book (BACC Author Group, 2008) is based on climate change considerations from Cusbach *et al.* (2001), which in turn is based on GCM forecasts using different emission scenarios. According to Cusbach *et al.*'s study the increase in greenhouse gases will lead to global warming during the 21st century. The estimated increase in global mean temperature is 1.4°C–5.8°C between 1990 and 2100. This range covers differences between the climate and models and various estimates of emissions. However, the results include some uncertainties but the range of temperature given is the most probable according to present knowledge.

10.2.2 Overall climate change in the Baltic Sea region

The resolution of today's global climate models remains coarse. Therefore, for the Baltic Sea region downscaling techniques are used in order to get more high-resolution information (see BACC Author Group, 2008 for details). Model-based estimates for climate change are produced as follows. Two different model runs are carried out: the control run and the scenario run. The control period is often chosen as the climatological normal period 1961–1990 and the horizon of scenario runs is 2100. The climate change estimate is obtained by computing the difference between these two runs.

Based on available regional model studies, a general warming of 3°C–5°C of atmospheric surface layer temperature is expected to take place in the Baltic Sea region by the end of the 21st century. On the seasonal timescale the largest warming is expected to take place in the eastern and northern parts of the Baltic Sea region during winter months and in the south during summer months. In connection with the general warming trend, spatiotemporal changes in precipitation are expected as well. Annual precipitation is predicted to increase especially in northern parts. On the seasonal level the increase in precipitation would take place mostly in wintertime. Regionally, southern areas would become dryer than northern ones, especially during summer. The increase in annual precipitation would lead to an increase in river runoff, and consequently river inflow is expected to increase in the northern part of catchments with a decrease in the southern part. In general, river inflow would decrease in summer and increase in winter.

10.2.3 Future changes in the Baltic Sea

Future scenarios for the Baltic Sea are based on calculations using different regional coupled atmospheric–ocean models. Details about the models can be found in BACC Author Group (2008) and in Meier (2006) and in the other papers mentioned here.

Sea surface temperature in the Baltic Sea is expected to be on average higher by 2°C–4°C during 2071–2100 than during the normal period 1961–1990 (BACC Author Group, 2008). The increase should be greatest in early summer (May–June) in the southern and central parts of the sea. Year-to-year variability is going to increase in the northern basins due to melting of ice.

The estimated increase in sea surface temperature shows up best in wintertime in the ice conditions. The first-order prediction is that an increase of mean atmospheric temperature by 1°C would delay the freezing date by 5 days, decrease the ice thickness by 5 cm–10 cm, decrease the ice extent by 45,000 km^2, and lead to ice breakup earlier by 5 days (Haapala and Leppäranta, 1997). Haapala *et al.* (2001a, b) estimated the average maximum annual ice extent to be reduced by 50%–80% by 2100 from the present value (190,000 km^2). Around 2100 there would be ice in the Bay of Bothnia and eastern Gulf of Finland every year, but the Sea of Bothnia, large areas of the Gulf of Finland, and the Gulf of Riga would be ice-free in normal years (Figure 10.4). The length of the ice season would decrease by as much as 1–2 months in northern parts and even 2–3 months in the central Baltic (Meier *et al.*, 2004b).

The reduction in maximum ice extent and the shortening of the ice season would have many consequences for Baltic Sea physics. Surface albedo would decrease and the sea would gain much more solar radiation thus contributing to a rise in sea temperature. The annual cycle of hydrography in the Baltic Sea would be modified, especially as a result of the weakening of spring stratification. Wind stress transfer to the water body and its influence on circulation would also be largely modified. In

Mean level ice thickness : 1 – 10 March

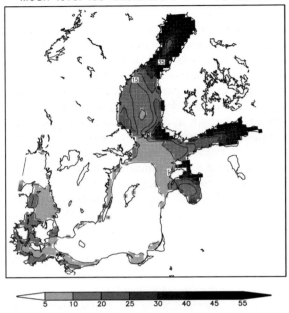

Mean level ice thickness : 1 – 10 March

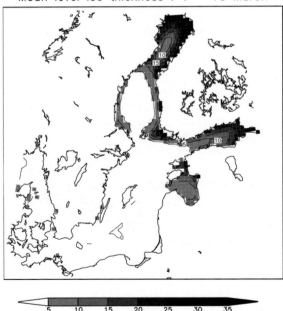

Figure 10.4. Mean thickness of level ice (cm) in the control simulation 1961–1990 (top) and at 2100 (bottom) based on a Baltic Sea ice–ocean model (Haapala *et al.*, 2001a).

addition, wind stress gained by the sea surface could increase in basins with open drift ice but further consequences are uncertain.

There are also indications that in the climate of the future the strength of wind is more likely to increase than decrease, but the magnitude of this change is still highly uncertain (BACC Author Group, 2008). Assuming that the number of extreme cases (storms) increases, high-wave conditions and extreme sea-level elevations are more probable than at present. Increasing wind speeds would in turn intensify mixing in the Baltic Sea and influence the distribution of nutrients. In summer, increased mixing may lead to intensification of cyanobacterial blooms and thus to anoxic conditions. In sediments, resuspension of contaminants would take place as a result of increased wave action.

The latest results (IPCC, 2007) predict global sea-level rise to be about 10 cm–50 cm by the year 2100 (Figure 10.5). This is somewhat smaller than previous estimates, and there is still some uncertainty in these numbers—especially concerning the development of polar ice sheets. In the Baltic Sea, future sea-level rise is not a straightforward process—it strongly depends on the sea area under investigation. Mean sea-level is determined by three factors: eustatic rise in global sea-level, land uplift (see Figure 3.1), and the water balance of the Baltic Sea. According to expected changes in these factors, scenarios for future sea-level change can be outlined. The results (BACC Author Group, 2008) show that expected sea-level rise is not uniform in the Baltic area. Many regions in which sea-level apparently retreats (land rise greater than eustatic rise in global sea level) at present would witness rising sea-levels in the future. An example is the Gulf of Finland. Land uplift would be overcome by the accelerated rise in global sea-level. Sea-level retreat would however still continue

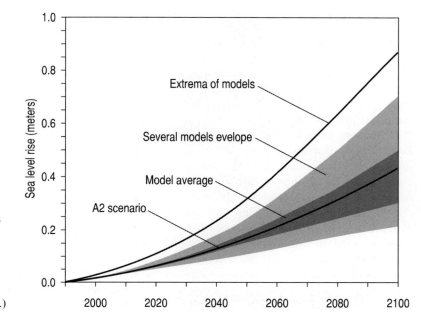

Figure 10.5. Forecast global sea-level changes in the Baltic until 2100. (Redrawn from BACC Author Group, 2008.)

in the Gulf of Bothnia. According to the most extreme scenarios, sea-level would start to increase in the Gulf of Bothnia after 2050 (*http://www.fimr.fi*). Extreme values in sea-level are expected to increase more than the mean sea-level, but this strongly depends on future wind conditions, which are still somewhat unclear.

The present net inflow of freshwater to the Baltic Sea is about $15,000\,\text{m}^3\,\text{s}^{-1}$ (see Section 4.1). River discharge is estimated to increase by as much as 54% during winters while summer inflows would decrease by 16% by 2100 (BACC Author Group, 2008). Annual river discharge shows an increase in northern catchments while in southernmost catchments the situation is opposite, and net flux would increase by some one-third. There would also be a higher chance of high-river-discharge events than at present.

There has even been discussion about whether the Baltic Sea could become a lake. Meier and Kauker (2003a) estimated that even if freshwater inflow increased by 100% the Baltic Sea would not become a freshwater body. There would still be a halocline, and thus wind-induced mixing would be restricted to the upper layer. If the change in freshwater input were to be within ±30%, the location of the halocline would not change much from present. Actually, an increase in river runoff could reduce surface layer salinity considerably and the salinity difference between the upper and lower layer would increase. This in turn would lead to reduced vertical mixing and consequent worsening of oxygen conditions in bottom water.

The salinity change is expected to vary between −45% and +4% according the recent regional scenario simulations (BACC Author Group, 2008). The largest negative change is not statistically significant. In the scenario with the largest negative change, sea surface salinity in the Bornholm Basin would be as low as it is now in the northern Sea of Bothnia. The Belt Sea front would shift northwards.

10.3 ENVIRONMENTAL STATE

Both natural and anthropogenic factors contribute to the environmental problems in the Baltic Sea. Euthrophication, pollutants, erosion by ship-induced waves, and the import of alien species by ships are anthropogenic. Oxygen depletion in deep basins is a natural phenomenon but amplified by eutrophication. There is a clear consensus that nutrient load leading to eutrophication should be drastically reduced in the Baltic Sea. Unfortunately, all the physical features of the Baltic Sea have the same impact: they make the sea very vulnerable to the effects of human activity.

The key question is to find the best strategy for nutrient load reduction. In this respect, recommendations have been made; for example, for the Gulf of Finland in the SEGUE project (Pitkänen and Tallberg, 2007). That is because the Gulf of Finland is among the most eutrophied areas in the Baltic Sea. External nutrient load relative to the surface area of the Gulf of Finland is two to three times that of the corresponding average for the whole Baltic Sea.

The results of present studies, based on the combined use of measurements and ecosystem modeling tools, clearly show that it is not possible to combat eutrophication in the Baltic Sea without considerable reduction of all major nutrient sources both at the coastline and inland in the large catchment area of the Baltic Sea. This seems to be a valid statement from ecological effects and cost–benefit analysis viewpoints.

Coupled hydrodynamic–ecological models can be used to carry out long-term simulations to study the effects of different loading scenarios. However, such work is not as straightforward as expected and the message from simulations is not always so reliable. That is because there are still large gaps in our understanding of nutrient dynamics in the Baltic. Even such basic information as the magnitude of nutrient load into the Baltic Sea from some sources is not known sufficiently well (Pitkänen and Tallberg, 2007). This means that monitoring should be more frequent and that more focus should be placed on quality assurance in laboratories as well.

Compared with other aquatic ecosystems only a relatively limited number of animal and plant species have adapted to live in the brackish Baltic waters. The small number of species in Baltic Sea foodwebs have important roles to play in the structure and dynamics of the Baltic Sea ecosystem. The disappearance of a single key species could have serious consequences in how the ecosystem functions. Other problems are harmful substances and risks associated with increased ship traffic.

10.4 FUTURE OF THE PHYSICAL OCEANOGRAPHY OF THE BALTIC SEA

The physical oceanography of the Baltic Sea has been investigated for almost 150 years, with some instrumental time-series records going back to the 1700s. It has been one of the most intensively studied basins in the world. Consequently, the physics of the Baltic Sea is well understood, and the mathematical models used can be regarded as good-quality models.

There are, however, a few critical items in which information based on observational data is limited, and hence trust in numerical simulations cannot be complete. First, deepwater circulation is mostly known for the resulting salinity distribution— not for circulation dynamics as such. The second item is diapycnal mixing, which is known to take place but for which the leading processes are not well understood. Third, hydrography and circulation under the ice-covered Baltic Sea suffers from a lack of observations: in particular, some features that are not closely connected to ice are unclear.

The evolution of the ice season is well known, and there are long-term observational records available. Satellite systems are excellent for the mapping of ice conditions, except for the thickness of ice, which is a major worldwide sea ice geophysics problem. Apart from the ice thickness the main Baltic Sea ice research question is the fast ice–drift ice boundary and the dynamics of this boundary. In its small and shallow basins the fast ice zone covers considerable areas and there is no satisfactory approach to put the physics of fast ice into large-scale sea ice models.

In this book the principal goal and the leading theme has been to present the physical oceanography of the Baltic Sea based on physical theory and observations. The theory consists of using general oceanography for a small and shallow basin, the major features of which are its brackish water as well as ice formation in wintertime. Field observations contain temperature and salinity characteristics, circulation, air–sea interaction, and ice conditions based on classical station data and satellite remote sensing. Mathematical models form a very powerful tool in physical oceanography, and a very large variety of problems have been successfully examined by model simulations. There is no modeling chapter in this book, but model results are analysed and discussed in most chapters in specific modeling sections. Having models in separate sections emphasizes the value of the fundamental theory and data—while these stay, new models replace old ones whose value is then purely historical.

The Baltic Sea is a heavily polluted sea area undergoing slow water renewal and having strong external and internal loading. In the future research of physical oceanography, "Save Our Sea" would be an appropriate slogan for our leading theme. For physicists this means the development of operational oceanographic systems, evaluation of the potential consequences of predicted climate change, working closely with environmental scientists and biologists to together develop survival strategies and to solve critical open science problems. It is in this that the physics of the Baltic Sea forms the fundamental basis of knowledge and understanding.

Appendix A

Useful constants and formulas

GENERAL PHYSICS

Avogadro's number	$N_A = 6.022 \times 10^{23}$
Planck's constant	$h = 6.6261 \times 10^{-34}$ J s
Stefan–Boltzmann constant	$\sigma = 5.6705 \times 10^{-8}$ W m^{-2} K^{-4}
Velocity of light in a vacuum	$c_0 = 2.9979 \times 10^8$ m s^{-1}
Celsius temperature reference	$0°C = 273.15$ K

EARTH AND SUN

Acceleration due to gravity on Earth	$g = 9.81$ m s^{-2}
Earth's rotation rate	$\Omega = 0.7292 \times 10^{-5}$ s^{-1}

The Sun–Earth distance (r)

$$\left(\frac{\tilde{r}}{r}\right)^2 = 1.000110 + 0.034221 \cos j + 0.001280 \sin j$$

$$+ 0.00719 \cos 2j + 0.000077 \sin 2j$$

where the tilde stands for averaging; and $j = (J - 1) \cdot 2\pi/365$, J is the Julian Day ($J = 1.0$ on January 1, at 00:00 hours).

Solar constant	$Q_{sc} = 1,370$ W m^{-2}
Optical band fraction	$\gamma \approx 0.48$ (range 0.44–0.50)

Appendix A: Useful constants and formulas

The solar zenith angle (Z)

$$\cos Z = \sin \phi \sin \delta + \cos \phi \cos \delta \cos \tau$$

$$\sin \delta = \sin \varepsilon \sin j$$

$$t_{\text{GMT}} = -\Lambda/15° + \Delta t$$

where ϕ is latitude, δ is declination, τ is hour angle (zero at solar noon), $\varepsilon = 23°27'$ is the inclination of ecliptic, t_{GMT} is local noon in GMT (Greenwich Mean Time), Λ is longitude, and Δt is time correction, $|\Delta t| < 30$ minutes and varies during the course of the year, available in astronomical tables. The solar altitude is $\frac{1}{2}\pi - Z$.

BALTIC SEA REFERENCES ($S \approx 7.4‰$)

Density of water	$\rho_w = 1{,}005\,\text{kg m}^{-3}$
Specific heat of water	$c_w = 4.15\,\text{kJ kg}^{-1}\,°\text{C}^{-1}$
Thermal emissivity of water	$\varepsilon = 0.97$
Latent heat of evaporation	$L_E = 2.5\,\text{MJ kg}^{-1}$
Latent heat of freezing	$L = 335\,\text{kJ kg}^{-1}$
Density of ice	$\rho_i = 917\,\text{kg m}^{-3}$
Specific heat of ice	$c_i = 2.1\,\text{kJ kg}^{-1}\,°\text{C}^{-1}$
Emissivity of ice	$\varepsilon = 0.97$

ATMOSPHERE

Standard pressure at sea-level	$p_0 = 1{,}013.25\,\text{mbar}$
Gas constant of air	$R_a = 287.04\,\text{J kg}^{-1}\,°\text{C}^{-1}$
Specific heat of air (constant pressure)	$c_p = 1.0\,\text{kJ kg}^{-1}\,°\text{C}^{-1}$
Density of air	$\rho_a = p_a/(R_a T)$ ($p_a = $ air pressure)
Atmospheric transmissivity	$T_{tr} \approx 0.9$

The saturation water vapor pressures (mbar) over a liquid water surface (e_w), over an ice surface (e_i), as functions of temperature (°C) are given by (see Gill, 1982)

$$\log_{10} e_w(T) = \frac{0.7859 + 0.03477T}{1 + 0.00412T}$$

$$\log_{10} e_i(T) = \log_{10} e_w(T) + 0.00422T \quad (T \leq 0°\text{C})$$

The relation between specific humidity (q) and water vapor pressure is

$$q = \frac{0.622 e}{p_a}$$

SEA ICE PHASE EQUATIONS

Denote by ρ and S the bulk density and salinity of a sea ice sample, by ρ_b and S_b the density and salinity of the brine, and by k the ratio of the mass of solid salts to the mass of the salts in brine. The brine volume is obtained by setting the salt mass in the bulk of ice equal to the salt masses in the brine and solid salt crystals,

$$\nu_b = (1 - \nu_a) \frac{\rho S}{\rho_b S_b (1+k)}, \quad \rho_b = 1,000 \cdot (1 + 0.8 S_b)$$

This is formulated using auxiliary functions $F_1(T)$ and $F_2(T)$ as

$$\nu_b = (1 - \nu_a) \frac{\rho_i S}{F_1(T) - \rho_i S F_2(T)}, \quad \rho_i = 917 - 0.1403 T$$

$$F_1(T) = \rho_b S_b (1+k) = a_0 + a_1 T + a_2 T^2 + a_3 T^3$$

$$F_2(T) = (1+C) \frac{\rho_b}{\rho_i} - C \frac{\rho_b}{\rho_{ss}} - 1 = b_0 + b_1 T + b_2 T^2 + b_3 T^3$$

The coefficients of the third-degree polynomials $F_1(T)$ and $F_2(T)$ for $-30°C \leq T \leq 0°C$ (Cox and Weeks, 1983; Leppäranta and Manninen, 1988) are

F_1	a_0	a_1	a_2	a_3
$0 \geq T \geq -2$	-4.122×10^{-2}	-1.841×10^1	5.840×10^{-1}	2.145×10^{-1}
$-2 \geq T \geq -22.9$	-4.732	-2.245×10^1	-6.397×10^{-1}	-1.074×10^{-2}
$-22.9 \geq T \geq -30$	9.899×10^3	1.309×10^3	-5.527×10^{-1}	-7.160×10^{-2}
F_2	b_0	b_1	b_2	b_3
$0 \geq T \geq -2$	9.031×10^{-2}	1.611×10^{-2}	1.229×10^{-4}	1.360×10^{-4}
$-2 \geq T \geq -22.9$	8.903×10^{-2}	-1.763×10^{-2}	-5.330×10^{-4}	-8.801×10^{-6}
$-22.9 \geq T \geq -30$	8.547	1.089	4.518×10^{-2}	5.819×10^{-4}

The $k \times 1,000$ of the mass of solid salts relative to the mass of salts in brine is given in the following table (temperature in °C). For $T \geq -2°C$, $k = 0$.

T	-2	-4	-6	-8	-10
k	0	0.554	1.050	1.400	55.277
T	-12	-14	-16	-18	-20
k	84.141	97.627	106.330	112.570	118.078
T	-22	-24	-26	-28	-30
k	123.090	509.787	1,312.69	2,065.83	2,685.71

Appendix B

Study problems

CHAPTER 2

1. What is the total watermass of the Baltic Sea?

2. List at least seven physical factors in the Baltic Sea that are clearly different from those of the World Ocean.

3. Let us assume that the salinity of water in the Baltic Sea is 6‰.

    a. The concentration of gold equals 10^{-11} in ocean water. How much gold is there in the Baltic Sea (in kilograms).
    b. If the seawater temperature is 10°C, how much energy is needed to produce 1 kg of salt?

CHAPTER 3

1. If the current speed is 5 cm/s, how should hydrographic measurements be carried out in a 100 km wide basin in order for the results to be comparable (on the synoptic timescale)? The speed of a research vessel is 10 knots and taking one CTD profile requires 30 minutes.

2. Calculate the speed of sound in the Gotland Basin at the 50 m depth, where the temperature is 3.1°C and salinity is 7.8‰. How does the speed of sound change as we go down to a 200 m depth, where the temperature is 4.5°C and salinity is 11.9‰.

3. How does the averaging of temperature profiles and salinity profiles (an average profile is calculated from several single profiles) influence the meaning and interpretation of a T–S diagram?

Appendix B: Study problems

4 Calculate the freezing point temperature and the temperature of maximum density in the Bay of Bothnia ($S = 3‰$), in the Åland Sea ($S = 6‰$), and in the Kattegat ($S = 19‰$).

CHAPTER 4

1 There is an infinite water basin with salinity S in the center of a basin and there are no currents. Let us assume that precipitation and evaporation are known. What is the surface salinity? (*Hint:* it might be that in the solution salinity depends on time.)

2 Calculate, using the Knudsen equations, the inflow and outflow in the Danish Straits when the salinity of outflowing water is 8.6‰ and that of inflowing water 15.0‰. The net discharge of freshwater equals $473 \, km^3/yr$. How is the result affected if freshwater flux is doubled?

3 Investigate the diffusion equation of heat

$$\frac{\partial T}{\partial t} = K \frac{\partial^2 T}{\partial z^2} \quad (K \approx 10^{-3} \, m^2 \, s^{-1})$$

with boundary conditions $T_0 = T_m + \Delta T \cos\left(\frac{2\pi t}{\tau} - \alpha\right)$ (surface) and $T \to T_m$ when $z \to 0$. What is the depth that daily and annual temperature cycles penetrates and what is the time lag at the corresponding depth compared with the sea surface?

4 Calculate the turbulent heat fluxes (latent heat and sensible heat) from the climatology of Utö island.

Month	T_s (°C)	T_a (°C)	U_a (m/s)	q_a	q_0
I	1.7	−2.1	7.9	0.002772	0.004305
II	0.7	−3.5	6.9	0.002556	0.004006
III	1.0	−1.7	6.5	0.002889	0.004094
IV	1.9	1.9	5.8	0.003711	0.004368
V	4.6	6.7	5.3	0.004957	0.005291
VI	11.6	12.6	5.3	0.007381	0.008537
VII	15.3	15.9	5.3	0.009264	0.010882
VIII	15.1	15.7	5.7	0.009033	0.010742
IX	12.2	11.9	6.7	0.007134	0.008884
X	9.1	7.9	7.6	0.005516	0.007217
XI	6.3	3.8	8.1	0.004148	0.005957
XII	3.6	0.4	8.3	0.003252	0.004930

CHAPTER 5

1. It takes two years before the influence of a Major Baltic Inflow is recognized in the northernmost Bay of Bothnia. What is the mean speed of the pulse?

2. How long is the inertial period at latitude 60°N? How much does the radius of a circle change when moving from the Bornholm Basin to the Bay of Bothnia when current velocity is 10 cm/s?

3. Estimate the orders of magnitude of the equation of motion when $U = 30$ cm/s, $T = 1$ day, $L = 25$ km, $H = 20$ m, $A_H = 500$ m^2/s, and $A_V = 0.05$ m^2/s.

4. What is the speed and direction of a wind-driven current at a 5 m and 25 m depth at latitude 60°N. when the surface current is 10 cm/s to the north and $A_V = 0.01$ m^2/s?

5. The current field $U = u\mathbf{i} + v\mathbf{j}$ has the form

$$u = a(y - y_0)$$
$$v = 2a(y - y_0); \quad a = 10^{-6} \, \text{s}^{-1}$$

   a. Sketch the current field with arrows ($\Delta x = \Delta y = 100$ km).
   b. Calculate $\nabla \cdot U$ and $\nabla \times U$.

7. There is a saltwater inflow to the Baltic Sea that lasts 7 days and the salinity of the inflowing water is 20‰.

   a. Calculate the Q_{87} index.
   b. How long should this pulse continue before it is ranked as "very strong"?

CHAPTER 6

1. Show how a standing wave $\eta(x, t) = 2A \cos(kx) \cos(\omega t)$ can be presented as a sum of two waves that are moving opposite to each other.

2. The period of an n-node seiche wave can be estimated using equation $T_n = \dfrac{2L}{n\sqrt{gH}}$. What are the approximations that led to the results (starting with linear wave theory)?

3. Solve, using a wave monogram (see Figure 6.12) for significant wave height, the following problems:

   a. What is the significant wave height when the wind blows for 5 hours at a speed of 5 m/s?
   b. The wind has blown for 18 h and fetch is 200 km. What is the factor that limits wave height: is it wind speed. duration of the wind, or fetch?

Appendix B: Study problems

c. The wind blows first for 9 hours at a speed of 10 m/s and then 3 hours at 20 m/s. What is the significant wave height after that?

CHAPTER 7

1. Snow-covered ice is floating in the water. Where the water boundary is located with respect to the snow/ice boundary when the depth of snow is h_s, the thickness of ice is h_i, the density of snow is ρ_s, the density of ice is ρ_i, and the density of water is ρ_w? Define the numerical value when $h_s = 15$ cm, $h_i = 50$ cm, $\rho_i = 0.9$ g cm^{-3}, $\rho_w = 1.0$ g cm^{-3}, and $\rho_s = 0.3$ g cm^{-3}.

2. How large must a 50 cm thick ice float be so that you can stand on it without getting your feet wet?

3. What will be the thickness of ice when the air temperature remains at $-15°$C for 20 days? How much would the ice thickness further increase if the same air temperature remained for 10 additional days?

4. A westerly wind is blowing at 15 m/s and water current speed is 7 cm/s to the south. Estimate the speed and direction of free-drifting ice.

CHAPTER 8

1. Let us assume that at the mouth of a river there is pronounced stratification. The volume of the mouth area is 5×10^8 m^3 and river discharge is 10 m^3 s^{-1}. The salinity at the bottom is 4.0‰ and 0.1‰ at the surface. What is a typical flushing time for the river mouth area?

CHAPTER 9

1. What is the euphotic depth when the attenuation coefficient of light is $c = 0.57$ m^{-1}? (*Hint:* at euphotic depth the light level has reduced to 1% of the value at the surface.)

References

Ågren, J. and Svensson, R. (2007) Postglacial land uplift model and system definition for the New Swedish Height System RH 2000. *Lantmäteriet, Rapportserie: Geodesi och Geografiska informationssystem*, **2007**(4), 124 pp. Gävle, Sweden.

Aitsam, A. and Elken, J. (1982) Synoptic scale variability of hydrophysical fields in the Baltic Proper on the basis of CTD measurements. In: J. C. J. Nihoul (ed.), *Hydrodynamics of semi-Enclosed Seas*, pp. 433–468. Elsevier, Amsterdam.

Aitsam, A. and Talpsepp, L. (1982) On one interpretation of the mesoscale variability in the Baltic Sea. *Oceanology*, **22**(3), 357–362.

Aitsam, A., Hansen, H. P., Elken, J., Kahru, M., Laanemets, J., Pajuste, M., Pavelson, J., and Talpsepp, L. (1984) Physical and chemical variability of the Baltic Sea: A joint experiment in the Gotland Basin. *Continental Shelf Research*, **3**, 291–310.

Alenius, P. (1978) On the use of CTD-profiling system on R.V. Aranda. *Proceedings of the 11th Conference of the Baltic Oceanographers, Rostock, April 24–27, 1978*, pp. 600–607.

Alenius, P. (1980) On currents, inertial and internal waves in a coastal zone of the Bothnian Sea. Department of Geophysics, University of Helsinki (Phil.Lic. thesis).

Alenius, P. (1985) Itämeren sisäiset aallot [Internal waves of the Baltic Sea]. In: *XII Geofysiikan päivät Helsingissä 14–15.5.1985*, pp. 25–30. The Geophysical Society of Finland, Helsinki.

Alenius, P. and Leppäranta, M. (1982) Statistical features of hydrography in the northern Baltic Sea. *Proceedings of the 13th Conference of the Baltic Oceanographers, Helsinki, August 24–27, 1982*, Vol. I, pp. 95–104.

Alenius, P. and Mälkki, P. (1978) Some results of the current measurement projects of the Pori–Rauma region. *Finnish Marine Research*, **244**, 52–63.

Alenius, P., Myrberg, K., and Nekrasov, A. (1998) The physical oceanography of the Gulf of Finland: A review. *Boreal Environment Research*, **3**(2), 97–125.

Alenius P., Nekrasov, A., and Myrberg, K. (2003) The baroclinic Rossby-radius in the Gulf of Finland. *Continental Shelf Research*, **23**, 563–573.

Alexandersson, H., Schmith, T., Iden, K., and Tuomenvirta, H. (1998) Long-term variations of the storm climate over NW Europe. *The Global Atmosphere and Ocean System*, **6**, 97–120.

References

Alford, M. and Pinkel, R. (2000) Patterns of turbulent and double diffusive phenomena: Observations from a rapidly profiling conductivity probe. *Journal of Physical Oceanography*, **30**, 833–854.

Alestalo, J., Heikkinen, O., and Tabuchi, T. (1986) Sea ice deformation in the Bay of Bothnia off Hailuoto in March 1986. *Bay of Bothnia Reports*, **5**, 51–63.

Al-Hamdani, Z. and Reker, J. (Eds.) (2007) *Towards Marine Landscapes in the Baltic Sea*. BALANCE Interim Report No. 10, Copenhagen. Available at *http://www.balance-eu.org/*

Allan, R. J. and Ansell, T. J. (2006) A new globally complete monthly historical gridded mean sea level pressure data set (HadSLP2): 1850–2004. *Journal of Climate*, **19**, 5816–5842.

Andersson, H. C. (2002) Influence of long-term regional and large-scale atmospheric circulation on the Baltic Sea level. *Tellus*, **54A**, 76–88.

Andersson, T. and Gustafsson, N. (1994) Coast of departure and coast of arrival: Two important concepts for the formation and structure of convective snowbands over seas and lakes. *Monthly Weather Review*, **122**, 1036–1049.

Andersson, T. and Michelson, D. (1999) The December 1998 snowfall over Sweden as seen by weather radars. *Polarfront*, **100/101**, 26–33.

Andersson, T. and Nilsson, S. (1990) Topographically induced convective snowbands over the Baltic Sea and their precipitation distribution. *Weather and Forecasting*, **5**, 299–312.

Andreas, E. L. (1998) The atmospheric boundary layer over polar marine surfaces. In: M. Leppäranta (ed.), *The Physics of Ice-covered Seas*, Vol. 2, pp. 715–773. Helsinki University Press, Helsinki.

Andrejev, O. and Sokolov, A. (1989) Numerical modelling of the water dynamics and passive pollutant transport in the Neva inlet. *Meteorologiya i Gidrologiya*, **12**, 75–85 [in Russian].

Andrejev, O., Myrberg, K., Mälkki, P., and Perttilä, M. (2002) Three-dimensional modelling of the Baltic main inflow in 1993. *Environmental and Chemical Physics*, **24**(3), 121–126.

Andrejev, O., Myrberg, K., Alenius, P., and Lundberg, P. A. (2004a) Mean circulation and water exchange in the Gulf of Finland: A study based on three-dimensional modelling. *Boreal Environment Research*, **9**, 9–16.

Andrejev, O., Myrberg, K., and Lundberg, P. A. (2004b) Age and renewal time of water masses in a semi-enclosed basin: Application to the Gulf of Finland. *Tellus*, **56A**, 548–558.

Anonymous (1999) Nordic Environmental Research Programme for 1993–1997: Final Report and Self-Evaluation. TemaNord Environment. TemaNord 1999: 548, 260 pp. Nordic Council of Ministers, Copenhagen.

Apel, J. R. (1987) *Principles of Ocean Physics*. Academic Press, 634 pp.

Armstrong, T., Roberts, B., and Swithinbank, C. (1966) *Illustrated Glossary of Snow and Ice*. Scott Polar Research Institute, Cambridge, U.K.

Arst, H. (2003) *Optical Properties and Remote Sensing of Multicomponental Water Bodies*. Springer/Praxis, Chichester, U.K., 231 pp.

Arst, H., Pozdnyakov, D., and Rozenstein, A. (1990) *Optical Remote Sensing in Oceanography*. Valgus, Tallinn [in Russian].

Arst, H., Erm, A., Leppäranta, M., and Reinart, A. (2006) Radiative characteristics of ice-covered freshwater and brackish water bodies. *Proceedings of the Estonian Academy of Sciences. Geology*, **55**(1), 3–23.

Askne, J., Leppäranta, M., and Thompson, T. (1992) Bothnian experiment in preparation for ERS-1, 1988 (BEPERS-88): An overview. *International Journal of Remote Sensing*, **13**, 2377–2398.

Assur, A. (1958) Composition of sea ice and its tensile strength. In: W. Thurston (ed.), *Arctic Sea Ice*, pp. 106–138. U.S. National Academy of Science/National Research Council, Publ. No. 598.

Axell, L. (1998) On the variability of the Baltic Sea deepwater mixing. *Journal of Geophysical Research*, **103**, 21667–21682.

BACC Author Group (2008) *Assessment of Climate Change for the Baltic Sea Basin*. Springer-Verlag, Berlin, 473 pp.

Barenblatt, G. (1996) *Scaling, Self-similarity and Intermediate Asymptotics* (Cambridge text in applied mathematics). Cambridge University Press, U.K., 386 pp.

Bergström, S. (1992) *The HBV Model: Its Structure and Applications*. SMHI Report RH No. 4, Norrköping.

Bergström, S. and Carlsson, B. (1994) River runoff to the Baltic Sea: 1950–1990. *Ambio*, **23**(4/5), 180–287.

Bergström, S., Alexandersson, H., Carlsson B., Josefsson, W., Karlsson. K. G., and Westring, G. (2001) Climate and hydrology of the Baltic Sea. In: F. Wulff et al. (eds.), *A System Analysis of the Baltic Sea* (Ecological Study No. 455, pp. 75–112). Springer-Verlag, Berlin.

Bin, C. (2002) On the modelling of sea ice thermodynamics and air–ice coupling in the Bohai Sea and the Baltic Sea. Contributions/Finnish Institute of Marine Research, 5 (Ph.D. thesis).

Blomquist, Å., Pilo, C., and Thompson, T. (1976) *Sea-ice '75* (a summary report). Winter Navigation Research Board Rep. No. 16(9), 28 pp., Norrköping.

Blumberg, A. F. and Mellor, G. L. (1987) A description of a three-dimensional coastal ocean circulation model. In: N. S. Heaps (ed.), *Three-dimensional Coastal Ocean Models*, pp. 1–16. American Geophysical Union, Washington, D.C.

Bock, K. H. (1971) Monatskarten des Salzgehalten der Ostsee, dargestellt für verschiedene Tiefenhorisonte [Monthly salinity maps of the Baltic Sea for different depths]. *Deutsche Hydrographische Zeitschrift*, Ergänzungshäft Reihe B, **12**, 1–147.

Bolin, B. and Rodhe, H. (1973) A note on the concepts of age distribution and transit time in natural reservoirs. *Tellus*, **25**(1), 58–62.

Bo Pedersen, F. (1993) Fronts in the Kattegat: The hydrodynamic regulating factor for biology. *Estuaries* **16**, 104–112.

Borenäs, K., Hietala, R., Laanearu, J., and Lundberg, P. A. (2007) Some estimates of the Baltic deep-water transport through the Stolpe trench. *Tellus*, **59A**, 238–248.

Brogmus, W. (1952) Eine Revision des Wasseraushaltes [A revision of the water exchange]. *Kieler Meeresforschung*, **9**(1), 15–42.

Broman, B., Hammarklint, T., Rannat, K., Soomere, T., and Valdmann, A. (2006) Trends and extremes of wave fields in the north-eastern part of the Baltic Proper. *Oceanologia* **48**(S), 165–184.

Brosin, H. J. (1996) *Zur Geschichte der Meeresforschung in der DDR* [On the history of marine research in GDR]. Institut für Ostseeforschung, Meereswissenschaftliche Berichte 17, Warnemünde, 212 pp.

Bryan, K. (1969) A numerical method for the study of the circulation of the World Ocean. *Journal of Computational Physics* **4**, 347–376.

Bunker, J. (1976) Computations of surface energy flux and annual air–sea interaction cycle of the North Atlantic Ocean. *Monthly Weather Review* **104**, 1122–1140.

Burchard, H., Lass, H.-U., Mohrholtz, V., Umlauf, L., Sellschopp, J., Fiekas, V., Bolding, K., and Arneborg, L. (2005) Dynamics of medium-intensity dense water plumes in the Arkona Basin, Western Baltic Sea. *Ocean Dynamics*, **55**(5/6), 391–402.

Bychkova I. and Viktorov, S. (1987) Use of satellite data for identification and classification of upwelling in the Baltic Sea. *Oceanology*, **27**(2), 158–162.

References

Bychkova, I., Viktorov, S., and Shumakher, D. (1988) A relationship between the large scale atmospheric circulation and the origin of coastal upwelling in the Baltic. *Meteorologiya i Gidrologiya*, **10**, 91–98 [in Russian].

Carlsson, M. (1997) Sea level and salinity variations in the Baltic Sea. An oceanographic study using historical data. Department of Oceanography, Gothenburg University (Ph.D. thesis).

Chen, C. T. and Millero, F. J. (1977) Speed of sound in seawater at high pressures. *Journal of the Acoustical Society of America*, **62**(5), 1129–1135.

Cox, G. F. N. and Weeks, W. F. (1983) Equations for determining the gas and brine volumes in sea ice samples. *Journal of Glaciology*, **29**, 306–316.

Cox, R. A. and Smith, N. D. (1959) The specific heat of sea water. *Proceedings of the Royal Society London A*, **252**, 51–62.

Crépon, M., Richez, C., and Chartier, M. (1984) Effects of coastline geometry on upwelling. *Journal of Physical Oceanography*, **14**(8), 1365–1382.

Csanady, G. T. (1982) *Circulation in the Coastal Ocean*. D. Reidel, Dordrecht, 279 pp.

Cusbach, U., Meehl, G. A., Boer, G. J., Stouffer, R. J., Dix, M., Noda, A., Senior, C. A., Raper, S., and Yap, K. S. (2001) Projections of future climate change. In: IPCC Climate Change 2001: The Scientific Basis. Contribution of Working Group I to the *Third Assessment Report of the Intergovernmental Panel on Climate Change*. Cambridge University Press, Cambridge, New York, pp. 525–582.

Curry, J. A. and Webster, P. J. (1999) *Thermodynamics of Atmospheres and Oceans*. Academic Press, San Diego, CA, 471 pp.

Cushman-Roisin, B. (1994) *Introduction to Geophysical Fluid Dynamics*. Prentice Hall, Englewood Cliffs, NJ.

Dahlin, H. (1976) Hydrokemisk balans för Bottenhavet och Bottenviken [Hydrochemical balance for the Bothnian Sea and Bay of Bothnia]. *Vannet in Norden*, **1**, 62–73 [in Swedish].

Dahlin, H., Fonselius, S., and Sjöberg, B. (1993) The changes of the hydrographic conditions in the Baltic proper due to 1993 major inflow to the Baltic Sea. *ICES Statutory Meeting*, Paper C.M. 1993/C:58, 15 pp.

Dahlström, B. (1977) Estimation of precipitation for the Baltic Sea: Preliminary result. Ad Hoc Meeting of the Pilot Study Group of Experts, Norrköping (mimeo.).

Dailidienė, I., Davulienė, L., Tilickis, B., Stankevičius, A., and Myrberg, K. (2006) Sea level variability at the Lithuanian coast of the Baltic Sea. *Boreal Environmental Research*, **11**, 109–121.

Danielssen, D., Edler, L., Fonselius, S., Hernroth, L., Ostrowski, M., Svendsen, E., and Talpsepp, L. (1997) Oceanographic variability in the Skagerrak and Northern Kattegat. May–June 1990. *ICES Journal of Marine Science*, **54**, 753–773.

DeCosmo, J., Katsaros, K. B., Smith, S. D., Anderson, R. J., Oost, W. A., Bumke, K., and Chadwick, H. (1996) Air–sea exchange of water vapour and sensible heat: The humidity exchange over the sea (HEXOS) results. *Journal of Geophysical Research*, **101**, 12001–12016.

Defant, A. (1961) *Physical Oceanography*. Pergamon Press, New York, 729 pp.

Deleersnijder, E., Campin, J.-M., and Delhes, E. J. M. (2001) The concept of age in marine modelling. I. Theory and preliminary model results. *Journal of Marine Systems*, **28**, 229–267.

Dera, J. (1992) *Marine Physics*. Elsevier, Amsterdam.

DHI Water Environment (2000) MIKE 21: Coastal hydraulics and oceanography. *DHI Software User Guide, Document and Reference Manual.* Danish Hydraulic Institute, Hørsholm, Denmark.

Dietrich, G. (1950) Die natürlichen Regionen von Nord- und Ostsee auf hydrographischer Grundlage [The natural regions in North Sea and Baltic Sea based on hydrographical basic laws]. *Kieler Meeresforschnung,* **7**, 35–69.

Dietrich, G. (1951) Oberflächenströmungen im Kattegat, im Sund und in der Beltsee [Surface currents in Kattegat, Öresund and Belt Sea]. *Deutsche Hydrographische Zeitschrift,* **4**, 129–150.

Dietrich, G. (ed.) (1972) *Upwelling in the Ocean and Its Consequences* (Geoforum Oxford 11), Elsevier Science/Pergamon, Frankfurt.

Dietrich, G., Kalle, K., and Ostapoff, F. (1963) *General Oceanography: An Introduction.* John Wiley & Sons, New York. [German version *Allgemeine Meereskunde* by G. Dietrich and K. Kalle in 1957 by Gebrüder Borntraeger, Berlin, 492 pp.]

Donner, J. (1995) *The Quaternary History of Scandinavia.* Cambridge University Press, 210 pp.

Döös, K., Meier, H. E. M., and Döscher, R. (2004) The Baltic haline conveyor belt or the overturning circulation and mixing in the Baltic. *Ambio,* **33**, 261–266.

Dybern, B. I. and Fonselius, S. (2001) International marine scientific activities in the Baltic Sea with special reference to Estonian participation. *Proceedings of the Estonian Academy of Sciences, Biology, Ecology,* **50**(3), 139–157.

Ehlin, U. (1981) Hydrology of the Baltic Sea. In: A. Voipio (ed.), *The Baltic Sea,* pp. 123–134. Elsevier, Amsterdam.

Ehlin, U. and Ambjörn, C. (1978) Bottniska vikens hydrografi och dynamic [The hydrography dynamics of the Bay of Bothnia]. *Finnish–Swedish Seminar of the Gulf of Bothnia, March 8–9, 1978, Vaasa* [in Swedish].

Ehn, J., Rasmus, K., Leppäranta, M., and Shirasawa, K. (2002) Parametrization of solar radiation in sea ice thermodynamics models. *Proceedings of the 16th IAHR Ice Symposium, Dunedin, New Zealand.*

Eilola, K. (1997) The development of a spring thermocline in low-saline seawater at temperatures below the temperature of maximum density with application to the Baltic Sea. *Journal of Geophysical Research,* **102**, 8657–8662.

Ekman, F.L. and Pettersson, O. (1893) Den svenska hydografiska expeditionen 1877 [The Swedish hydrogical expedition 1877]. *Kungliga Svenska Vetenskaps-Akademins Handlingar,* N.F. **25**(1), 163.

Ekman, M. (1988) The world's longest continued series of sea level observation. *Pure and Applied Geophysics,* **127**, 73–77.

Ekman, M. (1996) A common pattern for interannual and periodical sea level variations in the Baltic Sea and adjacent waters. *Geophysica,* **32**, 261–272.

Ekman, M. (1999) Climate change detected through the world's longest sea level series. *Global Planet Change,* **21**, 215–224.

Ekman, M. and Stigebrandt, A. (1990) Secular change of the seasonal variation in sea level and of the pole tide in the Baltic Sea. *Journal of Geophysical Research,* **95**, 5379–5383.

Ekman, V. (1905) On the influence of the earth's rotation on ocean currents. *Arkiv for Matematik, Astronomi och Fysik,* **2**(11), 1–52.

Elken, J. (1994) Numerical study of fronts between the Baltic sub-basins. *Proceedings of the 19th Conference of the Baltic Sea Oceanographers, Sopot,* Vol. 1, pp. 438–446.

Elken, J. (1996) *Deep Water Overflow, Circulation and Vertical Exchange in the Baltic Proper.* Estonian Marine Institute, Report Series 6, pp. 1–91, Tallinn, Estonia.

References

Elken, J. and Matthäus, W. (2008) Physical system description, Annex A1. In: The BACC Author Team (eds.), *Assessment of Climate Change for the Baltic Sea Basin.* Springer-Verlag, Berlin, pp. 379–386.

Elken, J., Pajuste, M., and Kõuts, T. (1988) On intrusive lenses and their role in mixing in the Baltic Sea. *Proceedings of 16th Conference of Baltic Oceanographers, Kiel*, pp. 367–376.

Elken, J., Talpsepp, L., Kõuts, T., and Pajuste, M. (1994) The role of mesoscale eddies and saline stratification in the generation of spring bloom heterogeneity in the southeastern Gotland Basin: An example from PEX '86. In: B. I. Dybern (ed.), *Patchiness in the Baltic Sea*, pp. 40–48. ICES Cooperative Research Report 201, Copenhagen.

Elken J., Raudsepp, U., and Lips, U. (2003) On the estuarine transport reversal in deep layers of the Gulf of Finland. *Journal of Sea Research*, **49**, 267–274.

Elken, J., Mälkki, P., Alenius, P., and Stipa, T. (2006) Large halocline variations in the Northern Baltic Proper and associated meso- and basin-scale processes. *Oceanologia*, **48(S)**, 91–117.

Engqvist, A. (1996) Long-term nutrient balances in the eutrophication of Himmerfjärden estuary. *Estuarine Coastal and Shelf Science*, **42**(4), 483–507.

Engqvist, A. and Andrejev, O. (2003) Water exchange of the Stockholm archipelago: A cascade framework modelling approach. *Journal of Sea Research*, **49**, 275–294.

Erm, A. and Soomere, T. (2004) Influence of fast ship waves on optical properties of sea water in Tallinn Bay, Baltic Sea. *Proceedings of the Estonian Academy of Sciences. Biology, Ecology*, **53**, 161–178.

Erm, A. and Soomere T. (2006) The impact of fast ferry traffic on underwater optics and sediment resuspension. *Oceanologia*, **48(S)**, 283–301.

Eronen, M. (1990) The evolution of the Baltic Sea. In: P. Alalammi (ed.), *Atlas of Finland, Folio 123–126: Geology*, pp. 15–18. National Board of Survey and Geographical Society of Finland, Helsinki.

Eronen, M. and Haila, H. (1990) The evolutionary stages of the Baltic Sea. In: P. Alalammi (ed.), *Atlas of Finland, Folio 123–126: Geology*, p. 13. National Board of Survey and Geographical Society of Finland, Helsinki.

Essen, H.-H. (1993) Ekman portion of surface currents, as measured by radar in different areas. *Deutsche Hydrographische Zeitschrift*, **45**(2), 57–85.

EU Water Framework Directive (2000) http://ec.europa.eu/environment/water/water-framework/objectives/implementation_en.htm

Feistel, R., Nausch, G., Matthäus, W., and Hagen, E. (2003) Temporal and spatial evolution of the Baltic deep water renewal in spring 2003. *Oceanologia*, **45**, 623–642.

Feistel, R., Nausch, G., Heene, T., Piechura, J., and Hagen, E. (2004) Evidence for a warm water inflow into the Baltic Proper in summer 2003. *Oceanologia*, **46**, 581–598.

Feistel, R., Nausch, G., and Hagen, E. (2006) Unusual Baltic inflow activity in 2002–2003 and varying deep-water properties. *Oceanologia*, **48(S)**, 21–35.

Feistel, R., Nausch, G., and Wasmund, N. (2008) *State and Evolution of the Baltic Sea, 1952–2005: A Detailed 50-year Survey of Meteorology and Climate, Physics, Chemistry, Biology, and Marine Environment.* John Wiley & Sons, Hoboken, NJ, 704 pp.

Feller, W. (1968) *An Introduction to Probability Theory and Its Applications*, Vol. 1. John Wiley & Sons, New York.

Fenger, J., Buch, E., and Jacobsen, P. R. (2001) Monitoring and impacts of sea level rise at Danish coasts and nearshore infrastructures. In: A. M. Jørgensen, J. Fenger, and K. Halsnaes (eds.) *Climate Change Research: Danish Contributions.* Danish Climate Centre, Copenhagen, pp. 237–254.

Fennel, W. and Neumann, T. (2004). *Introduction to the Modelling of Marine Ecosystems.* Elsevier Oceanography Series 72, Amsterdam, 297 pp.
Fennel, W. and Seifert, T. (1995) Kelvin wave controlled upwelling in the western Baltic. *Journal of Marine Systems*, **6**, 289–300.
Fennel, W. and Sturm, M. (1992) Dynamics of the western Baltic. *Journal of Marine Systems*, **3**, 183–205.
Fennel, W., Seifert, T., and Kayser, B. (1991) Rossby radii and phase speeds in the Baltic Sea. *Continental Shelf Research*, **11**(1), 23–36.
FMI (1982) Solar radiation measurements 1971–1980. *Meteorological Yearbook of Finland* (pp. 71–80, Part 4:1 Measurements of solar radiation 1971–1980). Finnish Meteorological Institute, Helsinki.
FMI (1991) Climatological Statistics in Finland 1961–1990. Supplement of the *Meteorological Yearbook of Finland*, Part 1, Vol. 1. Finnish Meteorological Institute, Helsinki.
Fischer, H. and Matthäus, W. (1996) The importance of the Drogden Sill in the Sound for major Baltic inflows. *Journal of Marine Systems*, **9**, 137–157.
Fonselius, S. (1962) Hydrography of the Baltic Sea deep basins. *Fishery Board of Sweden, Series Hydrography*, **13**, 1–41.
Fonselius, S. (1967) Hydrography of the Baltic Sea deep basins II. *Fishery Board of Sweden, Series Hydrography*, **20**, 1–31.
Fonselius, S. (1969) Hydrography of the Baltic Sea deep basins III. *Fishery Board of Sweden, Series Hydrography*, **23**, 1–97.
Fonselius, S. (1971) Om østerjöns och speciellt Botniska vikens hydrografi [About the hydrography of the Baltic Sea with a special focus to the Sea of Bothnian]. *Vatten*, **27**.
Fonselius, S. (1981) Oxygen and hydrogen sulphide conditions in the Baltic Sea. *Marine Pollution Bulletin*, **12**, 187–194.
Fonselius, S. (1995) *Västerhavets och østersjöns oceanografi* [Oceanography of the Baltic Sea, Kattegat and the Skagerrak] (report). SMHI, Norrköping, Sweden, 200 pp.
Fonselius, S. (2001) History of hydrographic research in Sweden. *Proceedings of the Estonian Academy of Sciences, Biology, Ecology*, **50**(2) 110–129.
Fonselius, S. and Valderrama, J. (2003) One hundred years of hydrographic measurements in the Baltic Sea. *Journal of Sea Research*, **49**, 229–241.
Franck, H., Matthäus, W., and Sammler, R. (1987) Major inflows of saline water into the Baltic Sea during the present century. *Gerlands Beiträge zur Geophysik*, **96**, 517–531.
Funkquist, L. (2001) HIROMB: An operational eddy-resolving model for the Baltic Sea. *Bulletin of the Maritime Institute in Gdansk*, **XXVIII**(2), 7–16.
Gammelsrød, T., Mork, M., and Røed, L. P. (1975). Upwelling possibilities at an ice edge: Homogeneous model. *Marine Science Communications*, **1**, 115–145.
Gidhagen, L. (1987) Coastal upwelling in the Baltic Sea: Satellite and in situ measurements of sea-surface temperatures indicating coastal upwelling. *Estuarine, Coastal and Shelf Science*, **24**, 449–362.
Gill, A. (1982) *Atmosphere–ocean Dynamics.* Academic Press, London, 662 pp.
Gill, A. E. and Clarke, A. J. (1974) Wind-induced upwelling, coastal currents and sea-level changes. *Deep Sea Research*, **21**, 325–345.
Goldstein, R., Osipenko, N., and Leppäranta, M. (2000) Classification of large-scale sea-ice structures based on remote sensing imagery. *Geophysica*, **36**(1/2), 95–109.
Gorringe, P. and Håkansson, B. (eds.) (2005) *Present Status of Operational Oceanography in the Baltic Sea* (a report from the EU-project PAPA Contract No:EVRI-CT-2002-20012PAPA NOW). SMHI, Norrköping, 126 pp.
Gothus, Olaus Magnus (1539) *Carta Marina.* [Available at National Land Survey, Finland].

Granqvist, G. (1926) Yleiskatsaus talven 1924–25 jääsuhteisiin [A general view on the ice situation in winter 1924–1925 at the Finnish coastal area]. *Merentutkimuslaitoksen Julkaisu/Havsforskningsinstitutets Skrift*, **44**, 48.

Granqvist, G. (1938a) Den baltiska isveckan 12–18. Febr. 1938 [The Baltic ice week, February 12–18, 1938]. *Terra*, **50**(4), 385–391.

Granqvist, G. (1938b) Zur Kenntnis der Temperatur und des Salzgehaltes des Baltischen Meeres und den Küsten Finnlands. *Merentutkimuslaitoksen Julkaisu/Havsforsknings-institutets Skrift*, **122**, 166.

Granskog, M., Leppäranta, M., Ehn, J., Kawamura, T., and Shirasawa, K. (2004) Sea ice structure and properties in Santala Bay, Baltic Sea. *Journal of Geophysical Research*, **109**, C02020.

Gustafsson, B. G. (2001) Quantification of water, salt, oxygen and nutrient exchange of the Baltic Sea from observations in the Arkona Basin. *Continental Shelf Research*, **21**, 1485–1500.

Gustafsson, B. G. and Andersson, H. C. (2001) Modeling the exchange of the Baltic Sea from the meridional atmospheric pressure difference across the North Sea. *Journal of Geophysical Research*, **106**(C9), 19731–19744.

Gustafsson, T. and Kullenberg, B. (1936) Untersuchungen von Trägheitströmungen in der Ostsee [Investigations of inertial currents in the Baltic Sea]. *Svenska Hydrografiska-Biologiska Kommitteens Skrifter, Ny Ser. Hydr.*, **13**, 1–28.

Haapala, J. (1994) Upwelling and its influence on nutrient concentration in the coastal area of the Hanko Peninsula, entrance of the Gulf of Finland. *Estuarine, Coastal and Shelf Science*, **38**, 507–521.

Haapala, J. (2000) On the modelling of ice-thickness distribution. *Journal of Glaciology*, **46**(154), 427–437.

Haapala, J. and Leppäranta, M. (1996) Simulating Baltic sea ice season with a coupled ice–ocean model. *Tellus*, **48A**(5), 622–643.

Haapala, J. and Leppäranta, M. (1997) The Baltic Sea ice season in changing climate. *Boreal Environment Research*, **2**, 93–108.

Haapala, J., Juottonen, A., Marnela, M., Leppäranta, M., and Tuomenvirta, H. (2001a) Modelling the variability of sea-ice conditions in the Baltic Sea under different climate conditions. *Annals of Glaciology*, **33**, 555–559.

Haapala, J., Meier, H. E. M., and Rinne, J. (2001b) Numerical investigations of future ice conditions in the Baltic Sea. *Ambio*, **30**, 237–244.

Hagen, E. and Feistel, R. (2001) Spreading of Baltic deep water: A case study for the winter 1997–1998. In: W. Matthäus and G. Nausch (eds.), *The Hydrographic-chemical State of the Western and Central Baltic Sea in 1999/2000 and during the 1990s* (Marine Science Report No. 45, pp. 99–133). Baltic Sea Research Institute, Warnemünde, Germany.

Håkansson, B., Broman, B., and Dahlin, H. (1993) The flow of water and salt in the Sound during the Baltic major inflow event in January 1993. *ICES Statutory Meeting*, Paper C.M. 1993/C:57, 26 pp.

Håkansson, B., Alenius, P., and Brydsten L. (1994) Fysisk miljö I Bottniska Viken [Physical environment in the Bay of Bothnia]. *Vatten*, **50**(3), 187–200 [in Swedish].

Häkkinen, S. (1980) Computations of sea level variations during December 1975 and 1 to 17 September 1977 using numerical models of the Baltic Sea. *Deutsche Hydrographische Zeitschrift*, **33**(4), 158–175.

Hankimo, J. (1964) Some computations of the energy exchange between the sea and the atmosphere in the Baltic area. *Ilmatieteellisen keskuslaitoksen toimituksia*, **57**, 1–26.

References

Hela, I. (1944) Über die Schwankungen des Wasserstandes in der Ostsee mit besonderer Berücksichtigung des Wasseraustausches durch die dänischen Gewässer [On the variations of the water level of the Baltic Sea with special consideration on the water exchange through the Danish waters]. *Annales Academiae Scientiarum Fennicae, Ser. A I, Mathematica-Physica*, **28**, 108.

Hela, I. (1976) Vertical velocity of the upwelling in the sea. *Societas Scientarium Fennica, Commentationes Physico-Mathematicae*, **46**(1), 9–24.

HELCOM (1993) First assessment of the state of the coastal waters of the Baltic Sea. *Baltic Sea Environmental Proceedings*, **54**, 1–166.

HELCOM (1996) Third periodic assessment of the state of the marine environment of the Baltic Sea, 1989–1993: Background document. *Baltic Sea Environment Proceedings*, **64B**, 1–252.

HELCOM (2000) Nutrient pollution to the Baltic Sea in 2000. *Baltic Sea Environment Proceedings*, **100**, 14.

HELCOM (2006) *Pollution Observation Log*. Helsinki Commission, Helsinki.

HELCOM (2007) Climate Change in the Baltic Sea area. *Baltic Sea Environment Proceedings*, **111**, 54.

Henning, D. (1988) Evaporation, water and heat balance of the Baltic Sea: Estimates of short- and long-term monthly totals. *Meteorologische Rundschau*, **41**, 33–53.

Herlevi, A. and Leppäranta, M. (1994) ERS-1 SAR over open water in the Baltic Sea. *EARSeL Advances in Remote Sensing*, **3**(2), 64–70.

Hibler, III W. D. (1979) A dynamic thermodynamic sea ice model. *Journal of Physical Oceanography*, **9**, 815–846.

Hietala, R., Lundberg, P. A., and Nilsson, J. A. U. (2007) A note on the deep-water inflow to the Bothnian Sea. *Journal of Marine Systems*, **68**, 255–264.

Højerslev, N. K. (1988) Surface water quality studies in the interior marine environment of Denmark. *Limnology and Oceanography*, **34**(8), 1630–1639.

Højerslev, N. K., Holt, N., and Aarup, T. (1996) Optical measurements in the North Sea–Baltic Sea transition zone. I. On the origin of the deep water in the Kattegat. *Continental Shelf Research*, **16**(10), 1329–1342.

Hollan, E. (1966) Das Spektrum des internen Bewegungsvorgänge der Westlichen Ostsee im Periodenbereich von 0.3 bis 60 Minuten [Spectrum of internal motions in the western Baltic Sea in periods from 0.3 to 60 minutes]. *Mitteilung der Institut für Meereskunde Universität Hamburg I* [in German].

Holthuijsen, L. H. (2007) *Waves in Oceanic and Coastal Waters*. Cambridge University Press, Cambridge, U.K.

Holton, J. R. (1979) *An Introduction to Dynamic Meteorology*, Second Edition (International Geophysics Series 23). Academic Press, New York, 391 pp.

Horstmann, U. (1983) Distribution patterns of temperature and water colour in the Baltic Sea as recorded in satellite images: Indicators for phytoplankton growth. *Berichte aus dem Institut für Meereskunde der Universität Kiel*, **106**, 147.

ICES (1984) Overall Report on the Baltic Open Sea Experiment (BOSEX) (edited by G. Kullenberg). *ICES Cooperative Research Report*, **127**.

ICES (1994) Patchiness in the Baltic Sea: Selected papers from a symposium held in Mariehamn June 3–4, 1991 (edited by B. I. Dybern). *ICES Cooperative Research Report*, **201**.

Ikävalko, J. (1997) Studies of nanoflagellate communities in the sea ice of the Baltic and the Greenland Sea (24 pp. + 6 articles), Walter and Andrée de Nottbeck Foundation (Ph.D. thesis).

References

IPCC (2007) *Climate Change 2007: Synthesis Report*. A Report from the Intergovernmental Panel on Climate Change, IPCC, Geneva, Switzerland, 104 pp.

Jakobsen, F. (1995) The major inflow to the Baltic Sea during 1993. *Journal of Marine Systems*, **6**, 227–240.

Jakobsen, F. and Ottavi, J. (1997) Transport through the contraction area in the Little Belt. *Estuarine, Coastal and Shelf Science*, **45**, 759–767.

Jakobsen, T. (1980) Mixing, exchange and transport processes. In: T. Melvasalo *et al.* (eds.), Assessment of the effects of pollution on natural resources of the Baltic Sea. *Baltic Sea Environment Proceedings*, **5B**, 114–128.

Jankowski, A. (2002) Variability of coastal water hydrodynamics in the southern Baltic: Hindcast modelling of an upwelling event along the Polish coast. *Oceanologia*, **44**(4), 395–418.

Janssen, F. (2002) Statistical analysis of multi-year variability of the hydrography in the North Sea and Baltic Sea, University of Hamburg (Ph.D. thesis) [in German].

Jerlov, N. (1976) *Optical Oceanography*. Elsevier, Amsterdam, 231 pp.

Jevrejeva, S., Drabkin, V. V., Kostjukov, J., Lebedev, A. A., Leppäranta, M., Mironov, Y. U., Schmelzer, N., and Sztobryn, M. (2002) *Ice Time Series of the Baltic Sea* (Report Series in Geophysics No. 44). Department of Geophysics, University of Helsinki.

Joffre, S. M. (1981) A theoretical and empirical study of the atmospheric boundary layer dynamics over a frozen sea (Report No. 21), Department of Meteorology, University of Helsinki, 75 pp. (Ph.D. thesis).

Joffre, S. M. (1984) *The Atmospheric Boundary Layer over the Bothnian Bay: A Review of Work on Momentum Transfer and Wind Structure*. Winter Navigation Research Board Report No. 40, Helsinki, 58 pp.

Johansson, M., Boman, H., Kahma, K. K., and Launiainen, J. (2001) Trends in sea level variability in the Baltic Sea. *Boreal Environmental Research*, **6**(3), 159–179.

Johansson, M., Kahma, K. K., Boman, H., and Launiainen, J. (2004) Scenarios for sea level on the Finnish coast. *Boreal Environment Research*, **9**(2), 153–166.

Jurva, R. (1937a). Über die Eisverhältnisse des Baltischen Meeres an den Küsten Finnlands [On the ice conditions in the Baltic Sea at the Finnish coastal area]. *Merentutkimuslaitoksen Julkaisu/Havsforskningsinstitutets Skrift*, **114** (Finnish Institute of Marine Research, Helsinki) [in German]

Jurva, R. (1937b) Laskelmia meriemme lämpövarastosta [Calculations on the heat storage of our seas]. *Suomi merella*, **12**, 162–191 [in Finnish]

Jurva, R. (1952) On the variations and changes of freezing in the Baltic during the last 120 years. *Fennia*, **75**, 17–24.

Kahma, K. K. (1981a) On the growth of wind waves in fetch-limited conditions. Report Series in Geophysics, University of Helsinki (Ph.D. thesis).

Kahma, K. K. (1981b) A study of the growth of the wave spectrum with fetch. *Journal of Physical Oceanography*, **11**(11), 1503–1515.

Kahma, K. K. (1986) On prediction of the fetch-limited wave spectrum in a steady wind. *Finnish Marine Research*, **253**, 52–78.

Kahma, K. K., Pettersson, H., and Tuomi, L. (2003) Scatter diagram wave statistics from the northern Baltic Sea. *Meri*, **49**, 15–32 (Report Series of the Finnish Institute of Marine Research).

Kahru, M., Håkansson, B., and Rud, O. (1995) Distributions of the sea-surface temperature fronts in the Baltic Sea as derived from satellite imagery. *Continental Shelf Research*, **15**(6), 663–679.

Kalas, M. (1993) Characteristics of sea level changes on the Polish Coast of the Baltic Sea in the last forty-five years. *International Workshop, Sea Level Changes and Water Management, 19–23 April 1993.* Noorsdwijerhout Nederlands, pp. 51–61.

Kalliosaari, S. and Seinä, A. (1987) Ice winters 1981–1985 along the Finnish coast. *Finnish Marine Research*, **254**, 5–63.

Kamenkovich, V. M., Koshlyakov, M. N., and Monin, A. S. (eds.) (1986) *Synoptic Eddies in the Ocean.* D. Reidel, Dordrecht, the Netherlands, 433 pp.

Kankaanpää, P. (1998) Distribution, morphology and structure of sea ice pressure ridges in the Baltic Sea. *Fennia*, **175**(2) (Ph.D. thesis).

Kask, J., Talpas, A., Kask, A., and Schwarzer, K. (2003) Geological setting of areas endangered by waves generated by fast ferries in Tallinn Bay. *Proceedings of the Estonian Academy of Sciences, Engineering*, **9**, 185–208.

Kawamura, T., Shirasawa, K., Ishikawa, N., Lindfors, A., Rasmus, K., Ehn, J., Leppäranta, M., Martma, T., and Vaikmäe, R. (2001). A time series of the sea ice structure in the Baltic Sea. *Annals of Glaciology*, **33**, 1–4.

Kielmann, J. (1978) Mesoscale eddies in the Baltic. *Proceedings of the 11th Conference of Baltic Oceanographers*, Vol. 2, pp. 729–755. Rostock, German Democratic Republic.

Kielmann J. (1981) Grundlagen und Anwendung eines numerischen Modells der geschichteten Ostsee [Fundaments and applications of numerical model for the stratified Baltic Sea]. *Berichte aus dem Institut für Meereskunde der Universität Kiel*, **87**(a, b) (Ph.D. thesis).

Killworth, P., Stainworth, D., Webbs, D., and Paterson, S. (1991) The development of a free-surface Bryan–Cox-Semtner–Killworth ocean model. *Journal of Physical Oceanography*, **21**, 1333–1348.

Kirk, J. T. O. (1994) *Light and Photosynthesis in Aquatic Ecosystems*, Second Edition, Cambridge University Press, Cambridge, U.K, 400 pp.

Kitaigorodskii, S. A. (1962) Applications of the theory of similarity to the analysis of wind generated wave motion as a stochastic process. *Izv. Akademia Nauk SSSR, Geophys. Ser.*, **1**, 105–117.

Kitaigorodskii, S. A. and Mälkki, P. (1979) Note on the parametrization of turbulent gas transfer across an air–water interface. *Finnish Marine Research*, **246**, 111–124.

Kitaigorodskii, S. A. and Miropolsky, Y. (1970) On the theory of the open-ocean active layer. *Izv. Atmospheric and Ocean Physics*, **6**(2), 178–188 (English edition).

Kleine, E. (1993) Numerical simulation of the recent (1993) major Baltic inflow. *ICES Statutory Meeeting*, Paper C.M. 1993/C:48, 12 pp.

Kleine, E. (1994) *Das operationelle Model des BSH für Nordsee und Ostsee, Konzeption und Übersicht* [Operational model of BSH for the North Sea and Baltic Sea, concepts and overview] (technical report). Bundesamt für Seeschifffahrt und Hydrographie, Hamburg. [in German]

Knudsen M. (1900) Ein Hydrographischer Lehrsatz [A hydrographical theorem]. *Annalen Maritimen Metorology*, **28**, 316–320 [in German].

Komen, G. J., Cavaleri, L., Donelan, M., Hasselmann, K., and Janssen, P. A. E. M. (1994) *Dynamics and Modelling of Ocean Waves*. Cambridge University Press, 532 pp.

Kononen, K. and Niemi, Å. (1986) Variation in phytoplankton and hydrography in the outer archipelago. *Finnish Marine Research*, **253**, 35–51.

Kononen, K., Kuparinen, J., Mäkelä, K., Laanemets, J., Pavelson, J., and Nõmmann, S. (1996) Initiation of cyanobacterial blooms in a frontal region at the entrance to the Gulf of Finland, Baltic Sea. *Limnology and Oceanography*, **41**, 98–112.

Kont, A., Ratas, U., and Puurmann, E. (1997) Sea-level rise impact on coastal areas of Estonia. *Climate Change*, **36**, 175–184.

References

Kortum, G. and Lehmann, A. (1997) A.v. Humboldts Forschnungsfahrt auf der Ostsee in Sommer 1834 [Expedition of A. von Humboldt to the Baltic Sea in summer 1834]. *Schr. Naturwiss. Ver. Schlesw.-Holst.*, **67**, 45–58 [in German].

Kostianoy, A. and Kosarev, A. N. (eds.) (2008) *The Black Sea Environment: The Handbook of Environmental Chemistry*, Vol. 5: *Water Pollution*, Part 5Q. Springer-Verlag, Berlin, 457 pp.

Kõuts, T. (1999) Processes of deep water renewal in the Baltic Sea. *Dissertationes Geophysicales Universitatis Tartuensis*, **10**, 50 pp. + 7 articles (Ph.D. thesis).

Kõuts, T. and Omstedt, A. (1993) Deep water exchange in the Baltic proper. *Tellus*, **45A**, 311–324.

Kõuts, T., Wang, K., and Leppäranta, M. (2007) On connection between mesoscale stress of geophysical sea ice models and local ship load. *Proceedings of the 10th International Symposium on Practical Design of Ships and other Floating Structures, Houston, Texas, October 1–5, 2007*.

Kowalewski, M. and Ostrowski, M. (2005) Coastal up- and downwelling in the southern Baltic. *Oceanologia*, **47**(4), 453–475.

Kowalik, Z. and Murty, T. S. (1993) *Numerical Modeling of Ocean Dynamics* (Advanced Series on Ocean Engineering, Vol. 5). World Scientific, Singapore.

Krauss, W. (1966) *Interne Wellen*. Gebrüder Borntraeger, 248 pp.

Krauss, W. (1972) Wind-generated internal waves and inertial-period motions. *Deutsche Hydrographische Zeitschrift*, **25**, 241–250.

Krauss, W. (1981) The erosion of the thermocline. *Journal of Physical Oceanography*, **11**, 415–433.

Krauss, W. (1990). The Institute of Marine Research in Kiel. In: W. Lenz and M. Deacon (eds.), Ocean sciences: Their history and relation to man. *Deutsche Hydrographische Zeitschrift, Ergänzungshäft Reihe*, **B22**, 131–140.

Krauss, W. and Brügge, B. (1991) Wind-produced water exchange between deep basins of the Baltic Sea. *Journal of Physical Oceanography*, **11**, 373–384.

Krauss, W. and Magaard, L. (1961) Zum Spektrum der internen Wellen der Ostsee. *Kieler Meeresforschung*, **17**(2), 137–147.

Krężel, A., Ostrowksi, M., and Szymelfenig M. (2005) Sea surface temperature distribution during upwelling along the Polish Baltic coast. *Oceanologia*, **47**(4), 415–432.

Krümmel, O. (1907) *Handbuch der Ozeanographie*, Vol. 1. J. Engelhorn, Stuttgart.

Kullenberg, G. (1981) Physical oceanography. In: A. Voipio (ed.), *The Baltic Sea*, pp. 135–181. Elsevier Oceanography Series No. 30, Amsterdam.

Kundu, P. K. and Cohen, I. M. (2004) *Fluid Mechanics*, Third Edition, Elsevier, Amsterdam.

Kutser, T., Metsamaa, L., Strömbeck, N., and Vahtmäe, E. (2006) Monitoring cyanobacterial blooms by satellite remote sensing. *Estuarine, Coastal and Shelf Science*, **67**, 303–312.

Kutser, T., Metsamaa, L., Vahtmäe, E., and Aps, R. (2007) Operative monitoring of the extent of dredging plumes in coastal ecosystems using MODIS satellite imagery. *Journal of Coastal Research*, **SI50**, 180–184.

Kuzin, V. I. and Tamsalu, R. E. (1974) Vetrovye techeniya v baroklinnom more poctoyannoy glubiny. In: *Chislennye metody rascheta okeanologicheskikh techeniy*, pp. 103–114, Novosibirsk [in Russian].

Kuzmina, N., Rudels, B., Stipa, T., and Zhurbas, V. (2005) The structure and driving mechanisms of the Baltic intrusions. *Journal of Physical Oceanography*, **35**, 1120–1136.

Kwiecień, K. (1987) Warunki klimatyczne. In: B. Augustowski (ed.), *Baltyk Poludniowy*, pp. 219–287. Gdanskie Towarzystwo Naukowe, Ossolineum, Wroclaw.

Laamanen M., Fleming, V., and Olsonen, R. (2004) Water transparency in the Baltic Sea between 1903 and 2004. HELCOM Indicator fact sheets 2004: http://www.helcom.fi/environment2/ifs/ifs2004

Laanearu, J. and Lundberg, P. A. (2003) Topographically constrained deep-water flows in the Baltic Sea. *Journal of Sea Research*, **49**(4), 257–265.

Laanearu, J., Lips, U., and Lundberg, P. A. (2000) On the application of hydraulic theory to the deep-water flow through the Irbe Strait. *Journal of Marine Systems*, **25**(3/4), 323–332.

Laanemets, J., Kononen, K. and Pavelson, J. (1997) Nutrient intrusions at the entrance area to the Gulf of Finland. *Boreal Environment Research*, **2**, 337–344.

Laanemets, J., Kononen, K., Pavelson, J., and Poutanen, E.-L. (2004) Vertical location of seasonal nutriclines in the western Gulf of Finland. *Journal of Marine Systems*, **52**, 1–13.

Lass, H.-U. and Matthäus, W. (1996) On temporal wind variations forcing salt water inflows into the Baltic Sea. *Tellus*, **48A**(5), 663–671.

Lass, H.-U. and Mohrholtz, V. (2003) On the dynamics and mixing of inflowing saltwater in the Arkona Sea. *Journal of Geophysical Research*, **108**(C2), 24/1–24/15.

Lass, H.-U. and Talpsepp, L. (1993) Observations of coastal jets in the Southern Baltic. *Continental Shelf Research*, **13**, 189–203.

Lass, H.-U., Prandke, H., and Liljebladh, B. (2003a). Dissipation in the Baltic proper during winter stratification. *Journal of Geophysical Research*, **108**(C6), 3187.

Lass, H.-U., Schmidt, T., and Seifert, T. (2003b) Hiddensee upwelling field measurements and modelling results. *ICES Cooperative Research Report*, **257**, 204–208.

Lass, H.-U., Mohrholtz V., and Seifert, T. (2005). On pathways and residence time of saltwater plumes in the Arkona Sea. *Journal of Geophysical Research*, **110**, C11019 (1–24).

Launiainen, J. (1979) Studies of energy exchange between the air and the sea surface on the coastal area of the Gulf of Finland. *Finnish Marine Research*, **246**, 110 pp. (Ph.D. thesis).

Launiainen, J. (1982) Variation of salinity at Finnish fixed hydrographic stations in the Gulf of Finland and river runoff the Baltic Sea. *Gulf of Finland Seminar, Leningrad, August 16–20, 1982*, 12 pp. (mimeo.).

Launiainen, J. and Vihma, T. (1990) Meteorological, ice and water exchange conditions. In: Second Periodic Assessment of the State of the Marine Environment of the Baltic Sea, 1984–1988: Background document. *Baltic Sea Environment Proceedings*, **35B**, 22–33.

Launiainen, J., Cheng, B., Uotila, J., and Vihma, T. (2001) Turbulent surface fluxes and air–ice coupling in the Baltic Air–Sea-Ice Study (BASIS). *Annals of Glaciology*, **33**, 237–242.

Lehmann, A. (1992) Ein dreidimensionales wirbeauflösendes Modell der Ostsee [A three-dimensional eddy-resolving model of the Baltic Sea]. *Berichte aus dem Institut für Meereskunde der Universität Kiel*, **231**, 104 (Ph.D. thesis).

Lehmann, A. (1993) The major Baltic inflow in November/December 1951: A model simulation. *ICES Statutory Meeting*, Paper C.M. 1993/C:48, 7 pp.

Lehmann, A. (1994a) The major Baltic inflow in 1993: A numerical model simulation. *ICES Statutory Meeting*, Paper C.M. 1994/Q:9, 17 pp.

Lehmann, A. (1994b) A model study of major Baltic inflows. *Proceedings of the 19th Conference of the Baltic Oceanographers, Sopot, August 29–September 1, 1994*, Vol. 1, pp. 410–421.

Lehmann A. (1995) A three-dimensional baroclinic eddy-resolving model of the Baltic Sea. *Tellus*, **47A**, 1013–1031.

Lehmann, A. and Hinrichsen, H.-H. (2000a) On the wind driven and thermohaline circulation of the Baltic Sea. *Physics and Chemistry of the Earth*, **B25**, 183–189.

Lehmann, A. and Hinrichsen. H.-H. (2000b) On the thermohaline variability of the Baltic Sea. *Journal of Marine Systems*, **25**, 333–357.

Lehmann, A. and Hinrichsen, H.-H. (2002) Water, heat and salt exchange between the deep basins of the Baltic Sea. *Boreal Environment Research*, **7**, 405–415.

Lehmann, A. and Myrberg, K. (2008). Upwelling in the Baltic Sea: A review. *Journal of Marine Systems* (in press).

Lehmann, A., Krauss, W., and Hinrichsen, H.-H. (2002) Effects of remote and local atmospheric forcing on circulation and upwelling in the Baltic Sea. *Tellus*, **54A**, 299–316.

Lehmann, A., Lorenz, P., and Jacob, D. (2004) Modelling the exceptional Baltic Sea inflow events in 2002–2003. *Geophysical Research Letters*, **31**(21).

Lentz, W. (1971) Monatskarte der Temperatur der Ostsee dargestellt für verschiedene Tiefenhorisonte [Monthly maps of the temperature of the Baltic Sea for different depths]. *Deutsche Hydrographische Zeitschrift, Ergänzungshäft Reihe B*, **11**, 1–148 [in German].

Leppäranta, M. (1981a) An ice drift model for the Baltic Sea. *Tellus*, **33**, 583–596.

Leppäranta, M. (1981b) On the structure and mechanics of pack ice in the Bothnian Bay. *Finnish Marine Research*, **248**, 3–86.

Leppäranta, M. (1983) A growth model for black ice, snow ice and snow thickness in subarctic basins. *Nordic Hydrology*, **14**(2): 59–70.

Leppäranta, M. (1987) Ice observations: Sea Ice '85 experiment. *Meri*, **15** (Report Series of the Finnish Institute of Marine Research).

Leppäranta, M. (1990) Observations of free ice drift and currents in the Bay of Bothnia. *Acta Regiae Societatis Scientiarum et Litterarum Gothoburgensis, Geophysica*, **3**, 84–98.

Leppäranta, M. (1993) A review of analytical models of sea-ice growth. *Atmosphere–Ocean*, **31**(1), 123–138.

Leppäranta, M. (ed.) (1998) *Physics of Ice-Covered Seas I–II*. University of Helsinki Press, Helsinki, 828 pp.

Leppäranta, M. (2005) *The Drift of Sea Ice*. Springer-Verlag, Heidelberg, 290 pp.

Leppäranta, M. and Hakala, R. (1992) Structure and strength of first-year sea ice ridges in the Baltic Sea. *Cold Regions Science and Technology*, **20**, 295–311.

Leppäranta, M. and Hibler III, W.D. (1985) The role of plastic ice interaction in marginal ice zone dynamics. *Journal of Geophysical Research*, **90**(C6), 11899–11909.

Leppäranta, M. and Lewis J. E. (2007) Observations of ice surface temperature and thickness in the Baltic Sea. *International Journal of Remote Sensing*, **28**(17), 3963–3977.

Leppäranta, M. and Manninen, T. (1988) *The Brine and Gas Content of Sea Ice with Attention to Low Salinities and High Temperatures*. Finnish Institute of Marine Research, Internal Report 2, Helsinki, 14 pp.

Leppäranta, M. and Peltola, J. (1986) *Satunnaiskulkumallin soveltuvuudesta Itämerelle* [On the applicability of a random walk model to the Baltic Sea]. Finnish Institute of Marine Research, Internal Report 4, Helsinki, 31 pp [in Finnish].

Leppäranta, M. and Wang, K. (2002) Sea ice dynamics in the Baltic Sea basins. *Proceedings of 15th IAHR Ice Symposium, Dunedin, New Zealand*.

Leppäranta, M. and Zhang, Z. (1992) Use of ERS-1 SAR data in numerical sea ice modeling. *Proceedings of the Central Symposium of the International Space Year Conference, Munich, Germany, March 30–April 4, 1992*, pp. 123–128 (ESA SP-341, July 1992).

Leppäranta, M., Kuittinen, R., and Askne, J. (1992) BEPERS pilot study: An experiment with X-band synthetic aperture radar over Baltic Sea ice. *Journal of Glaciology*, **38**(128), 23–35.

Leppäranta, M., Lensu, M., Kosloff, P., and Veitch, B. (1995) The life story of a first-year sea ice ridge. *Cold Regions Science and Technology*, **23**, 279–290.

Leppäranta, M., Sun, Y., and Haapala, J. (1998a) Comparisons of sea-ice velocity fields from ERS-1 SAR and a dynamic model. *Journal of Glaciology*, **44**(147), 248–262.

Lepparanta, M., Tikkanen, M., and Shemeikka, P. (1998b) Observations of ice and its sediments on the Baltic coast. *Nordic Hydrology*, **29**, 199–220.

Lepparanta, M., Reinart, A., Arst, H., Erm, A., Sipelgas, L., and Hussainov, M. (2003) Investigation of ice and water properties and under-ice light fields in fresh and brackish water bodies. *Nordic Hydrology*, **34**(3), 245–266.

Lessin, G. and Raudsepp, U. (2007) Modelling the spatial distribution of phytoplankton and inorganic nitrogen in Narva Bay, southeastern Gulf of Finland, in the biologically active period. *Ecological Modelling*, **201**(3/4), 348–358.

Lewis, J. E., Lepparanta, M., and Granberg, H. B. (1993) Statistical features of the sea ice surface topography in the Baltic Sea. *Tellus*, **45A**, 127–142.

Li, W. H. and Lam, S. H. (1964) *Principles of Fluid Mechamiscs*. Addison-Wesley, Reading, MA.

Lidberg, M. (2007) Geodetic reference frames in presence of crustal deformations. Chalmers University of Technology, Department of Radio and Space Science, Gothenburg, Sweden, (Ph.D. thesis).

Liljebladh, B. and Stigebrandt, A. (1996) Observations of the deepwater flow into the Baltic Sea. *Journal of Geophysical Research*, **101**(C4), 8895–8911.

Lindholm, T., Rönnberg, O., and Östman, T. (1989) Husöviken: en flada i Ålands skärgård [The Husö Bay: A flada in the Åland archipelago]. *Svensk botanisk tidskrift*, **83**, 143–147 [in Swedish].

Lisitzin, E. (1957) On the reducing influence of sea ice on the piling-up of water due to wind stress. *Commentationes Physico-Mathematicae/Societas Scientiarum Fennica*, **20**(6), 1–12.

Lisitzin, E. (1966a) Land uplift in Finland as a sea level problem. *Annales academiae scientiarum Fennicae*, **A III 90**, 237–239.

Lisitizin, E. (1966b) Mean sea level heights and elevation systems in Finland. *Societas Scientiarum Fennica, Commentationes Physico-Mathematica*, **32**(4).

Lisitzin, E. (1974) *Sea Level Changes* (Elsevier Oceanography Series, 286 pp.). Elsevier, Amsterdam.

Lundberg, P. A. (1983) On the mechanisms of the deep-water flow in the Bornholm Channel. *Tellus*, **35A**, 149–158.

Luyten, P. J., Jones, J. E, Proctor, R., Tabor, A., Tett, P., and Wild-Allen, K. (1999) *COHERENS—A coupled hydrodynamical-ecological model for regional and shelf seas: User Documentation* (MUMM Report). Management Unit of the Mathematical Models of the North Sea, Brussels, Belgium, 911 pp.

MacKenzie, B. R. and Schiedek, D. (2007) Daily ocean monitoring since the 1860s shows record warming of northern European seas. *Global Change Biology*, **13**(7), 1335-1347.

Magaard, L. and Rheinheimer, G. (eds.) (1974) *Meereskunde der Ostsee* [Marine science of the Baltic Sea]. Springer-Verlag, New York, 269 pp. [in German].

Malicki, J. and Wielbińska, D. (1992) Some aspects of the atmosphere's impact on the Baltic Sea waters. *Bulletin of the Sea Fisheries Institute*, **1**(125), 19–28.

Mälkki, P. (1978) Itämeri—maailman suurin murtovesiallas [The Baltic Sea—the largest brackish water basin in the World]. *Suomen luonto*, **37**(3/4), 109–112.

Mälkki, P. (2001) Oceanography in Finland 1918–2000. *Geophysica*, **37**, 225–250.

Mälkki, P. and Tamsalu, R. (1985) Physical features of the Baltic Sea. *Finnish Marine Research*, **252**, 110 pp.

Marshall, J., Hill, C., Perelman, L. and Adcroft, A. (1997a) Hydrostatic, quasi-hydrostatic, and nonhydrostatic ocean modelling. *Journal of Geophysical Research*, **102**(C3), 5733–5752.

Marshall, J., Adcroft, A., Hill, C., Perelman, L., and Heisey, C. (1997b) A finite-volume, incompressible Navier Stokes model for studies of the ocean on parallel computers. *Journal of Geophysical Research*, **102**(C3), 5753–5766.

Matciak, M., Urbański, J., Piekarek-Jankowska, H., and Szymelfenig, M. (2001) Presumable groundwater seepage influence on the upwelling events along the Hel Peninsula. *Oceanological Studies*, **30**(3/4), 125–132.

Matthäus, W. (1987) The History of the Conference of Baltic Oceanographers. *Beiträge für Meereskunde, Berlin*, **57**, 11–25.

Matthäus, W. (1993) Major inflows of highly saline water into the Baltic Sea: A review. *ICES Statutory Meeting*, Paper C.M. 1993/C:52, 16 pp.

Matthäus, W. (2006) The history of investigation of salt water inflows into the Baltic Sea: From the early beginning to recent results. *Marine Science Reports*, **95** (Baltic Sea Research Institute, Warnemünde, Germany).

Matthäus, W. and Franck, H. (1988) The seasonal nature of major Baltic inflows. *Kieler Meeresforschung*, Sonderheft **6**, 64–72.

Matthäus, W. and Franck, H. (1990) The water volume penetrating into the Baltic Sea in connection with major Baltic inflows. *Gerlands Beiträge zur Geophysik*, **99**, 377–386 (Leipzig).

Matthäus, W. and Franck, H. (1992) Characteristics of major Baltic inflows: A statistical analysis. *Continental Shelf Research*, **12**, 1375–1400.

Matthäus, W. and Lass, H.-U. (1995) The recent salt inflow into the Baltic Sea. *Journal of Physical Oceanography*, **25**, 280–286.

Matthäus, W. and Schinke, H. (1994) Mean atmospheric circulation patterns associated with major Baltic inflows. *Deutsche Hydrographische Zeitschrift*, **46**, 321–339.

Matthäus, W. and Schinke, H. (1999) The influence of river runoff on deep water conditions of the Baltic Sea. *Hydrobiologia*, **393**(1), 1–10.

Matthäus, W., Lass, H.-U., and Tiesel, R. (1993) The major Baltic inflow in 1993. *ICES Statutory Meeting*, Paper C.M. 1993/C:51, 16 pp.

Massel, S. R. (1989) *Hydrodynamics of Coastal Zones*. Elsevier, Amsterdam, 336 pp.

Maykut, G. A. and Untersteiner, N. (1971) Some results from a time-dependent thermodynamic model of sea ice. *Journal of Geophysical Research*, **76**, 1550–1575.

Meier, H. E. M. (1996) Ein regionales Modell der westlichen Ostsee mit offenen Randbedingungen und Datenassimilation [A regional model for the the western Baltic Sea with open boundary conditions and data-assimilation]. *Berichte aus dem Institut für Meereskunde der Universität Kiel*, **284**, 118 pp. (Ph.D. t hesis).

Meier, H. E. M. (1999) *First Results of Multi-year Simulations Using a 3D Baltic Sea Model*. SMHI Reports Oceanography No. 27, Norrköping, Sweden.

Meier, H. E. M (2005) Modeling the age of Baltic Sea water masses: Quantification and steady state sensitivity experiments. *Journal of Geophysical Research*, **110**(C2), C02006.

Meier, H. E. M. (2006) Baltic Sea climate in the late 21st century: A dynamical downscaling approach using two global models and two emission scenarios. *Climate Dynamics*, **27**, 39–68.

Meier, H. E. M. (2007) Modeling the pathways and ages of inflowing salt- and freshwater in the Baltic Sea. *Estuarine, Coastal and Shelf Science*, **74**(4), 610–627.

Meier, H. E. M. and Kauker, F. (2003a). Sensitivity of the Baltic Sea salinity to the freshwater supply. *Climate Research*, **24**, 231–242.

Meier, H. E. M. and Kauker, F. (2003b) Modeling decadal variability of the Baltic Sea: 2. Role of freshwater inflows and large-scale atmospheric circulation for salinity. *Journal of Geophysical Research*, **108**(C11), 3368.

Meier, H. E. M., Döscher, R., and Faxén, T. (2003) A multiprocessor coupled ice–ocean model for the Baltic Sea: Application to salt inflows. *Journal of Geophysical Research*, **108**(C8), 3273.

Meier, H. E. M.., Döscher, R., Broman, B., and Piechura, J. (2004a) The Major Baltic Inflow in January 2003 and preconditioning by smaller inflows in summer/autumn 2002: A model study. *Oceanologia*, **46**, 557–579.

Meier, H. E. M., Döscher, R., and Halkka, A. (2004b) Simulated distributions of Baltic sea-ice in warming climate and consequences for the winter habitat of Baltic ringed seal. *Ambio*, **33**, 249–256.

Meier, H. E. M., Feistel, R., Piechura, J., Arneborg, L., Burchard, H., Fiekas, V., Golenko, N., Kuzmina, N., Mohrholz, V., Nohr, C. *et al.* (2006) Ventilation of the Baltic Sea deep water: A brief review of present knowledge from observations and models. *Oceanologia*, **48(S)**, 133–164.

Meissner, T. and Wentz, F. J. (2004) The complex dielectric constant of pure and sea water from microwave satellite observations. *IEEE Transactions on Geoscience and Remote Sensing*, **42**(9), 1836–1849.

Meyer, H. A., Möbius, K., Karsten, G., and Hensen, V. (1873) Die Expedition zur physikalisch-chemischen und biologischen Untersuchung der Ostsee im Sommer 1871 auf S. M. Avisodampfer Pommerania. *Jahresber. Comm. Wiss. Unters. Dt. Meere Kiel. f. d. J. 1871*, **1**, 178 pp.

Michel, B. (1978) *Ice Mechanics*. Laval University Press, Québec, Canada.

Mietus, M. (1998) *The Climate of the Baltic Sea Basin*. World Meteorological Organization, Marine Meteorology and Related Oceanographic Activities, Rep. No 41, WMO/TD-No. 933, Geneva.

Mietus, M. and Owczarek, M. (1994) Charakterystyka warunkow meteorologicznych. In: *Warunki srodowiskowe polskiej strefy poludniowego Baltyku w 1993 roku*, pp. 9–26. Materiale Oddzialu Morskiego, IMGW, Gdynia.

Mikhailov, A. E. and Chernyshova, E. S. (1997) General water circulation. In: I. N. Davidan and O.-P. Savchuk (eds.), *"Baltica" Project*, Issue 5, Part 2, pp. 245–260. Hydrometeoizdat, St. Petersburg [in Russian].

Mikulski, Z. (1970) Inflow of river water to the Baltic Sea in the period 1951–1960. *Nordic Hydrology*, **4**, 216–227.

Mikulski, Z. (1972) The inflow of the river waters to the Baltic Sea in 1961–1970. *Proceedings of the 8th Conference Baltic Oceanographers, Copenhagen, October*.

Millero, F. and Kremling, I. (1976) The densities of the Baltic Sea deep waters. *Deep Sea Research*, **23**, 611–622.

Miropolsky, Y. (1981) *Dynamics of Internal Gravity Waves in the Oceans*. Hidrometeoizdat, Leningrad, 304 pp. [original in Russian, English translation in 2001 by Kluwer Academic, Dordrecht, the Netherlands].

Mörner, N.-A. (1995) The Baltic Ice Lake–Yoldia Sea transition. *Quaternary International*, **27**, 95–98.

Müller-Navarra, S., and Lange, W. (2004) Modelling tides in the Baltic Sea: A short note on the harmonic analysis of a one-year water-level time series. *Proceedings of the Sixth HIROMB Scientific Workshop, St. Petersburg*, pp. 16–20.

Multala, J., Oksama, M., Hautaniemi, H., Leppäranta, M., Haapala, J., Herlevi, A., Riska, K., and Lensu, M. (1996) An airborne electromagnetic system on a fixed wing aircraft for sea ice thickness mapping. *Cold Regions Science and Technology*, **24**(4), 355–373.

Murthy, R., Håkansson, B., and Alenius, P. (1993) *The Gulf of Bothnia Year 1991: Physical Transport Experiments*. SMHI Reports Oceanography RO 15, 127 pp., Norrköping.

Myrberg, K. (1991) *Simulating the Climatological Flow Field of the Northern Baltic Sea by a Two-layer Model.* ICES C.M.1991/C:12, 12 pp., La Rochelle, France.

Myrberg, K. (1997) Sensitivity tests of a two-layer hydrodynamic model in the Gulf of Finland with different atmospheric forcings. *Geophysica*, **33**(2), 69–98.

Myrberg, K. (1998) Analysing and modelling the physical processes of the Gulf of Finland in the Baltic Sea. *Monographs of the Boreal Environment Research*, **10**, 50 pp. + 5 articles, (Ph.D. thesis).

Myrberg, K. and Andrejev, O. (2003) Main upwelling regions in the Baltic Sea: A statistical analysis based on three-dimensional modelling. *Boreal Environment Research*, **8**, 97–112.

Myrberg, K. and Andrejev. O. (2006) Modelling of the circulation, water exchange and water age properties of the Gulf of Bothnia. *Oceanologia*, **48**(S), 55–74.

Myrberg, K., Koistinen, J., and Järvenoja, S. (1990) A study of non-forecast cyclogenesis in a polar air mass over the Baltic Sea. *Tellus*, **42A**, 165–173.

Myrberg, K., Leppäranta, M., and Kuosa, H. (2006) *Itämeren fysiikka, tila ja tulevaisuus* [The Physics, State and Future of the Baltic Sea]. University of Helsinki Press, Helsinki, 201 pp [in Finnish].

Myrberg, K., Lehmann, U., Raudsepp, U., Szymelfenig, M., Lips, I., Lips, U., Matciak, M., Kowalewski, M., Krężel, A., Burska, D. et al. (2008) Upwelling events, coastal offshore exchange, links to biogeochemical processes: Highlights from the Baltic Sea Science Congress, March 19–22, 2007 at Rostock University. *Oceanologia*, **50**(1), 95–113.

Neelov, I. A. (1982) A mathematical model of eddies in the ocean. *Okeanologiya*, **22**(6), 875–884 [in Russian].

Neumann, G. (1941) Eigenschwingungen der Ostsee. *Arch. Dtsch. Seewarte Marineobserv.*, **61**(4), 1–59.

Neumann, G. and Pierson, W. J., Jr. (1966) *Principles of Physical Oceanography*. Prentice-Hall, Englewood Cliffs, NJ.

Neumann, T., Fennel W., and Kremp C. (2002). Experimental simulations with an ecosystem model of the Baltic Sea: A nutrient load reduction experiment. *Global Biogeochemical Cycles*, **16**(3), doi: 10.1029/2001GB001450 (7, 1–19).

Niemi, Å. (1979) Blue-green algal blooms and N:P ratio in the Baltic Sea. *Acta Botanica Fennica*, **110**, 57–61.

Nihoul, J. and Djenidi, S. (1987) Perspective in three-dimensional modelling of the marine system. In: J. Nihoul and B. Jamart (eds.), *Three-dimensional Models of Marine and Estuarine Dynamics*, pp. 1–33, Elsevier Oceanography Series, Amsterdam.

Niiler, P. P. and Kraus, E. B. (1977) One-dimensional models of the upper ocean. In: E. B. Kraus (ed.), *Modelling and Prediction of the Upper Layers of the Ocean*, pp. 143–172. Pergamon Press, Oxford, U.K.

Niros, A., Vihma, T., and Launiainen, J. (2003) Marine meteorological conditions and air–sea exchange processes over the northern Baltic Sea in 1990s. *Geophysica*, **38**(1/2), 59–87.

Omstedt, A. (1985) On supercooling and ice formation in turbulent sea-water. *Journal of Glaciology*, **31**(109), 263–271.

Omstedt, A. (1990) Modelling the Baltic Sea as thirteen sub-basins with vertical resolution. *Tellus*, **A4**, 286–301.

Omstedt, A. and Svensson, U. (1984) Modelling supercooling and ice formation in a turbulent Ekman layer. *Journal of Geophysical Research*, **89**(C1), 735–744.

Omstedt, A., Sahlberg, J., and Svensson, U. (1983) Measured and numerically simulated autumn cooling in the Bay of Bothnia. *Tellus*, **35A**, 231–240.

Omstedt, A., Nyberg, L., and Leppäranta, M. (1994) *A Coupled Ice–ocean Model Supporting Winter Navigation in the Baltic Sea. Part 1. Ice Dynamics and Water Levels*. SMHI Reports in Oceanography No. 17, 17 pp., Norrköping, Sweden.

Omstedt, A., Meuller, L., and Nyberg, L. (1997) Interannual, seasonal and regional variations of precipitation and evaporation over the Baltic Sea. *Ambio*, **26**(8), 484–492.

Omstedt, A., Elken, J., Lehmann, A., and Piechura, J. (2004). Knowledge of the Baltic Sea physics gained during the BALTEX and related programmes. *Progress in Oceanography*, **63**, 1–28.

Ovsienko, S., Zatsepa, S., and Ivchenko, A. (1999) Study and modelling of behaviour and spreading of oil in cold water and ice conditions. *Proceedings of the 15th Conference on Port and Ocean Engineering under Arctic Conditions*, Vol. 2, pp. 848–857.

Pacanowski, R. C. and Griffies, S. M. (1998) *MOM 3.0 Manual*. NOAA/Geophysical Fluid Dynamics Laboratory, Princeton, NJ.

Paka, V. T. (1996) Thermohaline structure of the waters over the cross sections in the Slupsk Channel of the Baltic Sea in spring 1993. *Okeanologiya* (English Translation), **36**, 188–198.

Palmén, E. (1930) Untersuchungen über die Strömungen in den Finnland umgebenden Meeren [Investigations of currents in the seas surrounding Finland]. *Commentationes physico-mathematicae/Societas Scientiarium Fennica*, **12**, 93 [in German].

Palmén, E. and Söderman, D. (1966) Computation of the evaporation for the Baltic Sea from the heat flux of water vapour in the atmosphere. *Geophysica*, **8**(4), 261–280.

Palosuo, E. (1953) A treatise on severe ice conditions in the Central Baltic. *Merentutkimuslaitoksen Julkaisu/Havsforskningsinstitutets Skrift*, **156**, 130 (Helsinki).

Palosuo, E. (1961) *Crystal Structure of Brackish and Fresh-water Ice* (Snow and Ice Commission Publication 54, pp. 9–14). IASH, Gentbugge, Belgium.

Palosuo, E. (1963) The Gulf of Bothnia in winter. II. Freezing and ice forms. *Merentutkimuslaitoksen Julkaisu/Havsforskningsinstitutets Skrift*, **208** (Helsinki).

Palosuo, E. (1965) Frozen slush on lake ice. *Geophysica*, **9**(2), 131–147.

Palosuo, E. (1975) *Formation and Structure of Ice Ridges in the Baltic* (Winter Navigation Research Board, Rep. No. 12). Board of Navigation, Helsinki.

Paulson, C. A. (1970) The mathematical representation of windspeed and temperature profiles in the unstable atmospheric boundary layer. *Journal of Applied Meteorology*, **9**, 857–861.

Pavelson, J. (1988) Nature and some characteristics of thermohaline fronts in the Baltic Proper. *Proceedings of the 16th Conference of Baltic Oceanographers, Kiel*, Vol. 2, pp. 796–805.

Pavelson, J. (2005) Mesoscale physical processes and related impact on the summer nutrient fields and phytoplankton blooms in the western Gulf of Finland. Tallinn University of Technology, 38 pp. + 10 articles (Ph.D. thesis).

Pavelson, J., Laanemets, J., Kononen, K., and Nõmmann, S. (1997) Quasi-permanent density front at the entrance to the Gulf of Finland: Response to wind forcing. *Continental Shelf Research*, **17**, 253–265.

Pedlosky, J. (1979) *Geophysical Fluid Dynamics*. Springer-Verlag, New York, 624 pp.

Perovich, D. K. (1998) The optical properties of the sea ice. In: M. Leppäranta (ed.), *Physics of Ice-covered Seas*. Helsinki University Printing House, pp. 195–230.

Pettersson, H. (2004) Aaltohavaintoja Suomenlahdelta: suuntamittauksia 1990–1994. *Meri*, **44**, 37 (Report Series of the Finnish Institute of Marine Research) [in Finnish].

Pettersson, Heidi (2004) Wave growth in a narrow bay. *Contributions of the Finnish Institute of Marine Research*, **9**, 33 pp. + 4 articles, Helsinki (Ph.D. thesis).

Perttilä, M. and Ehlin, U. (1993) The year of the Gulf of Bothnia. *Aqua Fennica*, **23**(1), 3–4.

Philander, S. and Yoon, J. (1982) Eastern boundary currents and coastal upwelling. *Journal of Physical Oceanography*, **12**, 862–879.

References

Pickard, G. L. and Emery, W. J. (1990) *Descriptive Physical Oceanography*, Fifth Edition. Butterworth-Heinemann. Oxford, U.K.

Piechura, J. and Beszczyńska-Möller, A. (2004) Inflow waters in the deep regions of the southern Baltic Sea: Transport and transformations (corrected version). *Oceanologia*, **46**(1), 113–141.

Piechura, J., Walczowski, W., and Beszczyńska-Möller, A. (1997) On the structure and dynamics of the water in the Slupsk Furrow. *Oceanologia*, **39**(1), 35–54.

Pielke, R., Jr. (2004) What is climate change? *Issues in Science and Technology*, Summer 2004, 1–4.

Pitkänen, H. and Tallberg, P. (eds.) (2007) Searching efficient protection strategies for the eutrophied Gulf of Finland: The integrated use of experimental modelling tolls (SEGUE). *Finnish Environment*, **15**, 90 (Helsinki).

Polhausen, E. (1921) Der Wärmeaustausch zwischen festen Körpern und Flüssigkeiten mit kleiner Wärmeleitung. *Z. angen. Mathematik und Mechanik*, **1**, 115–121.

Pollard, R. T. (1970) On the generation by winds of inertial waves in the ocean. *Deep Sea Research*, **17**, 795–812.

Pond, S. and Pickard, G. L. (1983) *Introductory Dynamical Oceanography*, Second Edition. Elsevier, Amsterdam.

Pruszak, Z. and Zawadzka, E. (2005) Vulnerability of Poland's coast to sea-level rise. *Coastal Engineering Journal*, **47**(2/3), 131–155.

Rahm, L. A. and Svensson, U. (1986) Dispersion of marked fluid elements in a turbulent Ekman layer. *Journal of Physical Oceanography*, **16**, 2084–2096.

Ramsay, H. (1947) *I kamp med Östersjöns isar: en bok om Finlands vintersjöfart* [Fighting with Baltic Sea ice: A book of Finnish winter shipping]. Holger Schildts Förlag, Helsingfors [in Swedish].

Rannat. K. (2007) Long weakly nonlinear waves in geophysical applications. *Dissertationes Geophysicales Universitatis Tartuensis*, **21**, 112 pp. + 8 articles (Ph.D. thesis).

Rantajärvi, E. (ed.) (2003) Algaline in 2003: 10 years of innovative plankton monitoring and research and operational information service in the Baltic Sea. *Meri*, **48**, 55 (Report Series of the Finnish Institute of Marine Research).

Raschke, E., Meywerk, J., Warrach, K., Andrea, U., Bergström, S., Beyrich, F., Bosveld, F., Bumke, K., Fortelius, C., Graham, L. P. *et al.* (2001) The Baltic Sea Experiment (BALTEX): A European contribution to the investigation of the energy and water cycle over a large drainage basin. *Bulletin of the American Meteorological Society*, **82**(11), 2389–2413.

Rasmus, K., Ehn, J., Granskog, M., Lindfors, A., Pelkonen, A., Rasmus, S., Leppäranta, M., and Reinart, A. (2002) Optical measurements of sea ice in Tvärminne, Gulf of Finland. *Nordic Hydrology*, **33**(2/3), 207–226.

Raudsepp, U. (1998) Current dynamics of estuarine circulation in the lateral boundary layer. *Estuarine, Coastal and Shelf Science*, **47**, 715–730.

Raudsepp, U., Toompuu, A., and Kõuts, T. (1999) A stochastic model for the sea level in the Estonian coastal area. *Journal of Marine Systems*, **22**, 69–87.

Reed, R. (1977) On estimating insolation over the ocean. *Journal of Physical Oceanography*, **7**, 482–485.

Reinart, A. and Kutser, T. (2006) Comparison of different satellite sensors in detecting cyanobacteria bloom event in the Baltic Sea. *Remote Sensing of Environment*, **102**, 74–85.

Reissman, J. H. (2002) Integrale Eigenschaften von mesoskaligen Wirbelstrukturen in den tiefen Becken der Ostsee. *Marine Science Reports*, **52**, 3–149. Baltic Sea Research Institute, Warnemünde, Germany.

Rodhe, B. (1952) On the relation between air temperature and ice formation in the Baltic. *Geografiska Annaler*, **1/2**, 176–202.

Rodhe, J. (1999) The Baltic and North Seas. A process-oriented review of the physical oceanography. In: A. Robinson and K. Brink (ed.), *The Sea*, Vol. 11. John Wiley & Sons, New York.

Rodhe, J. and Winsor, P. (2002) On the influence of the freshwater supply on the Baltic Sea mean salinity. *Tellus*, **54A**(2), 175–186.

Rukhovets, L. A. (1982) Mathematical modelling of water exchange and tracer distribution in the Neva Bay. *Meteorologiya i Gidrologiya*, **5**, 78–87 [in Russian].

Rutgersson, A., Bumke, K., Clemens, M., Foltescu, V., Lindau, R., Michelson, D., and Omstedt, A. (2001) Precipitation estimates over the Baltic Sea: Present state of the art. *Nordic Hydrology*, **32**, 285–314.

Saloranta, T. (2000) Modeling the evolution of snow, snow ice and ice in the Baltic Sea. *Tellus*, **52A**, 93–108.

Samuelsson, M. (1996) Interannual salinity variations in the Baltic Sea during the period 1954–1990. *Continental Shelf Research*, **16**(11), 1463–1477.

Samuelsson, M. and Stigebrandt, A. (1996) Main characteristics of the long-term sea level variability in the Baltic Sea. *Tellus*, **48A**, 672–683.

Sandén, P. and Håkansson, B. (1996) Long-term trends in Secchi depth in the Baltic Sea. *Limnology and Oceanography*, **41**, 346–351.

Sarkisyan, A. S., Staśkiewicz, A., and Kowalik, Z. (1975) Diagnostic calculations of summer circulation in the Baltic Sea. *Okeanologia*, **15**(6), 1002–1009 [in Russian].

Sarkkula, J. (ed.) (1997) *Proceedings of the Final Seminar of the Gulf of Finland Year 1996, March 17–18*. Suomen Ympäristökeskus, Helsinki.

Savijärvi, H., Niemelä, S., and Tisler, P. (2005) Coastal winds and low-level jets: Simulations for sea gulfs. *Quarterly Journal of the Royal Meteorological Society*, Part B, **131**(606), 625–637.

Sayin, E. and Krauss, W. (1996) A numerical study of the water exchange through the Danish Straits. *Tellus*, Series A, **48**, 324–341.

Shaffer, G. (1979) Conservation calculations in natural coordinates (with an example from the Baltic). *Journal of Physical Oceanographers*, **9**, 847–855.

Schinke, H. and Matthäus, W. (1998) On the causes of major Baltic inflows: An analysis of long time series. *Continental Shelf Research*, **18**, 67–97.

Schulz, B. (1956) Hydrographische Untersuchungen in der Ostsee 1925–1938 mit dem Reichsforschungsdampfer „Poseidon". *Deutsche Hydrographische Zeitschrift*, Ergänzungschäft Reihe B, **1**, 87.

Schwerdtfeger, P. (1963) The thermal properties of sea ice. *Journal of Glaciology*, **4**, 789–807.

Seinä, A. and Kalliosaari, S. (1991) Ice winters 1986–1990 along the Finnish coast. *Finnish Marine Research*, **259**, 3–61.

Seinä, A. and Palosuo, E. (1993) The classification of the maximum annual extent of ice cover in the Baltic Sea 1720–1992. *Meri*, **20**, 5–20.

Seinä, A. and Peltola, J. (1991) Duration of the ice season and statistics of fast ice thickness along the Finnish coast 1961–1990. *Finnish Marine Research*, **258**, 46.

Shirasawa, K., Leppäranta, M., Saloranta, T., Polomoshnov, A., Surkov, G., and Kawamura, T. (2005) The thickness of landfast ice in the Sea of Okhotsk. *Cold Regions Science and Technology*, **42**, 25–40.

Shirokov, K. P. (1977) Vliyanie splochennosti na vetrovoj dreif l'dov. *Sb. Rab. Leningr. GMO*, **9**, 46–53.

Siegel, H., Gerth, M., Rudolff, R., and Tschersich, G. (1994) Dynamic features in the western Baltic Sea investigated using NOAA-AVHRR data. *Deutsche Hydrographische Zeitschrift*, **46**(3), 191–209.

Siegel, H., Gert, M., and Mutzke, A. (1999) Dynamics of the Oder River plume in the Southern Baltic Sea: Satellite data and numerical modelling. *Continental Shelf Research*, **18**, 1143–1159.

Simojoki, H. (1940) Über die Eisverhältnisse der Binnenseen Finnlands. *Mitteilungen des Meteorologischen Instituts der Universität Helsinki*, **43**, 194 pp. + appendices [in German]

Simojoki, H. (1949) Niederschlag und Verdunstung auf dem Baltischen Meer. *Fennia*, **71**(1), 1–25 [in German].

Simojoki, H. (1978) *The History of Geophysics in Finland 1828–1918*. Societas Scientiarum Fennica, Helsinki.

Simons, T. S. (1978) Wind-driven circulations in the southwest Baltic. *Tellus*, **30**, 272–283.

Sipelgas, L., Arst, H., Raudsepp, U., Kõuts, T., and Lindfors, A. (2004) Optical properties of coastal waters of northwestern Estonia: In situ measurements. *Boreal Environment Research*, **9**, 447–456.

Sipelgas, L., Raudsepp. U., and Kõuts, T. (2006) Operational monitoring of suspended matter distribution using MODIS images and numerical modelling. *Advanced Space Research*, **38**(10), 2182–2188.

Sjöberg, B. (ed.), (1992) *Hav och Kust*. Sveriges Nationalatlas Förlag, Almqvist & Wiksell International, Stockholm.

Sjöberg, L. E. and Fan, H. (1986) *Studies on the Secular Land Uplift and Long Periodic Variations of the Sea Level around the Coast of Sweden* (Research reports from the Department of Geodesy, Trita Geod 1003). Royal Inst. of Technology, Stockholm, Sweden.

Sjöblom, V. (1967) Meriveden kumpuaminen ja Porkkalan niemi [Upwelling of the sea water and the cape of Porkkala]. *Suomen Kalatalous*, **27**, 1–12.

Smed, J. (1990) Hydrographic investigations in the North Sea, the Kattegat and the Baltic before ICES. *Deutsche Hydrographische Zeitschrift*, Ergänzungsheft Reihe B, **22**, 357–366.

Smedman, A., Bumke, K., Högström, U., Rutgersson, A., Gryning, S-E., Batchvarova, E., Peters, G., Hennemuth, B., Tammelin, B., Hyvönen, R. *et al*. (2005) Precipitation and evaporation budgets over the Baltic Proper: Observations and modelling. *Journal of Atmospheric and Ocean Science*, **10**, 163–191.

SMHI and FIMR (1982) *An Ice Atlas for the Baltic Sea, Kattegat, Skagerrak and Lake Vänern*. Norrköping, Sjöfartsverket.

Smith, R. (1968) Upwelling. *Oceanography and Marine Biology: An Annual Review*, **6**, 11–46.

Sokolov, A., Andrejev, O., Wulff, F., and Rodriguez Medina, M. (1997) *The Data Assimilation System for Data Analysis in the Baltic Sea* (System Ecology Contributions, No. 3, 66 pp.). Stockholm University, Sweden.

Soomere, T. (1995) Generation of zonal flow and meridional anisotropy in two-layer weak geostrophic turbulence. *Physical Review Letters*, **75**, 2440–2443.

Soomere, T. (1996) Spectral evolution of two-layer weak geostrophic turbulence. Part I: Typical scenarios. *Nonlinear Proceedings in Geophysics*, **3**, 166–195.

Soomere, T. (2005). Fast ferry traffic as a qualitatively new forcing factor of environmental processes in non-tidal sea areas: A case study in Tallinn Bay, Baltic Sea. *Environmental Fluid Mechanics*, **5**(4), 293–323.

Soomere, T. (2006) Nonlinear ship wake waves as a model of rogue waves and a source of danger to coastal environment. *Oceanologia*, **48**(S), 185–202.

Soomere, T. and Keevallik, S. (2003) Directional and extreme wind properties in the Gulf of Finland. *Proceedings of Estonian Academy of Sciences, Engineering*, **9**(2), 73–90.

Soomere, T. and Rannat, K. (2003) An experimental study of wind waves and ship wakes in Tallinn Bay. *Proceedings of Estonian Academy of Sciences, Engineering*, **9**(3), 157–184.

Soomere, T., Elken, J., Kask, J., Keevallik, S., Kõuts, T., Metsaveer, J., and Peterson, P. (2003a) Fast ferries as a new key forcing factor in Tallinn Bay. *Proceedings of Estonian Academy of Sciences, Engineering*, **9**(3), 220–242.

Soomere, T., Rannat, K., Elken, J., and Myrberg K. (2003b) Natural and anthropogenic wave forcing in Tallinn Bay, Baltic Sea. In: C. A. Brebbia, D. Almorza and F. López-Aguayo (eds.), *Coastal Engineering*, Vol. VI, pp. 273–282. WIT Press, Southampton, Boston.

Soomere, T., Myrberg, K., Leppäranta, M., and Nekrasov, A. E. (2008a) Progress in physical oceanography of the Gulf of Finland. *Oceanologia*, **50**(3), 287–362.

Soomere, T., Behrens, A., Tuomi, L., and Nielsen, J. W. (2008b) Wave conditions in the Baltic Proper and in the Gulf of Finland during Windstorm Gudrun. *Natural Hazards and Earth Systems Sciences*, **8**(1), 37–46.

Soskin, I. M. (1963) *Mnogoletnie izmenenija gidrologi eskih harakteristik Baltijskogo morja* [Long-term Changes in the Hydrological Characteristics of the Baltic]. Hydrometeorological Press, Leningrad, 159 pp. [in Russian]

Stefan, F. (1891) Über die Theorie der Eisbildung, insbesondere über die Eisbildung in der Polarmeere. *Annalen der Physik und Chemie*, **42**, 269–286.

Stenius, S. (1904) Der osmotische Druck im Meerwasser [The osmotic pressure in sea water]. Mitteilung aus dem Laboratorium der Finnischen hydrographisch-biologischen Kommission. *Öfversigt af Finska Vetenskaps-Societetens förhandlingar*, XLVI 1903–1904, No. 6, pp. 1–16, Helsingfors.

Stigebrandt, A. (1983) A model for the exchange of salt and water between the Baltic and the Skagerrak. *Journal of Physical Oceanography*, **13**, 411–427.

Stigebrandt, A. (1985) A model for the seasonal pycnocline in rotating systems with application to the Baltic proper. *Journal of Physical Oceanography*, **15**, 1392–1404.

Stigebrandt, A. (1987) A model for vertical circulation of the Baltic deep water. *Journal of Physical Oceanography*, **17**, 1772–1785.

Stigebrandt A. (1999) Resistance to barotropic tidal flow in straits by baroclinic wave drag. *Journal of Physical Oceanography*, **29**, 191–197.

Stigebrandt, A. (2001) Physical oceanography of the Baltic Sea. In: F. Wulff, L. Rahm, and P. Larsson (Eds.), *A System Analysis of the Baltic Sea* (Ecological Studies, Vol. 148, pp. 19–74). Springer-Verlag, Berlin.

Stigebrandt, A. (2003) Regulation of vertical stratification, length of stagnation periods and oxygen conditions in the deeper deepwater of the Baltic proper. In: W. Fennel and B. Hentzsch (eds.), *Festschrift zum 65. Geburtstag von Wolfgang Matthäus* (Marine Science Report No. 54, pp. 69–80). Baltic Sea Research Institute, Warnemünde, Germany.

Stigebrandt, A., Lass, H.-U., Liljebladh, B., Alenius, P., Piechura, J., Hietala, R., and Beszczynska, A. (2002) DIAMIX: An experimental study of diapycnal deepwater mixing in the virtually tideless Baltic Sea. *Boreal Environment Research*, **7**, 363–369.

Stigge, H. J. (1993) Sea level changes and high-water probability on the German Baltic Coast. *International Workshop, Sea Level Changes and Water Management, April 19–23, Noordswijerhout, Nederlands*, pp. 19–29.

Stipa, T. (2004) Baroclinic adjustment in the Finnish coastal current. *Tellus*, **56A**(1), 79–87.

Stokes, G.G. (1847) On the theory of oscillatory waves. *Transactions of the Cambridge Philosophical Society*, **8**, 441–455.

References

Suursaar, Ü., Jaagus, J., and Kullas, T. (2006a) Past and future changes in sea level near the Estonian coast in relation to changes in wind climate. *Boreal Environment Research*, **11**, 123–142.

Suursaar, Ü., Kullas, K., Otsmann, M., Saaremäe, I., Kuik, J., and Merilain, M. (2006b) Hurricane Gudrun and modelling its hydrodynamic consequences in the Estonian coastal waters. *Boreal Environment Research*, **11**, 143–159.

Svansson, A. (1959) Some computations of water heights and currents in the Baltic. *Tellus*, **2**, 231–238.

Svansson, A. (1975) Interaction between the coastal zone and the open sea. *Merentutkimuslaitoksen Julkaisu/Havsforskningsinstitutets Skrift*, **239**, 11–28.

Svensson, U. (1979) The structure of the turbulent Ekman layer. *Tellus*, **31**, 340–350.

Svensson, U. (1998). *PROBE: An Instruction Manual* (Report Oceanography No.10, 48 pp.). Swedish Meteorological and Hydrological Institute, Norrköping, Sweden.

Sverdrup, H. (1938) On the process of upwelling. *Journal of Marine Research*, **1**, 155–164.

Talpsepp, L. (1983) Trapped topographic waves in the Baltic Sea. *Oceanology*, **23**(6), 928–931 [in Russian].

Talpsepp, L. (2006) Periodic variability of currents induced by topographically trapped waves in the coastal zone in the Gulf of Finland. *Oceanologia*, **48**(S), 75–90.

Tamsalu, R. and Myrberg, K. (1995) Ecosystem modelling in the Gulf of Finland. I. General features and the hydrodynamic prognostic model FINEST. *Estuarine, Coastal and Shelf Science*, **41**, 249–273.

Tamsalu, R., Mälkki, P., and Myrberg, K. (1997) Self-similarity concept in marine system modelling. *Geophysica*, **33**(2), 51–68.

Tennekes, H. and Lumley, J. L. (1972) *A First Course in Turbulence*. MIT Press, Cambridge, MA, 300 pp.

Terzieva, F. S., Rožkova, V. A., and Smirnovoy, A. I. (1992) *Gidrometeorologiya i gidrokhimiya morey SSSR. Tom III. Baltiyskoe more. Vypusk 1: gidrometeorologitseskie usloviya*. Gidrometeoizdat, Sankt Peterburg.

Thorade, H. (1909) Über die Kalifornischen Meeresströmung. Oberflächentemperaturen und Strömungen an der westküste Nordamerikas [The Californian Current, sea-surface temperature and currents at the west coast of the USA]. *Annalen der Hydrographie und Maritimen Meteorologie*, **37**, 17–34, 63–76.

Thorndike, A. S., Rothrock, D. A., Maykut, G. A., and Colony, R. (1975) The thickness distribution of sea ice. *Journal of Geophysical Research*, **80**, 4501–4513.

Tikkanen, M. and Oksanen, J. (2002) Late Weichselian and Holocene shore displacement history of the Baltic Sea in Finland. *Fennia*, **180**(1/2), 9–20.

Tinz, B. (1996) On the relation between annual maximum ice extent of the ice cover in the Baltic Sea and sea level pressure as well as temperature field. *Geophysica*, **32**(3), 319–341.

Tyrväinen, M. (1978) Upper layer observations and simulation using Kraus and Turner's model in the Gulf of Finland. *Nordic Hydrology*, **9**, 207–218.

UNESCO (1981) *The Practical Salinity Scale 1978 and the International Equation of State of Seawater 1980* (Tenth Report of the Joint Panel on Oceanographic Tables and Standards, Technical Papers in Marine Science 36). UNESCO, New York.

Uusitalo, S. (1960) The numerical calculation of wind effect on sea level elevations. *Tellus*, **12**, 427–435.

Vahtera, E., Laanemets, J., Pavelson, J., Huttunen, M., and Kononen, K. (2005) Effect of upwelling on the pelagic environment and bloom-forming cyanobacteria in the western Gulf of Finland, Baltic Sea. *Journal of Marine Systems*, **58**(1/2), 67–82.

Venkatesh, S., El-Tahan, H., Comfort, G., and Abdelnour, R. (1990) Modelling the behaviour of oil spills in ice-infested waters. *Atmosphere–Ocean*, **28**(3), 303–329.
Vermeer, M., Kakkuri, J., Mälkki, P., Boman, H., Kahma, K. K., and Leppäranta, M. (1988) Land uplift and sea level variability spectrum using fully measured monthly means of tide gauge recordings. *Finnish Marine Research*, **256**, 3–75.
Vestøl, O. (2006) Determination of postglacial land uplift in Fennoscandia from leveling, tide-gauges and continuous GPS stations using least squares collocation. *Journal of Geodesy*, **80**, 248–258, doi: 10.1007/s00190-006-0063-7.
Vihma, T. (1995) Atmosphere–surface interactions over polar oceans and heterogeneous surfaces. *Finnish Marine Research*, **264**, 41 pp. + 7 articles. Finnish Institute of Marine Research, Helsinki (Ph.D. thesis).
Voipio, A. (ed.) (1981) *The Baltic Sea*. Elsevier, Amsterdam, 418 pp.
Volkov, V. A., Johannessen, O. M., Borodachev, V. E., Voinov, G. N., Pettersson, L. H., Bobylev, L. P., and Kouraev, A. V. (2002) *Polar Seas Oceanography: An Integrated Case Study of the Kara Sea* (450 pp.). Springer/Praxis, Chichester, U.K.
Voltzinger, N. E., Zolnikov, A. V., Klevanny, K. A., and Preobrazhensky L. Yu. (1990) Calculation of hydrological regime of the Neva Bay. *Meteorologiya i Gidrologiya*, 70–77 [in Russian].
Vuorinen, I., Hänninen, J., and Kornilovs, G. (2003) Transfer-time function modelling between environmental variation and mesozooplankton in the Baltic Sea. *Progress in Oceanography*, **59**, 339–356.
Walin, G. (1972) Some observations of temperature fluctuations in the coastal region of the Baltic Sea. *Tellus*, **24**(3), 187–198.
WAMDI Group (1988) The WAM model: A third generation ocean wave prediction model. *Journal of Physical Oceanography*, **18**, 1775–1810.
Wang, D. W., Mitchell, D. A., Teague, W. J., Jarosz, E., and Hulbert, M. S. (2005) Extreme waves under hurricane Ivan. *Science*, **309**(5736), 896, doi: 10.1126/science.1112509.
Wang, K., Leppäranta, M., and Kõuts, T. (2003) A model for sea ice dynamics in the Gulf of Riga. *Proceedings of Estonian Academy of Sciences, Engineering*, **9**(2), 107–125.
Wang, K., Leppäranta, M., Gästgifvars, M., Vainio, J., and Wang, C. (2008) The drift and spreading of the Runner 4 oil spill and the ice conditions in the Gulf of Finland, winter 2006. *Estonian Journal of Earth Sciences*, **57**(3), 181–191.
Warren, S. G. (1982) Optical properties of snow. *Reviews in Geophysics and Space Physics*, **20**(1), 67–89.
Webb, D. J., Coward, A. C., de Cuevas, B. A., and Gwilliam, C. S. (1997) A multiprocessor ocean circulation model using message passing. *Journal of Atmospheric and. Oceanic Technology*, **14**, 175–183.
Webb, E. K. (1970) Profile relationships: The log-linear range and extension to strong stability. *Quarterly Journal of Royal Meteorological Society*, **96**, 67–90.
Weeks, W. F. (1998) Growth conditions and the structure and properties of sea ice. In: M. Leppäranta (ed.), *Physics of Ice-covered Seas*, Vol 1, pp. 25–104. Helsinki University Press.
Weeks, W. F., Gow, A. J., Kosloff, P., and Digby-Argus, S. (1990) *The Internal Structure, Composition and Properties of Brackish Ice from the Bay of Bothnia* (Monograph, 90/1, pp. 5–15). Cold Regions Research and Engineering Laboratory, Hanover, NH.
Welander, P. (1974) Two-layer exchange in an estuary basin with special reference to the Baltic Sea. *Journal of Physical Oceanography*, **84**(4), 542–546.
Winsor, P., Rodhe, B., and Omstedt, A. (2001) Baltic Sea ocean climate: An analysis of 100 yr of hydrographic data with focus on freshwater budget. *Climate Research*, **18**, 5–15.

Winterhalter, B., Flodén, T., Ignatius, H., Axberg, S., and Niemistö, L. (1981) Geology of the Baltic Sea. In: A. Voipio (ed.), *The Baltic Sea*, pp. 1–121. Elsevier, Amsterdam.

Witting, R. (1910) Suomen Kartasto. *Karttalehdet*, Nos. 6b, 7, 8, ja 9. Rannikkomeret [in Finnish].

Witting, R. (1911) Tidvatten i østersjön och Finska viken. *Fennia*, **29**(2), 84 pp.

Witting, R. (1912) Zusammenfassende Übersicht der Hydrographie des Bottnischen und Finnischen Meerbusens und der nördlichen Ostsee nach den Untersuchungen bis Ende 1910. *Finnländische hydrographisch-biologische Untersuchungen*, No. 7, 82 pp.

WMO (1970) *WMO Sea-ice Nomenclature, Terminology, Codes and Illustrated Glossary* (WMO/OMM/BMO 259, TP 145). World Meteorological Organization, Geneva.

Wübber, C. and Krauss, W. (1979) The two-dimensional seiches of the Baltic Sea. *Oceanologica Acta*, **2**(4), 435–446.

Wulff, F., Rahm, L., and Larsson, P. (eds.) (2001) *A System Analysis of the Baltic Sea: Ecological Studies*, Vol. 148. Springer-Verlag, Berlin.

Wyrtki, K. (1953) Die Dynamik der Wasserbewegungen im Fehmarnbelt I. *Kieler Meeresforschung*, **9**, 155–170.

Wyrtki, K. (1954) Die Dynamik der Wasserbewegungen im Fehmarnbelt II. *Kieler Meeresforschung*, **10**, 162–181.

Yen, Y.-C. (1981) *Review of Thermal Properties of Snow, Ice and Sea Ice* (CRREL Report 81/10, pp. 1–27). U.S. Army Cold Regions Research and Engineering Laboratory, Hanover, NH.

Zalesnyi, V. B., Tamsalu, R., and Kullas, T. (2004) Nonhydrostatic model of marine circulation. *Oceanology*, **44**(4), 461–471 [English edition].

Zeidler, R. B. (1997) Climate change vulnerability and response strategies for the coastal zone of Poland. *Climatic Change*, **36**(1/2), 151–173.

Zhang, Y. (2005) Surface water quality estimation using remote sensing in the Gulf of Finland and the Finnish Archipelago Sea. Helsinki University of Technology, Espoo, Finland, (Ph.D. thesis).

Zhang, Z. and Leppäranta, M. (1995) Modeling the influence of ice on sea level variations in the Baltic Sea. *Geophysica*, **31**(2), 31–46.

Zhurbas, V. and Paka, V. (1997) Mesoscale thermohaline variability in the Eastern Gotland Basin following the 1993 major Baltic inflow. *Journal of Geophysical Research*, **102**(C9), 20917–20926.

Zhurbas, V. and Paka, V. (1999) What drives thermohaline intrusions in the Baltic Sea? *Journal of Marine Systems*, **2**(1/4), 229–241.

Zhurbas, V., Oh, I., and Paka, V. (2003) Generation of cyclonic eddies in the Eastern Gotland Basin of the Baltic Sea following dense water inflows: Numerical experiments. *Journal of Marine Systems*, **38**, 323–336.

Zhurbas, V. M., Stipa, T., Mälkki, P., Paka, V. T., Kuzmina, N. P., and Sklyarov, V. E. (2004) Mesoscale variability of the upwelling in the southeastern Baltic Sea: IR images and numerical modeling. *Oceanology*, **44**(5), 619–628.

Zhurbas, V., Laanemets, J., and Vahtera, E. (2007) Integrated study of an upwelling event in the Gulf of Finland, July 1999. *Baltic Sea Science Congress, March 19–23, Rostock, Germany*. Abstract Volume, lectures—CBO Session, Topic B: Upwelling events, coastal offshore exchange, links to biogeochemical processes, 59 pp.

Zillman, J. W. (1972) *Study of Some Aspects of the Radiation and Heat Budgets of Southern Hemisphere Oceans* (Department of International Meteorological Studies Report No. 26). Canberra: Bureau of Meteorology.

Index

adiabatic 18, 113
Advanced Very High Resolution
    Radiometer (AVHRR) 280, 281
aerodynamic method 93
air–sea interface 105, 293
air temperature 25, 27, 28, 29, 32, 33, 70,
    88, 93, 94, 110, 111, 112, 115, 116,
    117, 118, 220, 223, 232, 234, 236,
    271, 278, 289, 290, 291, 293, 321,
    328
Åland Deep 17, 55
Åland Sea 5, 10, 15, 46, 53, 54, 55, 81, 82
albedo 29, 108–111, 115, 220, 223, 232,
    236, 331
Alg@line-project 23
Almagrundet 207, 209
amplitude
  annual cycle 32, 33
  heating 115
  sea-level 201, 272,
  temperature 106
  tidal 201, 204, 268
  wave 190–194, 196, 197, 201, 211, 212,
    215, 216
Anholt 49
anoxic 3, 90, 155, 158, 303, 304, 306, 311,
    312, 327, 333
attenuation of light 296, 298–303
Archipelago Sea 10, 15, 27, 46, 49, 53, 55,
    145, 148, 152, 183, 265, 267, 308

Arctic front 18
airmass
  Arctic 26, 40
  continental 25
  descending 91
  maritime 271
  polar 10
Arkona Basin 15, 43, 46, 74, 84, 95, 97,
    129, 145, 150, 154, 158, 160, 161,
    169, 171, 172, 178, 179, 184–186,
    216
atmospheric boundary layer 27, 40, 93,
    212, 203, 251, 293, 321
atmospheric forcing 67, 84, 105, 117, 178,
    186, 234, 238, 255, 275, 278, 321,
    330
atmospheric pressure 27, 40, 56, 57, 60, 84,
    94, 170, 179, 203, 213, 321

Baltic Ice Code 19, 318, 320
Baltic Ice Week 19
Baltic herring 5, 327
Baltic Monitoring Program (BMP) 21
Baltic Operational Observing System
    (BOOS) 24, 318
Baltic salmon 5, 327
Baltic Experiment for ERS-1 (BEERS) 24
Baltic Sea Action Plan 327
Baltic Sea Experiment (BALTEX) 23, 24,
    89, 90, 176, 177, 330

Index

Baltic Sea Index (BSI) 179, 180
baroclinic 33, 127, 128, 129, 134, 135, 140, 150, 151, 153–155, 157, 171, 172, 177, 184, 187, 201, 263, 277, 282, 291, 321
baroclinic instability 33, 153, 182, 201, 291
barotropic 64, 127, 128, 129, 134, 140, 141, 160, 177, 178, 180, 182, 187, 201, 274, 275
Bay of Bothnia 10, 11, 15, 42, 45, 46, 53, 55, 70, 71, 74, 81–84, 86, 87 , 92, 95, 96–98, 107, 121, 121, 128, 140, 144, 151, 179, 186, 207, 213, 221, 225, 236, 239, 240, 242, 243, 245, 248, 249, 252, 253, 265, 266, 268–270, 281, 284, 285, 291, 313, 316, 317, 331
bedrock 4, 10
Beer's law 296
Before Present (BP) 10–13
Belt Sea 15, 46, 49, 62, 64, 74, 84, 145, 148, 150, 161, 167, 169, 179, 184, 186, 201, 265, 282, 307, 334
Belt Sea front 62, 334
Billingen Gateway 10
Bornholm Basin 7, 43, 46, 49, 51, 74, 75, 84, 86, 87, 95, 97, 129, 145, 150, 154, 160, 161, 169, 171, 172, 178, 183, 184, 200, 201, 334
Bornholm Channel 129, 145, 169, 178, 179
Bornholm Deep 17, 145, 171, 179, 185
bottom
  boundary layer 103, 106, 117, 138, 148, 150, 186,
  currents 138, 145, 148, 152, 153, 178, 181, 182, 277
  friction 127, 133, 139, 145, 274, 275
  layer (lower layer) 46, 62, 64, 70, 74, 76, 78, 84, 138, 141, 148, 150, 151, 160, 169, 172, 173, 186, 187, 200, 267, 304, 309, 312, 315
  topography 42, 44, 45, 48, 50, 52, 54, 55, 62, 64, 151–153, 158, 161, 175, 177, 194, 201, 216, 226, 243, 283, 284, 308, 314
boundary conditions 102, 103, 105, 106, 107, 121, 123, 134, 138, 175, 177, 186, 193, 197, 199, 232, 237, 268, 285, 287

boundary layer 21, 27, 40, 93, 107, 110, 112, 121, 122, 123, 138, 148, 193, 249, 251
Boussinesq approximation 132
box model 124–130, 272
brackish ice 8, 24, 219, 220, 225, 231, 329
brackish water 1, 3, 4, 6, 8–10, 15, 40–42, 56–61, 72, 74, 88, 103, 221, 225, 303, 312, 326, 328, 336
brash ice 239, 241, 244, 257
brine pocket 227–231, 302
Brunt–Väisälä frequency 216
bulk formula 93, 111, 112
bulk modulus 56, 57
buoy 25, 142, 146, 147, 177, 208, 313, 318, 323
buoyancy 119–121, 122, 145, 150, 151, 182, 212, 231, 282, 284

capillary wave 191, 203, 205, 212
chlorophyll a 296–300, 302, 303, 319
Clausius–Clapeyron equation 116
climate change 5, 6, 10, 23, 24, 84, 172, 176, 220, 315, 328, 330, 331
climate model 257–259
climate scenario 259, 330, 331, 333, 334
cloudiness 29–32, 108, 110
coastal
  current 144, 182, 276, 277, 282
  ice zone 87, 223, 238, 243–245, 256, 283, 284, 286, 288, 302
  jet 158, 180, 200, 265, 277
  trapped waves 157, 200
  zone 3, 15, 23, 40, 42, 72, 75, 84, 118, 121, 128, 135, 186, 201, 205, 261–264, 278, 280, 281, 283, 284, 292, 300, 301, 307, 326
Colored Dissolved Organic Matter (CDOM) 296–302
compact ice 240, 246, 247, 248, 252, 253, 254, 257, 283, 321
compressibility 56, 57, 60
Conductivity–Temperature–Depth (CTD) 7, 124, 153
congelation ice 226, 227, 231, 232, 234, 236–238,
conservation of mass 90, 100, 127, 132, 174, 251, 275
conservation of momentum 132, 174

continuity equation 132, 133, 193, 274
continuum 246, 248
convection 4, 33, 57, 58, 61, 67, 70, 72, 90, 92, 145, 157, 183, 220, 222, 291, 303, 329
Coriolis acceleration 127, 132–138, 140, 193, 195, 196, 215, 248, 252, 253, 288
coupled ice–ocean model 256, 257
Curonian Lagoon 51, 262
cyanobacteria 282, 320, 333
cyclone 27, 33, 40, 91, 161, 200, 216, 291, 315

Darss Sill 11, 49, 100, 101, 105, 150, 159–161, 164, 166, 170–172, 179, 186, 282
dead water 214
deepwater circulation 22, 55, 58, 64, 81, 102, 128, 132, 134, 135, 145, 148–158, 170, 172, 186, 303, 327, 335
deepwater waves 128, 134, 157, 189, 191, 192, 203, 205, 206, 215
deformation, ice 223, 239, 244, 245, 247, 248, 251, 255, 259
deformation, Rossby radius 134, 136, 199, 263
deformed ice 239, 241, 243, 246, 251, 259
density stratification 127
depression 5, 10, 55
detrainment 124, 161
diagnostic model 22, 175, 177
Diapycnal Mixing (DIAMIX) 23, 154, 157
dicothermal layer 16, 69–72,
discharge 15, 61, 62, 90, 95, 97–99, 106, 125, 127, 161, 167, 171, 178, 183, 186, 267, 332, 333
dispersion relation 191, 193, 194, 197, 198, 203, 214, 215
dissipation 120, 157, 203, 206
diurnal tides 201, 202, 268
divergence 155, 212, 275, 276
double diffusion 155,157
downwelling 29, 108, 109, 135, 180, 276–278, 289, 298, 302
drag coefficient 111, 112, 249, 251, 278, 288
drainage basin 14–16, 25, 33, 36, 37, 86, 87, 94–97, 226, 300, 326

drift 135, 138, 139, 144, 180–182, 212, 220, 225, 226, 252, 253, 257, 259, 286, 287, 288, 292, 309, 310, 321, 333, 335 drift ice 19, 72, 90, 220, 223, 227, 240, 241, 243–246, 248, 249, 251, 256, 257, 262, 283, 284, 286, 287, 289, 313, 317, 320, 321, 333, 335
Drogden Channel 49
Drogden Sill 74, 105, 148, 160, 161, 166, 169, 171, 183, 186

Eastern Gotland Basin 7, 46, 49, 51, 75, 76, 78, 86, 87, 95, 97, 145, 150, 155, 156, 157, 169, 172, 178, 183, 186, 207
ecology 24, 295, 320, 323, 328
eigenoscillation 189, 192
Ekenäs Sill 53
Ekman
  depth 138, 139
  drift 180, 182, 212
  layer 122, 138, 139, 140
  number 134
  pumping 267, 276
  spiral 138, 139, 144, 276
  transport 138, 139, 276, 277, 282
energy balance 20, 69, 72, 94, 206, 255, 320
energy spectrum 135, 202, 205, 206, 215
entrainment 64, 118, 119, 124, 145, 148, 171, 178, 186
equation of motion 132, 133, 136, 140, 141, 175, 214, 247, 248, 252, 253, 255, 274, 287
equation of state 56–60, 132, 174, 175
erosion 14, 44, 45, 118, 190, 307, 308, 314, 315, 334
estuarine circulation 46, 65, 127, 151, 263, 267
Eulerian 185
European Centre for Medium-range Weather Forecasting (ECMWF) 176
European Remote Sensing Satellite-1 (ERS-1) 213, 250, 265, 266
eustatic 4, 5, 10, 11, 14, 42, 268–270, 333
eutectic point 228, 229
eutrophication 4, 23, 263, 295, 300, 304, 310–312, 327, 329, 334

Index

evaporation 85, 90, 91, 93–95, 98, 99, 101, 103, 111, 174, 291

Fårö Channel 51
Fårö Deep 51, 150, 173,
fast ice 19, 219, 223, 224, 225, 227, 238, 239, 241, 243, 244, 248, 262, 263, 283–289, 321, 335
Fehmarn Belt 49
fetch 21, 203, 205–207, 209, 210, 239, 283, 284
Finngrundet 55, 115, 151
frazil ice 226, 227, 231, 232, 235, 241, 283
free drift 139, 225, 246, 248, 252, 253, 255, 287, 288
free wave 192, 194, 195, 198
freezing date 118, 121, 220, 316, 317, 331
freezing-degree-days 234, 235, 238
freezing point temperature 59, 105, 220, 221, 229, 230, 235
freshwater 1, 3, 4, 10, 15, 53, 59–61, 65, 67, 71, 72, 86, 90, 98, 100–104, 127, 128, 134, 135, 173–175, 186, 221, 225–227, 230, 231, 240, 252, 263, 267, 272, 321, 326, 327, 334
freshwater budget 3, 15, 90, 96, 98, 99
friction 33, 92, 112, 127, 134–141, 192, 193, 195, 203, 225, 245–250, 252, 253, 263, 265, 274–276, 287, 288
friction velocity, 112
front 25–27, 32, 33, 61, 62, 64, 84, 91, 125, 150, 154, 155, 161, 221, 263–265, 278, 281, 290, 334
Froude number, 128, 134, 193

Gdansk Bay 15, 21, 46, 49, 51, 95, 97, 154, 281, 301
geostrophic adjustment 154, 158, 282
geostrophic flow 140, 141, 161, 194, 199, 291
global radiation 29
Gotland 21, 42, 51, 196, 281
Gotland Deep 17, 65, 73, 145, 150, 154, 155, 167, 168, 169, 172, 173, 178, 179, 183, 185, 186, 216, 304, 306
Gotland Sea 14, 15, 19, 43, 45, 46, 49–53, 55, 59, 65, 74, 75, 84–86, 93, 105, 125, 129, 135, 144, 151, 154, 161, 171, 182, 183, 186, 195, 196, 200, 201, 223, 237, 259, 265, 301, 303, 304
Gotska Sandön 51
gravity 127, 132, 134, 145, 155, 169, 191, 193, 201, 203, 212, 215
Great Belt 11, 14, 15, 49, 74, 141, 142, 161, 169, 179, 307
gray-body law 109
group velocity 191, 206, 216
growth of ice 103, 220, 223, 225, 227, 231, 232, 234–238, 241, 255, 259, 283, 284
Gulf of Bothnia 10, 11, 13, 15, 21, 23, 27, 42, 43, 46, 51, 53, 54, 59, 65, 70, 75, 76, 78, 83, 95, 124, 137, 144–146, 151, 176, 178, 180, 182, 183, 185, 195, 196, 221, 227, 243, 252, 270, 272, 281, 291, 304, 313, 314, 334
Gulf of Finland 15, 17, 19, 21, 23, 27, 37, 42, 43, 46, 51–53, 65, 74–76, 78, 84, 86, 95–98, 106, 119, 124, 139, 144, 145, 151, 154, 169, 176, 178, 180–183, 195, 196, 200, 201, 207, 211, 216, 221, 223, 227, 228, 231, 252, 253, 263–265, 270, 272, 273, 280–283, 290–292, 300–302, 304, 306–310, 313–316, 322, 331, 333, 334
Gulf of Riga 15, 23, 27, 43, 46, 51, 52, 70, 74, 76, 78, 84, 95–98, 124, 144, 145, 150, 151, 178, 195, 221, 223, 252, 253, 270, 281, 314, 315, 317, 322, 331
Gulf Stream 144, 153, 270, 281

Hailuoto 55, 284, 313
haline conveyor belt 145
halocline 16, 61, 74, 65, 67, 70, 72, 74, 75, 81, 86, 90, 105, 113, 114, 117, 139, 145, 148, 150, 151, 154, 155, 169, 172, 178, 183, 185, 186, 200, 216, 220–223, 265, 277, 303 306, 334
Hamrare Strait 49, 171
heat budget 27, 84, 89, 105, 107, 113, 114, 129, 255, 328
High Resolution Limited Area Model (HIRLAM) 25, 176, 211, 321
Hiiumaa 51–53

hindcast 184, 320
High Resolution Operational Model for the Baltic Sea (HIROMB) 176, 320, 321
Hoburg Bank 51
Holmöarna 55, 151
human impact 4, 172, 174, 296, 306, 310
humidity 28, 29, 37, 93, 94, 111, 112, 115, 116, 278, 293, 321
hummock 223, 240, 241, 251
hydraulic control 53, 127
hydrodynamic–ecological model 23, 335
hydrography 16–18, 20, 22, 42, 60, 84, 95, 265, 317, 329, 331, 335
hydrostatic 132, 140, 151, 175, 176, 193, 214

ice
 algae 227, 231
 compactness 223, 240, 244, 245, 248, 251–253, 257, 321
 crystal 219, 226–229, 231, 234, 235,
 extent 14, 87, 88, 211, 222, 239, 245, 314, 315, 316, 331, 332
 drift 139, 239, 245, 248, 249, 251, 252, 253, 255, 256, 257, 259, 287, 288, 320, 321
 forecasting 21, 254–256, 318, 321
 pressure 220, 241, 246, 248, 256, 257, 313
 rheology 249, 250, 253, 255, 287
 salinity 225, 227–231
 season 14, 15, 19, 24, 88, 93, 94, 109, 219, 220, 223, 225, 226, 231, 258, 259, 295, 313, 315–318, 334
 state 244, 251, 255, 257, 287
 thickness 27, 107, 109, 223, 225–227, 231–239, 241, 243–248, 251, 254, 257, 259, 283, 284, 313, 316, 321, 322, 332, 335
 type 223, 226, 240, 241, 244, 245, 257
inertial
 oscillations 19, 133–136, 191
 period 123, 124, 135, 137, 157, 276
 waves, 276
internal circulation 20, 128, 132
internal friction 133, 225, 245, 246, 247, 249, 250, 252, 253, 288
internal stress 247, 253, 287
internal waves 23, 119, 128, 134, 151, 157, 189, 191, 193, 212–216, 277, 280

International Baltic Year (IBY) 21
International Panel on Climate Change (IPCC) 42, 269, 270, 328, 330, 333
intra-continental 3, 4, 14
Irbe Strait 52, 128, 150, 264

k–ε model 120, 121
Kattegat 2, 11, 14 16, 23, 42, 45, 46, 48, 49, 61, 61, 64, 72, 76, 78, 84, 86, 88, 94 98, 100 102, 128, 145, 150, 158, 160, 179, 183, 186, 201, 202, 204, 272, 282
Kelvin wave 135, 157, 196, 198, 199, 265, 278, 280
Kiel Bight 15, 49
Kihnu 52
Kihti Strait 55, 265
kinematics 220, 248, 257
kinematic viscosity 132
Kopparstenarna 51
Korsör 142

Lagrangian 185, 309
Lågskär Deep 54
land–sea breeze 289–292
landfast ice 19, 227, 238–241, 243, 244, 262, 283, 284, 286
Landskrona Deep 49
Landsort Deep 42, 51, 57, 145, 173
land uplift 4, 5, 10–12, 14–16, 42, 43, 268–270, 326, 333
Langmuir circulation 212
latent heat 27, 94, 106, 107, 111, 113–116, 220, 232, 234, 240, 252
Lawica Slupska 51
lead 220, 223, 239, 241, 252, 257, 258
level ice 241, 244
lightship 17, 18, 142, 143
Little Belt 14, 15, 49, 161
Luleå Deep 17, 57

main inflow period 159, 160, 161, 167, 169, 165
Major Baltic Inflow (MBI) 132, 158, 159–162, 164–168, 170–174, 183, 184, 304, 306, 311
Manning formula 127
mareograph 203

marine optics 18, 20, 108, 303
Märket 55
Markov chain 125
mean circulation 4, 72, 125, 126, 135, 142–144, 175–177, 178, 179–182, 264
Mecklenburg Bight 15, 49
Medium Resolution Imaging Spectrometer Instrument (MERIS) 303
melting of ice 90, 93, 101, 103, 104, 109, 220, 223, 232, 235–237, 251, 255, 258, 259, 286, 302, 316, 317
Merian formula 195, 196
mesoscale 132, 135, 145, 146, 151–153, 155, 158, 182
meteoric ice 226, 227
Midsjö Bank 49, 51
mixed layer 67, 70–72, 103, 113, 118, 119, 120, 123, 147, 157, 183, 276, 277
mixing length 104, 112, 132
Moderate Resolution Imaging Spectroradiometer (MODIS) 303, 320
momentum equation 132, 140, 174, 193, 214, 248–250, 252, 255, 287
Monin–Obukhov length 112
monitoring 4, 9, 21, 23–25, 68, 314, 318, 319, 327, 335
Monte Carlo 125, 257, 309
morphology 3, 15, 42, 125, 238, 265, 326

Navier–Stokes equation 132, 189, 232
net radiation 29, 114, 115, 223
new ice 103, 223, 227, 232
North Atlantic 10, 20, 25, 27, 40, 61, 161, 174, 207, 315, 322
North Atlantic Oscillation (NAO) 25, 40, 84, 85, 88, 174, 179, 180, 268, 269, 272, 273
North Sea 4, 10, 11, 15, 17, 20, 42, 46, 48, 61, 62, 64, 65, 67, 69, 75, 84, 89, 90, 95, 98, 100, 102, 105, 114, 145, 158, 159, 160, 167, 168, 222, 269, 273, 275, 312, 321, 327
Northern Gotland Basin 15, 46, 49, 51, 54, 75, 80, 81, 145, 150, 151, 169, 207, 208, 236, 322
Northern Quark 53, 55, 125, 128, 151, 240, 264, 313

Norrköping Deep 51, 145
Norwegian Coastal Current 46, 62
nutrient loading 4, 76, 334

oil spill 257, 309–311, 318, 320, 321
oil transport 5, 306, 307, 327
Öland 51
operational 11, 21, 23–25, 176, 211, 244, 256, 275, 296, 318–323, 336
Optically Active Substances (OAS) 296, 297, 298, 303
Öresund (Sound) 10, 14, 15, 42, 46, 49, 62, 64, 74, 101, 145, 148, 169, 171, 179, 308, 309
orographic 33, 37
osmotic pressure 59, 60
overmixing 127, 128
oxygen 4, 53, 60, 64, 75, 76, 80, 148, 155, 158, 166–169, 171, 172, 174, 183, 227, 296, 303–306, 311, 312, 319, 327, 334

Paldiski Deep 53
pancake ice 241, 244
Pärnu Bay 252
phase change 107, 113, 230, 232
phase diagram 219, 228, 229
phase velocity 190–192, 200, 215, 309
phytoplankton 282
Photosynthetically Active Radiation (PAR) 298, 299, 302
Poincaré wave 196–200
Pojo Bay 53
polar air 10, 25, 26
polar front 25–27, 32, 33, 84
post-period 159, 160, 161, 171
potential density 58
potential temperature 58, 111
Practical Salinity Unit (PSU) 56
precipitation 28, 33, 36, 37, 86, 90–92, 94–96, 98, 99, 101, 103, 107, 113, 115, 174, 184, 226, 237, 238, 289, 291, 331
precursory period 159–161, 167
pressure 17, 25–27, 56–61, 84, 86, 94, 100, 108, 111, 116, 133, 134, 140, 141, 151, 155, 162, 170, 171, 174, 179, 184, 193, 194, 203, 212, 220,

239–241, 243, 248–250, 252, 256, 257, 268, 281–286, 289–291, 312–314, 317, 321
pressure gradient 40, 127, 133, 134, 140, 141, 155, 171, 184, 193, 194, 212, 247, 249, 250, 254
primary production 6, 231, 303
Princeton Ocean Model (POM) 176
PROBE model 22, 119, 122
pycnocline 61, 118, 212, 215, 216, 264

radiation balance 27, 29, 32, 67, 107, 110, 115, 223, 236
rafted ice 239, 241, 246
random walk 102, 125,126
relict species 13
remote sensing 6, 21, 24, 240, 244, 248, 257, 296, 298, 303, 319, 320, 329, 336
Reynolds stress 132
rheology 247, 248, 253, 255, 287
Richardson number 112
ridged ice 5, 239, 240, 242–244, 246, 254
Ristna 51
river plume 263, 265, 266, 303
river runoff 67, 80, 85, 86, 90, 96, 135, 167, 173, 174, 176, 178, 181, 182, 321, 331, 334
Rossby number 134
Rossby radius 128, 134, 152, 181, 199, 263, 277, 282
Rossby wave 191, 199, 200
roughness length 112
Ruhnu 32
runoff 95, 96, 98, 104, 181, 184, 300

Saaremaa 51, 52, 207, 323
salt budget 6, 89, 101, 106, 304
sea ice ecology 24
sea ice sediment 219, 231, 240, 252
Sea of Bothnia 10, 15, 21, 45, 46, 53, 55, 74, 81, 82, 83, 84, 86, 87, 95–98, 114, 115, 125, 128, 144, 148, 151, 178, 182, 201, 202, 207, 213, 221, 252, 259, 265, 281, 334
sea surface temperature 27, 29, 33, 69, 75, 82–85, 93, 94, 106, 109, 113, 115,

117, 118, 271, 275, 289, 291, 293, 319, 331
Secchi depth 108, 299, 300, 302
secondary halocline 67
sedimentation 44, 45, 312
self-similarity 121, 123, 124, 243
seiche 135, 192, 195–197, 268, 272, 314, 315
semidiurnal tides 201, 202
semi-enclosed 3, 53, 100, 195, 220, 252, 275
sensible heat 107, 111, 113–116 , 220
shallow water wave 134, 189, 191–194, 198, 205, 206, 214, 308
Ship of Opportunity (SOOP) 319
ship wake 308, 309
significant wave height 206–211, 322, 323
sill 15, 43, 45, 46, 49, 51–55, 61, 64, 65, 74, 75, 81, 127, 128, 145, 148, 150, 151, 158, 159, 161, 164, 166, 167, 169, 173, 185, 265, 267
sill depth 43, 49, 51, 52, 55, 65, 125, 128, 150, 151, 169, 265, 267
Sjælland 49
Skagerrak 15, 46, 48, 61–64, 100, 125, 183, 201, 268
Skagerrak Front 62, 64
skeleton layer 226, 231
slush 226, 234, 237, 238, 257
snow 29, 33, 93, 95, 98, 104, 108, 110, 113, 115, 220, 223, 225–227, 231–238, 245, 291, 292, 296, 302
snow ice 226, 227, 231, 234, 236–238
solar constant 108
solar elevation 29
solar radiation 27, 29, 61, 70, 84, 105–111, 113–115, 119, 220, 223, 232, 345, 236, 255, 286, 296–298, 331
Southern Quark 53, 55, 129, 147
specific heat 58, 60, 107, 111, 232
specific humidity 93, 94, 111
speed of sound 60
stagnation period 148, 150, 158, 159, 167, 172–174, 184, 304, 306, 311
Stefan's law 232, 236, 238
Stokes drift 212
strain rate 132, 247, 248, 288
Stolpe Channel 49, 51, 129, 145, 150, 154, 155, 161, 169, 171, 178, 179, 183, 201
superimposed ice 226, 227, 232, 234, 236

surface circulation 125, 126, 132, 135, 142–145, 176, 178
surface temperature 27, 29, 33, 69, 72, 75, 76, 78, 82–85, 93, 94, 105, 106, 110, 111, 114, 115, 116, 117, 118, 120, 169, 220, 232, 234, 236, 255, 259, 265, 271, 275, 282, 289, 292, 293, 319, 320, 331
surface wave 191, 192, 212, 205, 206, 211, 214, 215, 319, 320
suspended matter 231, 296–298, 300, 304
Suursaari (Gogland) 33

Temperature–Salinity (TS) 60, 61, 67, 68
temperature of maximum density 56–59, 69, 70, 72, 120, 221
terrestrial radiation 27, 29, 33, 107, 115
theoretical mean sea-level 268
thermal conduction 232
thermal expansion 269, 271
thermal radiation 107, 110, 116
thermal wind 141
thermocline 22, 60, 61, 67, 70, 72, 114, 119, 123, 124, 138, 141, 147, 200, 216, 221, 265, 276, 278, 282
thermohaline circulation 72, 175, 176, 178
thermohaline intrusion 155
topographic eddy 200
topographic steering 152
topographic wave 199–201
transmissivity 108
turbulent exchange coefficient
turbulent flux 27, 93, 107, 111, 113, 115, 223, 232
turbulent kinetic energy 120, 157

Ulvö Deep 17, 55
upper layer 58, 64, 67, 69–72, 74, 75, 80, 81, 86, 90, 103–105, 113, 114, 117, 120, 127, 135, 139, 145, 147, 148, 178, 179, 186, 200, 216, 264
upwelling 16, 71, 108, 135, 139, 180, 262–265, 275–283, 289, 293, 298
Utö 27, 28, 67, 152, 227, 233, 236

Väinameri Sea 52
vertical diffusion coefficient 102, 106
vertical friction 133, 134, 138, 274
vertical mixing 6, 67, 104, 145, 155, 157, 264, 267, 304
vertical velocity 93, 111, 132, 133, 138, 174, 193, 194
viscosity 60, 132, 138, 139, 209, 214, 247, 257, 276
von Kármán's constant 112
vorticity 24, 154, 203, 291

water balance 11, 70, 90, 94, 135, 268, 269
water budget 15, 89, 96, 99, 182, 183
water exchange 1, 11, 17, 20, 22, 43, 46, 49, 51, 55, 64, 65, 85, 89, 90, 98, 100, 103, 122, 127, 128, 134, 148, 150, 158, 178, 180–183, 220, 262, 265, 270, 312
water renewal 3, 14, 15, 53, 64, 101, 102, 158, 169, 185, 303, 304
water type 60, 61, 74
warm inflow 169–172, 185
wavelength 190–192, 195, 201, 203, 205, 215, 231, 272, 285, 296, 298, 299, 302
wave height 191, 205–212, 309, 314, 315, 320, 322, 323
wave number 190, 191, 200, 214
wave period 190, 191, 195–197, 200, 201, 203, 205, 207–209, 215, 216, 309, 322
wave spectrum 21, 205–207, 209, 215, 268, 272, 320
wave vector 190, 193, 194, 214
Western Gotland Basin 15, 46, 49, 51, 75, 145, 150, 151, 178, 179, 186, 209
wind-generated wave 6, 189–192, 203, 206, 263, 320
wind direction 28, 38, 40, 139, 144, 159, 161, 178, 212, 252, 256, 275–277, 288
wind speed 28 38, 40, 111, 139, 144, 161, 206, 207, 209, 210, 248, 249, 252, 255, 256, 278, 284, 289
winter navigation 17, 21, 256
winter thermocline 72, 221